Also in the Variorum Collected Studies Series:

R.B. SERJEANT
Customary and Shari'ah Law in Arabian Society

R.B. SERJEANT
Studies in Arabian History and Civilisation

PAOLO M. COSTA
Studies in Arabian Architecture

WAEL B. HALLAQ
Law and Legal Theory in Classical and Medieval Islam

SHAUL SHAKED
From Zoroastrian Iran to Islam: Studies in Religious History and Intercultural Contacts

ETAN KOHLBERG
Belief and Law in Imāmī Shī'ism

GEORGE MAKDISI
Religion, Law and Learning in Classical Islam

M.J. KISTER
Studies in Jāhiliyya and Early Islam

WILFERD MADELUNG
Religious and Ethnic Movements in Medieval Islam

ROGER M. SAVORY
Studies on the History of Ṣafawid Iran

DAVID AYALON
Islam and the Abode of War: Military Slaves and the Adversaries of Islam

FRANZ ROSENTHAL
Muslim Intellectual and Social History

GEORGES VAJDA
Etudes de théologie et de philosophie arabo-islamiques à l'époque classique

MAJID FAKHRY
Philosophy, Dogma and the Impact of Greek Thought in Islam

MARIE-THÉRÈSE d'ALVERNY
La connaissance de l'Islam dans l'Occident médiéval

COLLECTED STUDIES SERIES

Farmers and Fishermen in Arabia

Professor R.B. Serjeant

R. B. Serjeant

Farmers and Fishermen in Arabia

Studies in
Customary Law and Practice

Edited by
G. Rex Smith

VARIORUM
1995

Published by VARIORUM
Ashgate Publishing Limited
Gower House, Croft Road,
Aldershot, Hampshire GU11 3HR
Great Britain

Ashgate Publishing Company
Old Post Road,
Brookfield, Vermont 05036
USA

ISBN 0-86078-491-6

British Library CIP Data
Serjeant, R.B.
Farmers and Fishermen in Arabia:
Studies in Customary Law and Practice.
(Variorum Collected Studies Series; CS 494)
I. Title II. Smith, G. Rex III. Series
340.5253

US Library of Congress CIP Data
Serjeant, R.B. (Robert Bertram)
Farmers and Fishermen in Arabia: Studies in Customary Law and Practice / R.B. Serjeant: edited by G. Rex Smith
p. cm. — (Collected Studies Series: CS494)
ISBN 0-86078-491-6 (hardback: acid-free paper)
1. Agricultural laws and legislation—Arabian Peninsula.
2. Fishery law and legislation—Arabian Peninsula. 3. Customary law—Arabian Peninsula. 4. Agriculture—Arabian Peninsula.
5. Fisheries—Arabian Peninsula. I. Smith, G. Rex (Gerald Rex).
II. Title. III. Series: Collected Studies: CS494.
KMC768.S47 1995 95–3348
343.53'07692—dc20 CIP
[345.3037692]

The paper used in this publication meets the minimum requirements of the American National Standard for Information Sciences - Permanence of Paper for Printed Library Materials, ANSI Z39.48-1984. ∞ ™

Printed by Galliard (Printers) Ltd
Great Yarmouth, Norfolk, Great Britain

COLLECTED STUDIES SERIES CS494

CONTENTS

Introduction

I	Fisher-folk and fish-traps in al-Baḥrain *Bulletin of the School of Oriental and African Studies 31. University of London, 1968*	486–514
II	Customary law among the fishermen of al-Shiḥr *Middle East Studies and Libraries: a Felicitation Volume for Professor J.D. Pearson, ed. B.C. Bloomfield. London: Mansell, 1980*	193–203
III	Star calendars and an almanac from South-West Arabia *Anthropos 49. Sankt Augustin: Anthropos Redaktion, 1954*	433–459
IV	A Socotran star calendar *A Miscellany of Middle Eastern Articles: In Memoriam Thomas Muir Johnstone, ed. A.K. Irvine, R.B. Serjeant and G. Rex Smith. Harlow: Longman, 1988*	94–100
V	Ḥaḍramawt to Zanzibar: the pilot-poem of the Nākhudhā Sa'īd Bā Ṭāyi' of al-Ḥāmī *Paideuma 28. Stuttgart: Franz Steiner Verlag, 1982*	109–127
VI	A *maqāmah* on palm-protection (*shirāḥah*) *Journal of Near Eastern Studies 40. University of Chicago, 1981*	307–322
VII	The cultivation of cereals in mediaeval Yemen *Arabian Studies 1, ed. R.B. Serjeant and R.L. Bidwell. London: C. Hurst & Co., 1984*	25–74
VIII	Some irrigation systems in Ḥaḍramawt *Bulletin of the School of Oriental and African Studies 27. University of London, 1964*	33–76
IX	Observations on irrigation in South-West Arabia *Proceedings of the Seminar for Arabian Studies 18. London, 1988*	145–153

X Customary irrigation law among the Baḥārnah of al-Baḥrayn 471–496
Bahrain through the Ages: the History, ed. Shaikh ʻAbdullāh bin Khālid al-Khalīfa and M. Rice. London and New York: Kegan Paul International, 1993

XI A Yemeni agricultural poem 407–427
with Ḥusayn ʻAbdullāh al-ʻAmrī
Studia Arabica et Islamica: Festschrift for Iḥsan ʻAbbās, ed. Wadād al-Qāḍī. American University of Beirut, 1981

Index 1–8

This volume contains x + 276 pages

INTRODUCTION

This volume of articles is dedicated to memory of their author, Professor R.B. Serjeant. After a long, distinguished career as an Arabist, culminating in his appointment as Sir Thomas Adams's Professor of Arabic in the University of Cambridge, he died suddenly in Denhead near St Andrews in April, 1993.

Bob Serjeant was born in 1915 and read Arabic and Hebrew at the University of Edinburgh where he came under the early influence of the late Richard Bell, a Quran scholar of some note. From Edinburgh, he travelled to Trinity College, Cambridge, where, supervised by Professor C.A. Storey, he gained a PhD degree about the time of the outbreak of the Second World War. Though clearly influenced by both, Serjeant always attributed his precision and meticulousness of scholarship to Storey. The years of the War saw Serjeant in the Aden Protectorate, a period which marked the beginning of his major scholarly effort in Arabian Studies, and working back in Britain with the BBC, editing the Arabic Listener. Academic posts at the School of Oriental and African Studies followed with periods of fieldwork in Arabia, mostly southern Arabia. In 1964, he returned to Cambridge where he became Director of the Middle East Centre and in 1970 he was appointed to the Sir Thomas Adams's chair of Arabic which he held until he retired in 1981. He and Marion, his wife, went to Denhead near St Andrews and there Bob Serjeant continued his academic work until his death.

The fruits of his labours were many. A comprehensive Serjeant bibliography can be found to the year 1983 by J.D. Pearson, 'Published works of Robert Bertram Serjeant' in Robin Bidwell and G. Rex Smith (eds.), *Arabian and Islamic Studies* (London & New York 1983), 268–79. This will be updated in the forthcoming *New Arabian Studies* 3 (1995), where a 1983–93 supplementary bibliography will appear. Among the books were *The Portuguese off the South Arabian Coast* (Oxford, 1963), *Ṣan'ā', an Arabian Islamic City* (London, 1983), which he edited with Ronald Lewcock and a great deal of which he wrote himself. Some of his articles have already formed volumes of the Variorum series: thirteen were published as *Studies in Arabian History and Civilisation* (1981) and a further eighteen as *Customary and Shari'ah Law in Arabian Society* (1991). This present volume of eleven articles is the third and it is hoped that a fourth will be assembled at a later date.

* * *

The eleven articles of this volume fall naturally into well definable groups. Fishing and fishermen (I & II), star-calendars (III & IV) (equally important to mariners and agriculturalists), navigation (V), agriculture (VI, VII & XI) and finally irrigation (VIII, IX and X). The golden thread which runs through almost all is the centuries-old customary law practised by fishermen, farmers and mariners alike in the Arabian Peninsula. To the information in his field notebooks on customary law and the precise documentation of such trades and professions, Serjeant joined the evidence of the vast array of Arabic texts with which he was conversant. It was in this marriage of the contents of his field notebooks with the literature he found in the library that Serjeant's enormous scholarly talents and strength lay.

The eleven articles span a broad chronological range (1954–1993). Re-reading these contributions, I am struck by their continuing relevance and their freshness. They come in a variety of shapes and sizes in different journals. It is not surprising then that different systems of transliteration and house rules have been employed. Very little updating of sources has been necessary. Cross references to other articles in the volume have been added and printing and other typographical errors have been corrected.

* * *

My thanks are due to Mrs Marion Serjeant who has been – as she always is – immensely kind and helpful during the process of putting this volume together. She, I know, would wish to join me in expressing our thanks to all those who have given permission for these articles to be reprinted in this way and who are recorded in detail in the Acknowledgements. Our final pleasant duty is to thank Dr John Smedley and his staff for all the help and advice which were so willingly and courteously given throughout work on the volume.

G. REX SMITH

Department of Middle Eastern Studies
University of Manchester
December 1994

ACKNOWLEDGEMENTS

The editor and publisher would like to thank the following institutions and individuals, who have given permission to reproduce the articles in this volume: School of Oriental and African Studies, University of London, and Oxford University Press (I & VIII); Mansell Publishing, Cassell plc, London (II); Herr A. Quack, Editor, *Anthropos*, Sankt Augustin (III); Longman Group, Harlow (IV); Franz Steiner Verlag Wiesbaden GmbH, Stuttgart (V); The University of Chicago Press (VI); C. Hurst & Co, London (VII); Dr Geraldine King, Seminar for Arabian Studies, London (IX); Michael Rice and the Ministry of Information, the State of Bahrain (X); Professor Munir Basashur, Chairman University Publications Committee, American University of Beirut (XI).

PUBLISHER'S NOTE

The articles in this volume, as in all others in the Collected Studies Series, have not been given a new, continuous pagination. In order to avoid confusion, and to facilitate their use where these same studies have been referred to elsewhere, the original pagination has been maintained wherever possible.

Each article has been given a Roman number in order of appearance, as listed in the Contents. This number is repeated on each page and quoted in the index entries.

Amendments are marked by an asterisk and noted in an Addenda at the end of articles I, IV, VII, IX and XI.

I

FISHER-FOLK AND FISH-TRAPS IN AL-BAḤRAIN

For assistance in every kind of way to obtain the information presented here I am indebted to Shaikh 'Abd al-'Azīz b. Muḥammad b. 'Abdullāh Āl Khalīfah, and particularly for introducing me to his friend the *nawkhudhā* 'Abdullāh b. Khamīs al-Shurūqī of al-Ḥidd village, situated on the south-east tip of al-Muḥarraq island, the northern island of the Baḥrain group. Al-Shurūqī entertained us to a *fuwālih* and took us carefully over his *ḥadrah*, explaining the technicalities involved in siting it, etc. The aerial photographs reproduced here were made available through the kindness of the station commander Group Captain Sanderson, R.A.F., and through the exertions of Flight-Lieutenant W. T. Wenham and his staff. I am also indebted to Eric and Beverley Fitzsimmons of the British Council for help in this and other matters.

In most parts of Arabia, as on the western bank of the Red Sea also, fishing is a despised occupation, and has been so from remote times. Al-Jāḥiẓ [1] in the early 'Abbāsid period alludes to the ' low ' nature of the fisherman's profession as people considered it in Baṣrah and Ubullah, but in al-Baḥrain to-day, on the contrary, no such sentiment obtains.

The marked social cleavage comes between the Arabs of tribal descent in al-Baḥrain and the Shī'ah Baḥārnah of the islands, as also of course of the area once included in al-Baḥrain on the mainland, Qaṭīf and al-Ḥasā'/al-Aḥsā'. These latter were considered inferior to the tribal element, and despite the improvement in their condition commented upon by Belgrave [2] bitter remarks against the Sunnīs may still occasionally be heard, just as indeed, one finds in the *Dīwān* of al-Karkhī [3] a sense of grievance among the Shī'ah Mi'dān of southern Iraq. Al-Shurūqī stated that people of noble birth were called Bin Ajwād or Ajwadī, or *qabīlī*, i.e. tribal—a usage not only identical with S. M. Salīm's [4] *Ajāwīd* of the Iraqi marshes, but with the *ajwād mujjād* of al-Jāḥiẓ.[5] Al-Shurūqī said, however, that one speaks of *al-ghanī wa 'l-ḍa'īf* in the sense of ' rich and poor ', while the equivalent of the south-west Arabian modern and classical *ḍa'īf* class is the Baisar (pl. Bayāsir) group. Jalāl al-Ḥanafī [6] reports Baisar as Kuwaitī usage for *man lā aṣl lahu*, persons of low class, and considers

[1] Dr Wadī'ah Najm has supplied this reference to a passage, not apparently in the printed text of Ṭaha al-Ḥājirī, from a MS. Cf. al-Khaṭīb al-Baghdādī, *al-Bukhalā'*, Baghdad, 1964, 186.

[2] Sir Charles Belgrave, *Personal column*, London, 1960, 100.

[3] *Dīwān al-Karkhī* (i.e. 'Abbūd al-Karkhī), ed. Ḥusain Ḥātim al-Karkhī, Baghdad, 1956, in the poem entitled *al-Majrashah*.

[4] *Marsh dwellers of the Euphrates Delta*, London, 1962, 76.

[5] *Al-Bukhalā'*, ed. Ṭaha al-Ḥājirī, Cairo, 1948, 144.

[6] *Mu'jam al-alfāẓ al-Kuwaitīyah*, Baghdad, 1964, 56.

the word to be Persian, but it is current in the Trucial States and Oman where it is an ancient term.[7]

In al-Baḥrain it is no *'aib* (disgrace or impugnment of honour) certainly, to be a *ṭawwāsh*, pearl-merchant and buyer-cum-financier—in fact he is accounted the most honourable (*ashraf*) of a respected group comprising the *ghawwāṣ*, i.e. anyone engaged in the pearling industry, a class to which the *nawkhudhā*, captain of a pearling outfit, belongs as well as the fisherman (*ṣaiyād samak*). Of course, said the *nawkhudhā* al-Shurūqī, if you happen to be of low origin (*aṣl-ak dānī*) then you are of low origin, but fishing in no way demeans you, and it is no *'aib* to engage in fishing or to sell fish. On the other hand, there was and *is* no marriage between the tribes and people of other classes who practise the types of profession or craft detailed below. Even if these persons should be extremely wealthy the tribesmen would not give them their daughters in marriage—this being the criterion of social standing in all parts of Arabia that I know.

In al-Baḥrain 60 years ago, said al-Shurūqī, it was an *'aib* to open a shop (*dukkān*), to be a tailor (*dirzī*), barber (*muḥassin*), carpenter (*najjār*, pl. *najājīr*), butcher (*jazzār*), to engage in agricultural work, or even to work in an office, to wear a coat or a sleeveless jacket [8] (*ṣudairīyah*) over the coat. Now (1963) all this has gone, he said. Nevertheless it is still *'aib* for an Arab in al-Baḥrain, even if he be poor, to carry hens or eggs, or to sell them, though Arabs seem to eat them without repugnance.[9] The 'Ajam, Persians, I was told, do deal in hens, eggs, and the like. This taboo is the legacy, in my view—based upon the literature of Arabic and upon field-work—of the sentiment of tribes-folk (not necessarily migrating Bedouin) that poultry and their products are unclean, perhaps only to be eaten by the *ḥaḍar*.

When at variance, the Baisar group, and the *baḥāḥīr* (sing. *baḥḥār*), i.e. sea-faring folk, take their cases for decision to a respected tribesman who acts as conciliator (*muṣliḥ*) between them. If unsuccessful in reaching an agreement they repair to the court (*maḥkamah*), but as a court case costs money, whereas the *muṣliḥ* charges nothing, and since the court imposes a decision whereas the view of the *muṣliḥ* which they seek is accepted by *consent* of the two parties, it has a greater appeal to them as a means of settling disputes. In fact it is said that not only does the *muṣliḥ* receive no fee, but he has to produce a *fuwālih*, a tray (*ṣīnīyah*, also known as *qudū'*), upon which are set fruits, sweets, and biscuits, and he entertains the parties to the dispute with coffee, so that to act

[7] Bertram Thomas, *Arabia Felix*, London, 1938, 12 ; al-Hamdānī, *Ṣifat jazīrat al-'Arab*, ed. D. H. Müller, Leiden, 1884–91, 52, and *al-Iklīl erstes Buch*, ed. O. Löfgren, Uppsala, 1965, II, 74, who describes them as the ancient inhabitants of Raisūt (pl. al-Bayāsirah). In *Alarms and excursions in Arabia*, London, 1931, 152, Thomas describes those in Oman as half-caste and fair-skinned. Al-Jāḥiẓ, *Rasā'il*, ed. 'Abd al-Salām Hārūn, Cairo, 1964, II, 298, and *Ḥayawān*, ed. 'Abd al-Salām Hārūn, Cairo, 1938–45, I, 311–12, mentions a type of poultry ' between al-Baiḍā' (white type) and al-Sindī ' known as Baisarī.

[8] Al-Ḥanafī, op. cit., 216, says it is worn by the *kandarī* or water-carrier. Shaikh 'Abd al-'Azīz spells it *sidairī* (with *sīn*), pl. *sidairīyah*, from *sidar/sadur* ' chest '.

[9] cf. my remarks on the Arab attitude to poultry in *BSOAS*, XXIV, 1, 1961, 144.

in this capacity costs him money. As al-S͟hurūqī said to me, ‘ We apportion [10] in inheritance (*irt͟h*) cases, partnership in houses, and other types of case ’. In the court at al-Baḥrain then newly formed, Belgrave [11] reports that in the very numerous cases of boundaries, fish-traps, and water rights the court used to appoint an arbitrator and each of the parties a representative. As the *nawk͟hud͟hā* described it, the *maḥkamah* selects two men, the *muddaʿī* two men, and the *muddaʿā ʿalaih* two men to form a committee (*lajnah*) to resolve their dispute. These men are *baḥāḥīr* and the *baḥāḥīr* in general are drawn from all classes in al-Baḥrain. According to Belgrave, because of the kudos gained as mediators, busy men used to be glad to give time to help the court, though this involved very often visits to distant fish-traps, but nowadays they are less willing to undertake such duties.

The S͟hīʿah of al-Baḥrain, called al-Baḥārnah, form about half the population, and, as remarked, are regarded as a class inferior to the Arabs. It has been supposed, though proof is not yet forthcoming, that they are descended from converts from the original population of Christians (Aramaeans ?), Jews, and Majūs inhabiting the island and cultivated coastal provinces of eastern Arabia at the time of the Arab conquest. The Arabs of al-Baḥrain are conscious of the distinct dialect of the Baḥārnah, and of such peculiarities in their names as the diminutive forms ʿAlwawh for ʿAlī, K͟halfawh for K͟halīfah, ʿAbdawh and Ḥamdawh (عَلْوَوْه، خَلْفَوْه عَبْدَوْه حَمْدَوْه), etc. In the latter half of the seventh/thirteenth century Ibn al-Mujāwir [12] reports that the island (here called Uwāl) contains 360 villages, all Imāmī (i.e. S͟hīʿī) in *mad͟hhab* except one, their diet being dates and fish. It may be significant that like the ancient Jews they have taboos regarding the eating of certain categories of fish. ‘ Anything which has not got scales is prohibited ’ (*kull shī mā fīh ṣafad* [13] *ḥarām*). So they do not eat crab (*saraṭān*), shark (*jarjūr*), the *kim* (كِمّ c͟himm ?), resembling the *kanʿad*, which is stupid and blind (*balīd wa-aʿmā*), or the octopus (*k͟hat͟ht͟hūq*) which changes colour according to the colour of the ground (*yitlawwan bi-lawn al-arḍ*), but the *k͟hat͟ht͟hūq* is used as bait (*ṭiʿim*) for fish. They do not eat turtles, known as *sulḥāfah* and *ḥamasah* (pl. *ḥamas*), but they do appear to eat the *rūbiyān* [14] or prawn. Confirming that the S͟hīʿah will not touch shark's flesh, Lorimer [15] adds that it is eaten by the Arabs who consider it to have aphrodisiac properties—in fact Sunnīs appear to eat all the above-mentioned

[10] *Naqsim*, v.n. *qismah*.

[11] op. cit., 33.

[12] *Descriptio Arabiae meridionalis . . . Taʾrīk͟h al-mustabṣir*, ed. O. Löfgren, Leiden, 1951–4, 301. Cf. Ibn Mājid, in G. Ferrand, *Instructions nautiques*, Paris, 1921–8, I, 69b, who says the population consists of Arab tribes (*qabāʾil min al-ʿArab*), and a number of merchants—perhaps by the latter the Baḥārnah are intended.

[13] *Ṣafad* is explained as *fulūs* ‘ scales ’.

[14] Al-Jāḥiẓ, *al-Ḥayawān*, Cairo ed., I, 297, IV, 102, VI, 79, *irbiyān* (sing. *ah*). In Aden it is still called *rūbiyān* (crawfish *Panulirus* spp.).

[15] J. G. Lorimer, *Gazetteer of the Persian Gulf, 'Omān and central Arabia*, Calcutta, 1908 and 1915, 2308. Cf. Appendix E, ‘ Fisheries of the Persian Gulf ’, for some notes on *ḥaḍrahs*, 2316.

fish. Nobody in the Gulf, however, eats oysters.[16] For the Baḥārnah the Christian is unclean (*al-Naṣrānī nijis*) and hence they break the coffee-cup from which a Christian has drunk, as Belgrave describes.[17]

While the Baḥārnah, in respect of these fish-taboos, appear to be following normal Shī'ah law, it is to be recalled that southern Iraq was the very cradle of Shī'ism and that taboos local to the Gulf may have become incorporated into the sect's legal system, though deriving ultimately from non-Muslim custom going back to a remote age.

To the posterity of the Prophet the Shī'ah give, theoretically at least, the *khums yad* in accordance with the Qur'anic verse (VIII, 42), *Wa-'lamū anna-mā ghanimtum min shai'in fa-anna lillāhi khumusa-hu wa-li 'l-rasūli wa-li-dhi 'l-qurbā wa 'l-yatāmā* .... Shī'īs (not in this case Baḥrain Shī'ah) inform me that the 'fifth' would only be paid to those of the Prophet's descendants engaged in some religious office, and the 'fifth' which is subscribed by the members of the sect is used for general welfare purposes as required by the latter part of the Qur'anic verse, that which goes *lillāh* being meant for the poor, whereas *al-qurbā* they interpret as the Prophet's house.[18] It may be for reason of this that the Shī'ah Waqf Department in al-Baḥrain holds a sizeable number of properties, including fish-traps.

Fish-traps in al-Baḥrain

The long shallow shore waters of the Persian–Arabian Gulf, with sandy beaches extending distances of a mile or two under the sea before meeting deeper water, are specially suited for the catching of fish in permanent fish-traps (*ḥadrah*), or tidal weirs as they have been described, such as may be seen all the way from Iraq along the Arabian coast. At high tide these traps are largely submerged, but as the water recedes the fish are left stranded within their fences. These fish-traps are not found, for example, on the South Arabian coasts, from Perim to al-Maṣna'ah on the borders of Mahrah territory, to my certain knowledge, for there are few beaches there like those so general to the Gulf. Nor can I report having seen them during flights over the Yemenite coast. Possibly weather conditions also are unfavourable, for Val T. Hinds [19]

[16] Belgrave, op. cit., 47.

[17] ibid., 56.

[18] This Shī'ah tradition may have its roots in ancient Arabian practice. For the entitlement of the Hāshimites to the *khums* through their kinship with the Prophet, cf. the traditions assembled by Abū 'Ubaidah, *Kitāb al-amwāl*, Cairo, n.d., 332 f., and Qudāmah b. Ja'far in A. Ben Shemesh, *Taxation in Islam*, II, Leiden, 1965, 51–2, unfortunately very incorrectly translated at this point. The methods by which the revenues of certain shrines of saints in the tribal districts of South Yemen are disposed of to-day seem to be closely analogous to the Shī'ah traditional view on the treatment of the *khums*. Cf. al-Wāqidī, *al-Maghāzī* (ed. Marsden Jones), OUP, 1966, II, 693 f.

For historical data on local Shī'ah a useful source is the treatise of the well-known merchant and bookshop proprietor of al-Baḥrain Muḥammad 'Alī Tājir, *Kitāb muntaẓam al-durrain fī a'yān al-Qaṭīf wa 'l-Aḥsā' wa 'l-Baḥrain*, consisting of Sunnī and Shī'ah biographies, still in MS form. Cf. also 'Alī b. Ḥasan al-Bilādī al-Baḥrānī, *Anwār al-badrain fī tarājim 'ulamā' al-Qaṭīf wa 'l-Aḥsā' wa 'l-Baḥrain*, al-Najaf, 1960.

[19] *Impressions of a fishing industry—Dubai—Trucial States—Arabian Gulf*, report, July 1964, to the Ministry of Agriculture and Fisheries, Federal Government of South Arabia.

has pointed out that they are rarely seen on the eastern section of the Trucial Coast where the north-westerly winds build up high waves during the rough weather period. On the other hand they are to be found along the East African coastal marine shelf near, for instance, Mombasa and Dar es Salaam. It may well be discovered that they were introduced there from Arabia which has so deeply imbued the coastal maritime and fishing peoples with its own culture that one finds practically all the parts of the African coasting dhow, and even some of the fish-names in Swahili are Arabian. The Swahili name for a fish-trap is, however, *jarife* (جريفة), pronounced in the Ḥaḍramī way, according to Hinds, a bottom set tangle-net, though the Ḥaḍramī *jarīf* is not a fish-trap, but a drag-net with two arms, operated from the shore, but it might be said to resemble the fish-trap in shape quite closely. It is puzzling to find a Ḥaḍramī name given by Swahilis to a fish-trap which I postulate may have reached Africa from the Gulf, though of course East African Muslim communities show evidence of far deeper cultural influence from Ḥaḍramawt than from Oman.

At a first glance the fish-trap (*ḥaḍrah/ḥaẓrah*) looks a simple enough affair, but one soon discovers that considerable expertise is required for siting and constructing it, and it must be concluded that these traps underwent a lengthy period of evolution before developing to their present shape. *Ḥaḍrahs* are subject to certain customary laws (only figuring incidentally here), they are heritable, can constitute a *waqf* property, and they are financed in their operation by methods very similar to those of share-cropping in agriculture.

Al-Ḥidd village

The village of al-Ḥidd lies on the south-east extremity of Muḥarraq island, the word *ḥidd* meaning a sand-spit. South of al-Ḥidd is the sweet water spring in the sea (*kawkab*, pl. *kawākib*) [20] known as Umm al-Sawālī. The spring or springs are so called because there they wash out the *sawālī* (sing. *sāliyah*, a landing-net with two sticks used as handles).[21] There are other sea-springs around the coasts of al-Muḥarraq as is well known, such as Yizīrat/Jazīrat il-Sāyih. A mile or so to the south of Umm al-Sawālī is al-Qaṣāṣīr.[22] A *qaṣṣār* was described to me as a rock in the sea (*ṣakhrah fi 'l-baḥr*) which at high tide looks like a castle. Ibn Mājid [23] says of al-Baḥrain, ' The most astonishing thing in it is a place called al-Qaṣāṣīr. A man dives into the salt sea with a water-skin [24] and fills it with sweet water while he is submerged (*gharqān*) '.[25]

[20] *Bahrain Trade Directory*, 1383/1963, 114.

[21] Jalāl al-Ḥanafī, op. cit., 167 cites this word but in the sense of a net, as also in Qaṭar (Dr. T. M. Johnstone). Cf. I. MacIvor, ' Notes on sea-fishing in the Persian Gulf ', *Administrative Report of the Persian Gulf Political Residency and Muscat Political Agency* for 1880–1, 55, where it is described as a circular cast-net with lead weights round the circumference.

[22] cf. Ibn Mājid, loc. cit., and George Rentz, ' Pearling in the Persian Gulf ', in W. J. Fischel (ed.), *Semitic and Oriental studies : a volume presented to William Popper* (University of California Publications in Semitic Philology, XI), Berkeley, 1951, 402.

[23] loc. cit.

[24] According to the *Bahrain Trade Directory*, 114, the large water-skin in which the pearling dhows used to take water from these sea-springs was called *farzih*.

[25] C. de Landberg, *Glossaire daṯînois*, Leiden, 1920–42, *gharqān* ' noyé '.

I

The village appears to be largely concerned with the fishing industry, gear for which lies about the sea-front. Just north of al-Shurūqī's *ḥaḍrah* can be seen (plate I) a flat sanded rectangular platform on the sea-front with a low wall, where the fishermen sit outside in the evenings. It is called *barāḥah* a word also in use for a town square in al-Ḥasā'.[26] Al-Muḥarraq, of which al-Ḥidd is a part, has somewhat of a reputation for turbulence, and it was so during the political riots of the early spring of 1965 after which I last visited the island, but I do not know if al-Ḥidd was specifically involved.

The sea-front of al-Ḥidd faces due west, and the fish-traps run in a general westerly direction from the shore. They lie within *al-baṭīn*,[27] the side of al-Muḥarraq sheltered from the *Shimāl* wind, not *al-ẓahr/ẓahar*, the side exposed to the full *Shimāl* gale which Dickson [28] describes as a north-west wind. The term *al-ẓahar* was explained to me as *maḍrab al-mawj, maḍrab al-rīḥ*, the place where the wind and waves strike. Of course the *Shimāl/Shamāl* blowing over al-Muḥarraq would, to judge from plate I, blow into the long arm of the *ḥaḍrahs*.

From the sand-spit known as al-Ḥidd, then, the beach running down from the village is *al-barr*, turning into a flat sandy area which at low tide (*thabur*) [29] is uncovered, but at high tide (*sajī*= *saqī*) is under water. Upon this latter part of the beach, known as *al-baṭḥ*, grows both *mash'ūrah* [30] seaweed, and *chinn/kinn* which is a seaweed rather like a light green wool. The fish feed on the *chinn* which is collected also for use as bait. The long stringy *mash'ūrah* is employed as bait for the shark (*jarjūr*).

The ḥaḍrah of 'Abdullāh b. Khamīs al-Shurūqī at al-Ḥidd

The siting of the *ḥaḍrah* and its shape are determined by the nature of the terrain. Al-Shurūqī's *ḥaḍrah* (fig. 1), called Ibzūr, may be described as open to the north and bounded on the east by the shore of al-Ḥidd. The whole is constructed of palm-*jarīds* from the palm plantations of al-Baḥrain, the only suitable material locally available. The long arm (*al-madd*, pl. *mudūd*) extends up to the land (*al-barr*), running roughly from east to west. The other short

[26] F. S. Vidal, *The oasis of al-Hasa*, Dhahran (?), 1955, 215. In Ḥaḍramawt *barāḥ* means a breeze, and at al-Shiḥr a *mibraḥ* is a space upon which pebbles have been laid, used for drying fish.

[27] *Baṭīn* or *baṭn* and *ẓahr* are widely used in the Persian Gulf and Arabian coasts in general in this sense, as may be perceived by a study of such authors on navigation as Ibn Mājid and 'Īsā al-Quṭāmī, *Dalīl al-muḥtār fī 'ilm al-biḥār*, third ed., Kuwait, 1383/1964, *passim*. I have noted also that *baṭn al-shirā'* is the inside of a sail which the wind is filling, and the outside is the *ẓahr*.

[28] H. R. P. Dickson, *The Arab of the desert*, London, 1949, 248.

[29] One says, *al-baḥr thābir* ' the tide is receding '. The verb is *thabara*. Cf. al-Quṭāmī, op. cit., 247 ; al-Ḥanafī, op. cit., 68. Vidal, op. cit., 216, gives *thabr* as a gathering canal in irrigation at al-Ḥasā'.

[30] *Mash'ūrah* is a seaweed growing in long strands which contains at best only a few fish. Sometimes the *lazzāq* fish is found in it. The *lazzāq* was described as like the *kan'ad* (*sīr*/seer fish, *Cybium guttatum*) but with a head like the heel of a shoe which sticks to the deck—presumably a sort of sucker ?

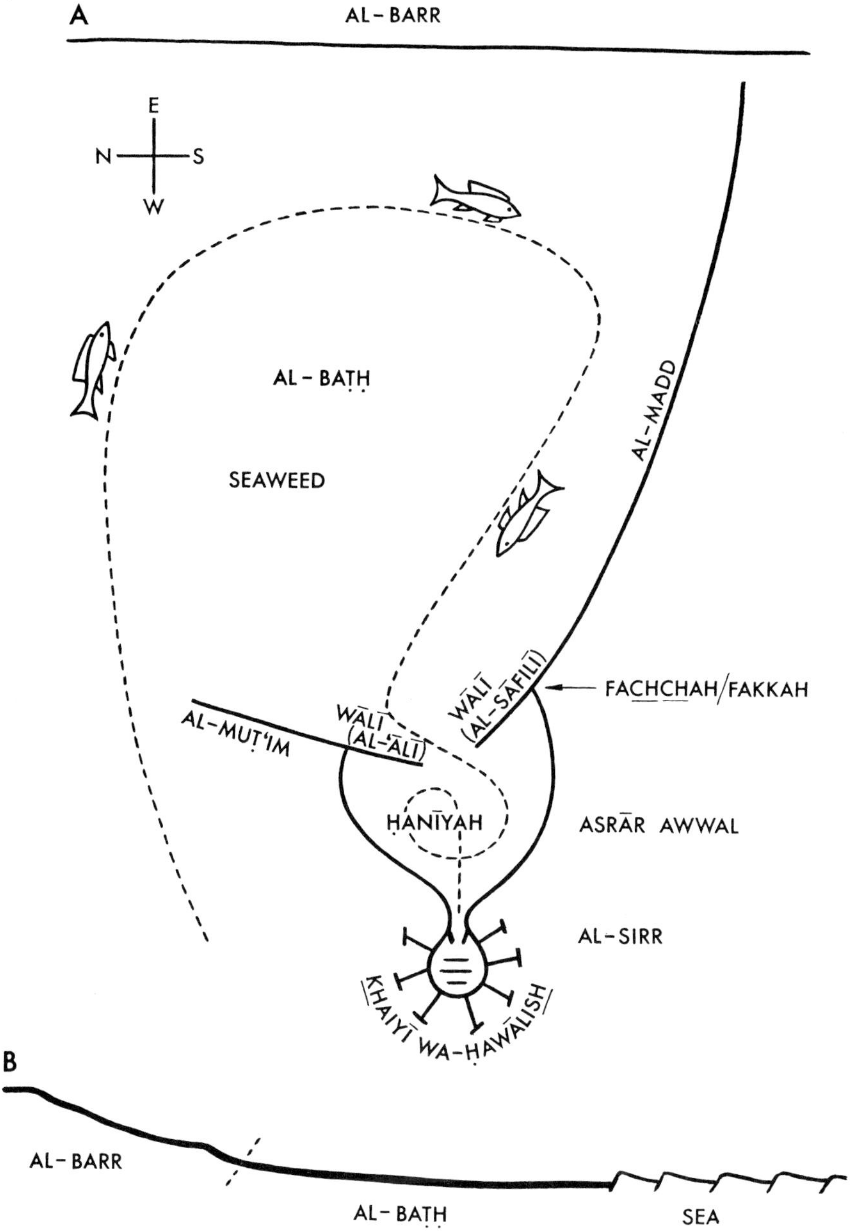

FIG. 1. Plan of Ibzūr fish-trap.

A. The dotted line indicates the movement of the fish feeding upon the seaweed of the flat sea-bed uncovered at low tide. Water and wind (*al-hawā*) bring them to the *madd* arm. There is dust on the water because of the wind. The *madd* arm causes them to change direction, and as the tide ebbs they are diverted by stages from one point to another, until, perforce (*bi 'l-jabar*) they enter the final *sirr* chamber, the opening into which is very narrow.

B. Profile of the shore to show the parts known as *al-barr* and *al-baṭh*.

(*qaṣīr*) arm called *al-muṭ'im* runs approximately north to south. These arms are linked by another fence enclosing a space called *al-ḥanīyah* 'the curved area', and the parts of the arms *madd* and *muṭ'im* forming the top of the *ḥanīyah* section on the east side are known as *wālī* (pl. *awlayā*), distinguished in the case of this particular *ḥaḍrah* by the specific epithets *al-sāfilī* 'the southern' and *al-'ālī* 'the northern'.[31] There appears to be a sort of join of the *madd* arm and the fence of the *ḥanīyah* known as *fa<u>chch</u>ah* (*fakkah* 'jaw' ?). The fence enclosing the *ḥanīyah* area was named to me as *asrār awwal*, and the actual chamber within which the fish are trapped is *al-sirr*. On the outside the *sirr* is held in position and strengthened by guy-ropes (*<u>kh</u>aiyah*, pl. *<u>kh</u>aiyī*) attached to pegs (*ḥāli<u>sh</u>*, pl. *ḥawāli<u>sh</u>*) stuck in the flat ground of the area *al-baṭḥ*. On the inside ropes run from one side to the other of the *sirr* as shown in fig. 1, known as *ṣuwar* (pl. *ṣuwārah*)[32] *al-ḥaḍrah*, a term also used for the beams running from side to side of a dhow and bearing the deck. Upon one of these is hung the gear (*mā'ūn*), i.e. the fish-gaff and the basket in which the fish are collected. The fish are taken out of the *sirr* by hand, with the *sāliyah*[33] a kind of landing-net rather like a large string shopping bag with two wooden spars at the top. It is used in the manner of a scoop, the top being drawn together but not very closely. The fish are then taken out of the *sāliyah* and placed in a *jirāb*, a bag made of palm-leaf (*<u>kh</u>ūṣ*). These bags are carried away on a pole borne over the shoulders.

The height of the fence of the Ibzūr *ḥaḍrah* is about 9 feet, and the *jarīds* are linked together by a horizontal tie near the top also made of *jarīds* and known as *<u>sh</u>idād*, those below it, four in all, being made of rope (*ḥabl*). The two drawings (figs. 2 and 3) illustrate a sort of ornamental tuft of coir (*qumbār*) at the top of the poles of the point of the *ḥaḍrah*, which I had not specially noted myself, known as the *kadīl* (pl. of *kidlih*) and so called because they are thought to resemble the Baḥrainī girl's fringe of hair over her forehead (fig. 4). In places where there is rock near the surface of the ground of the fish-trap so that *jarīds* cannot be fixed simply by planting them firmly in the sand, the arms have sometimes to be built of rock instead.

Construction of the ḥaḍrah

The *ḥaḍrah* (pl. *ḥuḍūr*) or *ḥaẓrah*[34] as, according to the canons of Classical Arabic, it would be more correctly spelled, is to be connected with such purely

[31] cf. my *Portuguese off the South Arabian coast*, Oxford, 1963, 195 *passim*. In al-Baḥrain one says, *Fulān 'ālī 'annā* 'So-and-so is north of us' and *Nās 'allaw* (*yi'allī*) 'People who have gone northwards'. To go southwards is *dabbar*, and the south is *al-sāfil*.

[32] T. M. Johnstone and J. Muir, 'Some nautical terms in the Kuwaiti dialect of Arabic', *BSOAS*, XXVII, 2, 1964, spell this with *sīn*, but cite Lorimer's spelling with *ṣād*. For al-Baḥrain my own observation was probably checked with al-<u>Sh</u>urūqī at the time, but I have noted from south-west Arabia that the distinction between these two letters is often unclear. One says of a ship, *ṣuwar al-maḥmal*. The word *maḥmal* itself in the sense of a vessel, ship, was linked by my informant with Qur'ān XVII, 72. لَقَدْ كَرَّمْنَا بَنِى آدَمَ وَحَمَلْنَاهُمْ فِى ٱلْبَرِّ وَٱلْبَحْرِ.

[33] cf. p. 490.

[34] T. M. Johnstone has recorded a plural *hẓār* for al-Baḥrain. Cf. al-Ḥanafī, op. cit., 96.

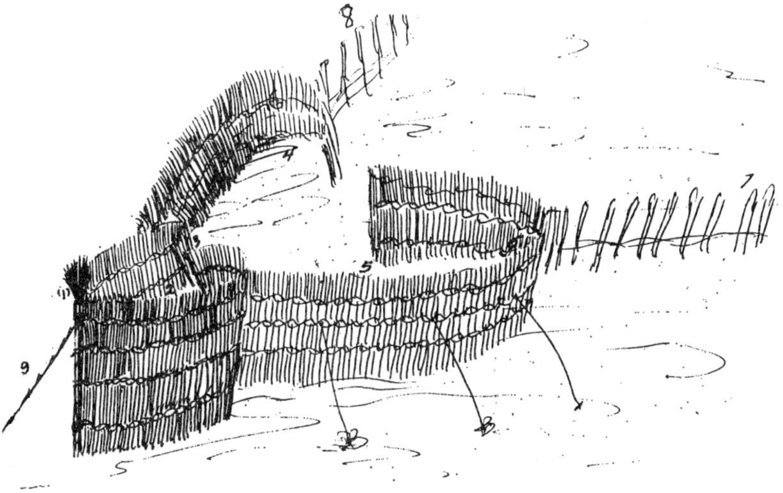

[By courtesy of Shaikh 'Abd al-'Azīz Āl Khalīfah

FIG. 2. Sketch of fish-trap.

1. The *kadīl* (sing. *kidlih/kidlah*) like a girl's fringe ; cf. fig. 4.
2. *Sirr/sir.*
3. *Sirr/sir awwal*, the opening.
4, 5. *Ḥanīyah* (pl. *ḥanāyā*) curved arm.
6. *Maqran*, the corner of the *ḥanāyā*.
7. *Madd* arm, longer than shown in sketch.
8. *Muṭ'im/miṭ'im* arm, about half the length of the *madd*.
9. *Khīyih/khaiyah* (pl. *khaiyī*), guy-ropes.

[By courtesy of Shaikh 'Abd al-'Azīz Āl Khalīfah

FIG. 3. Sketch of Sitrah type of fish-trap showing the *kadīl* and the cable known as *mijdab/mijdhab* used to anchor the *madd* arm.

I

[*By courtesy of Shaikh 'Abd al-'Azīz Āl Khalīfah*

FIG. 4. Baḥrainī girl of Sitrah with a fringe (*kidlih*).

agricultural terms as *ḥiḍār*, a fence of palm-*sa'f* and *jarīd*, *taḥḍīr* being the action of constructing such a fence, which I noted from documents in the Shī'ah Waqf Department.[35]

Ḥaḍrahs are constructed by experts—the *bannā'ī al-ḥaḍrah*, or fish-trap builder. There is a sort of foreman (*ustādh*) of the builders who receives a wage of 25 rupees (1963) at each water (*kull māyah*), i.e. every low tide that he works, at which time alone can he carry on the construction of the trap, the site being left exposed. The rest of the trap builders receive 12 rupees a time. The various parts of the *ḥaḍrah* are built according to careful measurement, the *bannā'ī* building according to a measure (*yibnī bi 'l-miqyās*). The materials employed are palm-sticks (*jarīd al-nakhīl*) stripped of their leaf (*sa'f*), and rope (*ḥabl*) of coir, coco-nut fibre (*qumbār*) or the fibre (*līf*) of the date-palm. Each year the *ḥaḍrah* must be re-made because barnacles (*al-naw*) grow on the *jarīds* and they become black—which frightens the fish so that they are scared (*yitwaḥḥash*). Al-Shurūqī's trap is reconstructed each autumn, the work taking about three days. It is, of course, the man who rents the *ḥaḍrah* who bears the expense of reconstructing it. The trap builder also receives *ruyūq* (i.e. *faṭūr*), breakfast, or *ghadā*, lunch, or perhaps both. Provision of a meal as part of a labourer's wages is also general custom in south-west Arabia.

In al-Baṣrah the *ḥaḍrah*/*ḥaẓrah*, there known as *gaṭ'ah*/*qaṭ'ah* or *mailān*, is made of *jarīds* and marsh-reed (*qaṣab*) which grows to a great size. In former

[35] Agricultural terms peculiar to the island occur in *Qānūn miyāh al-nakhīl*, al-Baḥrain, 1379/1960, which I am examining.

times the Kuwait *ḥaḍrahs* were made of Iraqi reed or cane (*qaṣab*) [36] but now wire is used for them, just as it is used for fish-pots in many parts of the Gulf. Names for the traps differ from place to place (*al-asāmī alladhīn 'inda-hum ghair shakl*).

The pattern of the fish-traps shows considerable variation depending on the nature of the terrain, the currents, etc., but the basic features of their construction appear to be the same. The Sitrah type (figs. 3 and 5) does not contain

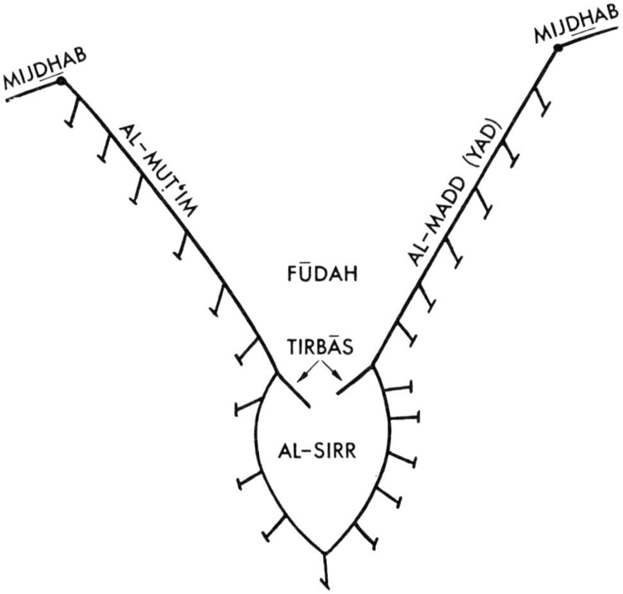

FIG. 5. Plan of fish-trap at Sitrah.

Additional technical terms are :
(1) the *fūdah*, the area enclosed by the *madd* and *muṭ'im/miṭ'im* arms ;
(2) *tirbās/ṭirbās* (*tarābīs/ṭarābīs*), the turned-in parts of the *ḥanāyā* between which lies the *sirr/sir awwal*.

the vestibules known as *ḥanāyā*, because the current is very strong there, but only the *sirr* and the two *madds*. The Kuwait Museum displays a diagram or model of a type of *ḥaẓrah* which I believe is also to be found in al-Baḥrain though I do not remember actually seeing it there. A long arm (*al-yad*) from the shore leads into the actual trap (fig. 7), but there is an entrance on each side of the *yad* to the enclosed antechamber (*ḥawsh*) to the *sirr* where the fish will be left stranded at low tide. The enclosing palisade of the trap is called *al-minshab* [37] (vocalization uncertain). A plan of this type sketched in al-Baḥrain is shown in fig. 6.

[36] Al-Ḥanafī, op. cit., 97. Quṭāmī, op. cit., 235, gives a photograph of the Kuwait form of *ḥaẓrah*.
[37] cf. al-Ḥanafī, op. cit., 365.

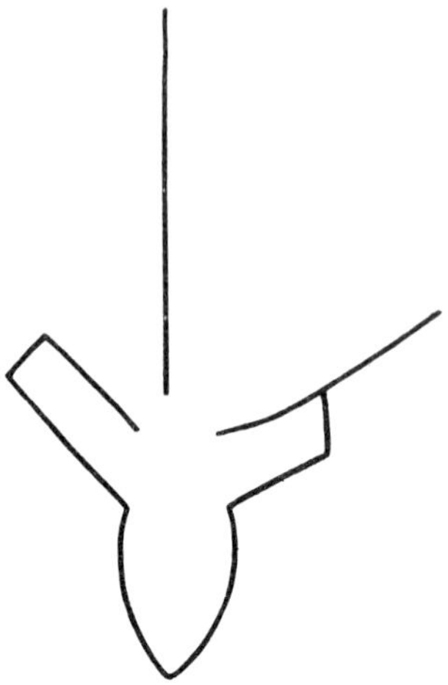

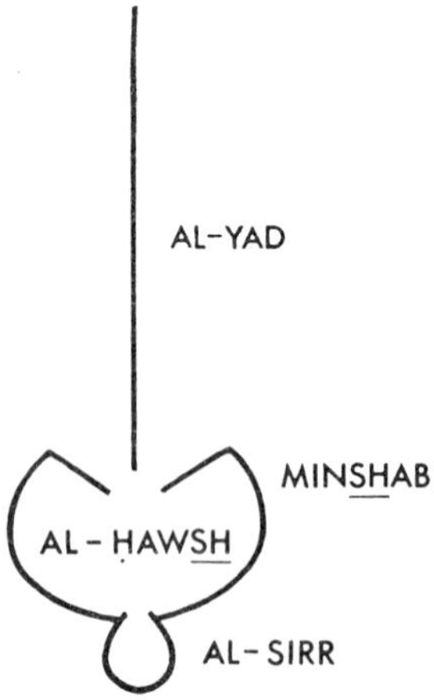

FIG. 6. Another pattern of *ḥaḍrah*.

FIG. 7. A Kuwait type of *ḥaḍrah*. The *minshab* is the enclosing palisade of *al-ḥawsh*.

This pattern is remarkably similar in appearance to the Malay *kelong* or off-shore palisade fish-traps at Singapore, though the Persian Gulf traps seem to be sited on tidal beaches only so far as I am aware. The Malay *penajur* corresponding to the Arabic *yad* ' runs at an angle to the current carefully determined by the *pawang* or *kelong* specialist '.[38] The chief difference between the two is that of materials, long poles being available in Malaya, so that the traps can be sited in much deeper water.

Deeds relating to the fish-trap Ibzūr at al-Ḥidd

'Abdullāh al-Shurūqī, in answer to my inquiry about property deeds relating to his *ḥaḍrah,* Ibzūr, brought me the documents to copy, in their little rusty iron cylinder in which they were kept for safety. The *ḥuḍūr*, he said, are *amlāk,* possessions, property. For Kuwait, Jalāl al-Ḥanafī[39] states, ' The fishermen owners of the *ḥuẓūr* have places off the coasts of al-Kuwait special to them, considered as their property (*milk*) in conformity with official deeds and documents. No person can put up a *ḥaẓrah* except on the coast which is owned by him '. Before accepting this statement as it stands I should like to see the documents of which he speaks, for I suspect the matter of ownership may be far less simple than he says.

The first of the deeds that follow is a plain declaration of ownership, the second a transfer from female heritors to a single male heritor, the third and most interesting, establishes the rights of the fish-trap owner to the unimpeded access of the fish to his trap. It would be possible to arrive at a code of customary law governing the operation of the traps by collecting such documents,

[38] *Straits Times Annual* (Singapore), 1966, 82–4, illus.

[39] op. cit., 97.

but in all probability there are persons in the seaboard community regarded as experts, retaining in their memory the cases which form the necessary precedents.

The original spellings and grammar have been retained in all three documents, but the vocalization has been added.

Document 1

بسم الله الرحمن الرحيم   بسم الله وحده وَلصَّلاة والسلام علا من لا نبىّ
بعده. وبعد اقول، وانا أُمْبارَكْ[40] بن سَيْف بن فَضالة، بان الحَضْرة المُسّماه
اِبْزُور[41] لعبد الله ولدى وعيال سيف ولدى، ولا عليهم تِتِبِّعَات[42] لاحد، وانا
مِسْتَارِثَتْها[43] أُمَّتِى من جدّى، وانا اِسْتَرَثْتُها من امى من سنة بن مذكور
سنة نَصُّور، شهد بذالك شيخ عثمان بن جامع، وشهد بذالك محمد بن عبد
الرحيم، وشهد بذالك عبد الحَىّ، وشهد بذالك على بن طَوْق، وشهد بذالك
أَرْحَمَه بن محمد بن فضالة، وشهد بذالك خَلَف[44] بن ثانى إن ما عليها
تِتْبِعَّات لاحد. ولله[45] خير شاهد، وَلّله على ما نقول وكيل، والفقير على
بن عبد الله
مطَّوع السَلَطَه شاهد
على منطوق[46] الورقه   Seal-impression
سنه ١٢٧٣

' In the name of God, the Compassionate, the Merciful. In the name of God alone, with blessings and peace upon him after whom there is no prophet.

Herewith I, being Umbārak b. Saif b. Faḍālah, declare that the fish-trap (*ḥaḍrah*) called Ibzūr belongs to 'Abdullāh my son, and to the children of my son Saif, they not being subject to any (rights) pertaining to any other person. My mother obtained it by inheritance from my grandfather, and I have inherited it from my mother, from the year of Bin Ma<u>dh</u>kūr, the year of Naṣṣūr.

[40] Umbārak as elsewhere in Arabia is a popular form of Mubārak.

[41] A derivation suggested for Ibzūr is from *zūr*, i.e. *quwwah*, as in *a<u>kh</u>a<u>dh</u>-ah bi 'l-zūr* ' he took it by force '. The name seems to me non-Arabic but I can see no ready derivation from Persian.

[42] Explained as Class. Ar. *tatabbu'āt*, i.e. *mā ḥawlah ḥaqq 'inda-hum*, and, *mā ḥad yitba'-hum bi 'l-ḥaqq aw mu'āraḍah*. One says, *Al-būlīs tatabba' al-qaḍīyah*.

[43] *Mistāri<u>th</u>ah*, i.e. *ḥaṣṣalat bi 'l-ir<u>th</u>*. Literally the text runs, ' and I, my mother inherited it '.

[44] Names like <u>Kh</u>alaf and <u>Kh</u>alfān mean 'Awaḍ (a name very common in south-west Arabia). *<u>Th</u>ānī* means Monday, and other names for days of the week used as proper names are *Liḥdān*, Sunday, *Rabī'ah*, Wednesday, *<u>Kh</u>amīs*, Thursday, *Jum'ah*, Friday, *Sabt*, Saturday. Month-names used as proper names are Ramaḍān, Rajab, <u>Sh</u>a'bān, 'Ā<u>sh</u>ūr, and Ṣafar, though of course the last-named month is unlucky. In al-Baḥrain they consider Wednesday unlucky (*yitashā'amūn bi-yawm al-Arbi'ā'*), but they say, *Al-Rubū' li 'l-dawā* ' Wednesday is medicine day '. On it they take such medicines as *sanā Makkī*, also known as *'i<u>sh</u>riq/'i<u>sh</u>rij*.

[45] For *Wallāh*, the *alif* omitted as in *wal-ṣalāh, supra*.

[46] *Manṭūq*, explained as ' contents '.

The shaikh ‘Uthmān b. Jāmi‘ bears testimony to this, as do Muḥammad b. ‘Abd al-Raḥīm, ‘Abd al-Ḥaiy, ‘Alī b. Ṭawq, Arḥamah b. Muḥammad b. Faḍālah, and Khalaf b. Thānī, namely that it is not subject to (rights) belonging to any other person.

God is the best witness, and God is testimony to what we say, and the *faqīr* ‘Alī b. ‘Abdullāh the *muṭawwa‘* of the Silaṭah is witness to what is stated in the document. Year 1273 (1856–7).’

Document 2

بسم الله الرحمن الرحيم

الباعث لتحرير هذه الورقة هو انه قد حضر الرجل المكرّم راشد بن عمرو الفَضاله ورَضِيعة[47] وَسبِيكة بنت عمرو الفضاله، وشهدا شهادة ثابتة لله تعالى، بأن أمهما مريم بنتُ مُبْارك ووخالتهما مَوْزه بنت سيف حضرا فى حال صحّتهما ورشادتهما، وباعا بيعاً صحيحاً شرعياً، بايجاب وقبول، حصّتهما من الحضرة الموروثة من أمهما المسمّات[47] فاطمة بنت مبارك، اعنى الحضرة الكاينة فى الحِدّ فى ارض[48] المسمّات بالبَطِين، اسمها إبْزُور، بثمن معلوم القدر — وَذالك خمسة تَوَامِين — على اخيهما مبارك بن سيف، والمشترى اشترى منهما، ومسلّم الثمن المذكور بيديهما بالتمام والكمال. فصارت الحضرة المحدودة المذكورة كلها ملكاً لمبارك بن سيف المذكور يتصرف فيها كسائر املاكه. جرا الاستشهاد الشرعى الثالث فى غرة ذى القعدة الحرام سنة ١٢٨٨.

نعم — صحيح راشد بن عمرو الفضاله

Seal-impression راشد بن عمرو بن سلطان الواثق بالرحمن

‘ In the name of God, the Compassionate, the Merciful.

The occasion for recording this document in writing is that the honourable gentleman, Rāshid b. ‘Amr al-Faḍālah appeared with (his) sister Sabīkah bint ‘Amr al-Faḍālah, and the two (of them) bore true witness to God Exalted, that their mother, Miryam bint Mubārak,[49] and their maternal aunt, Mawzah [50]

[47] My notes suggest that the correct reading here is *wa-raḍī‘atu-hu* ‘ and his sister ’. One says, *Fulānah dhī raḍī‘atī* ‘ So-and-so is my sister ’. Shaikh ‘Abd al-‘Azīz informs me that a foster-sister is *ikhtī min al-riḍā‘*.

[48] *Sic*, incorrectly, for *al-arḍ*. The *tā’* of *musammāt* is incorrect, here and previously, for *tā’ marbūṭah*.

[49] Though these documents were checked over with al-Shurūqī with care, I am inclined to think that one should read ‘ Miryam bint Saif ’. This would make the situation simple, but if it is correct to take the names as they stand, the only solution could be that Fāṭimah bint Mubārak married twice, firstly Saif by whom she had Mawzah, and later Mubārak by whom she had Miryam. This, however, seems to me a far less likely explanation than a simple error in transcription of the names.

[50] This lady is called ‘ Banana ’ because the banana is nice ! Perhaps it might also be a reference to the colour of her skin though I made no inquiries on this point. In other parts of Arabia I have come across Mawzah as a slave-name. In al-Baḥrain it was said that one calls a slave a nice name for *tabarruk*, but this is common practice in general.

bint Saif appeared, sound of health and adult, and sold by valid legal (*sh*ar'ī) sale, with assent and acceptance, their share of the fish-trap inherited from their mother, Fāṭimah bint Mubārak by name ; I mean the fish-trap at al-Ḥidd, on (the) ground known as al-Baṭīn,[51] which is called Ibzūr, for a price of a fixed amount, namely five *tūmāns*, to their brother Mubārak b. Saif. The purchaser has bought it from them, delivering the aforesaid price into their hands, entire and complete. The aforesaid demarcated fish-trap has become in its entirety property possessed by Mubārak b. Saif the afore-mentioned, to dispose of it as the rest of his properties.

The legal attestation (to the above) took place on the third (day) of the opening of Dhu 'l-Qa'dah al-Ḥarām of the year 1288 (14 January 1872).

Yes [52]—signature of Rāshid b. 'Amr al-Faḍālah
(seal).

Rāshid b. 'Amr b. Sulṭān, trusting in the Merciful.'

Document 3

بسم الله الرحمن الرحيم

ثَبَتَ لَدَيَّ ما ذُكِرَ فى هذه
الورقه وانا عيسى
بن على آل خَليفه

Seal-impression

مُوجِبه[53] انه لما ان عبد الله بن يلَيِّيل استحدث الحضرة التى فى بحر الحِدّ من القِبْله من المكرّم المعظّم شيخنا الشيخ عيسى بن على آل خليفه ، والحال[54] انها مُجَانَبَة[55] لحضرة مبارك بن فضالة ، وعلى مَجْرَى بحرها ، اشتكى مبارك المذكور ضررها لحضرته. فتوسّط بينهما المكرم السيد عبد الله بن ابراهيم بالصلح، وطلب من المذكور إعدامها. فاعطاه ذلك وسمحت نفسه بتركها على ان يسلّم له مبارك ما أنفقه عليها ، وتَبْقَى كما كانت ارضا قبل ذلك ، ولا يبنيها هو ولا غَيْرَه حتى لا يَخْفَى.[56]

حرّر فى ١٥ ربيع سنة ١٣٣٩
يشهد على ما حرر محمد
بن بيات
Seal-impression

صحيح عبد الله
يلييل
Thumb-print

جرى بحضرة
عبد اللطيف
بن محمود

يشهد على ما حرر
محمد بن محمود
Seal-impression

يشهد على هذه
خليفه بن حَبْتُور
Seal-impression

[51] Al-Baṭīn, cf. p. 491, n. 27, the sheltered side of al-Ḥidd.
[52] This seems to be a stereotyped formula in documents, as *Na'am wa-ana fulān b. fulān*.
[53] Explained as *alladhī awjaba 'l-kitābah*.
[54] Explained as *al-ḥaqīqah*.
[55] So pronounced by al-Shurūqī.
[56] The last sentence seems to be in *saj'*.

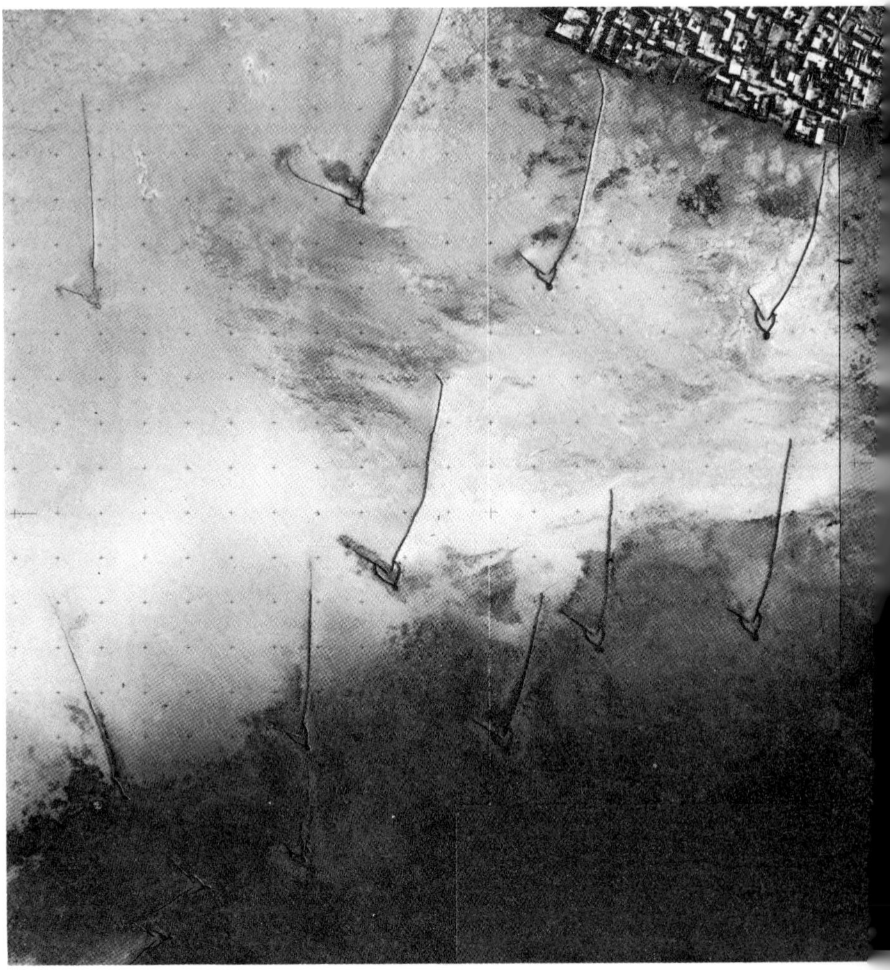

FISH-TRAPS AT AL-ḤIDD

Al-Ḥidd village the shore of which faces westwards, is set on a peninsula, a flat sand-spit a little above sea-level. In the *saṭwah* or sea-beach area pertaining to the village on the sheltered western side of al-Ḥidd, lie the traps extending from in front of the village westwards. The directions taken by the tide as it recedes can be perceived from the siting of the traps, i.e. it ebbs in a western, then a generally southern direction, but takes a sharp turn to the east and round the southern tip of the peninsula where a deeper channel seems to have been scored in the shallows round al-Ḥidd.

Al-Shurūqī's fish-trap, Ibzūr, is indicated by a white square. To the west of it lie the two other traps—al-Wusṭā 'the Middle (trap)', and al-Baḥrīyah 'the Sea (trap)'.

PLATE I

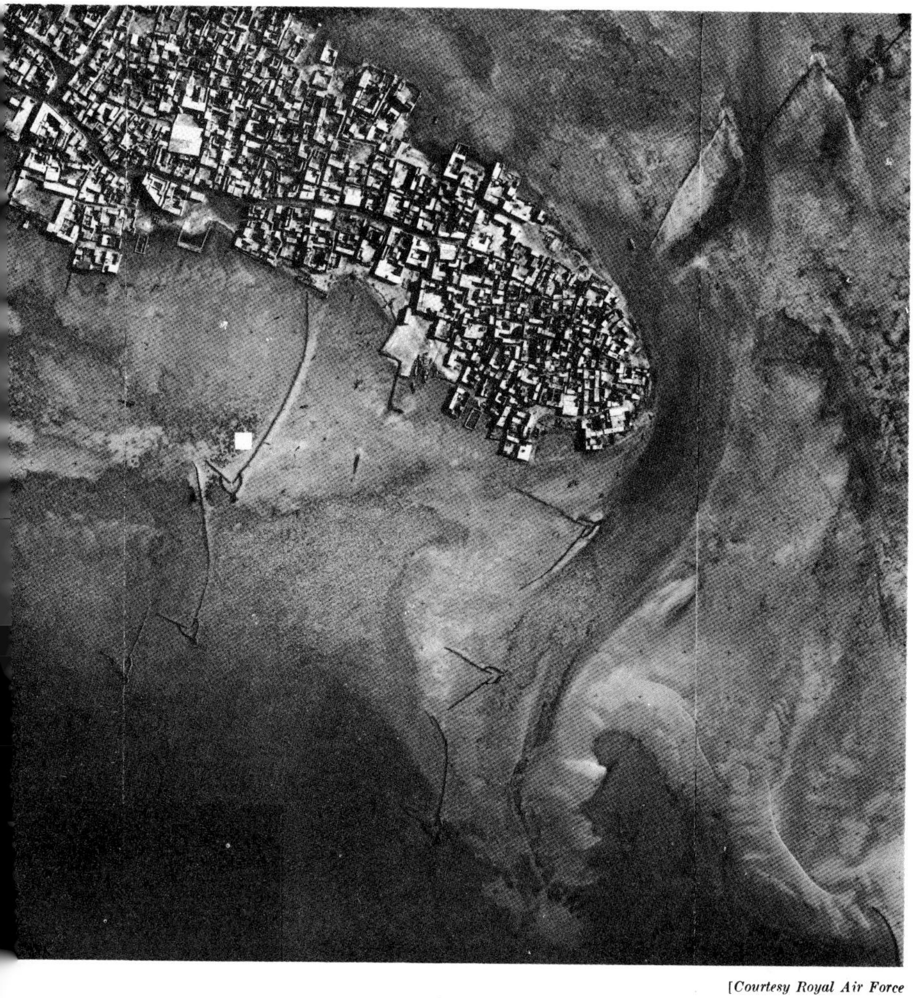

[*Courtesy Royal Air Force*

E
N ✛ S
W

جملة الدراهم التى سلمها مبارك
لعبد الله يلييل ست ماية ربيه وتسعين ربيه

‘ In the name of God, the Compassionate, the Merciful.

That to which reference is made in this document is confirmed by me, I being ‘Īsā (seal) b. ‘Alī Āl Khalīfah.

The reason for it is that when ‘Abdullāh b. Yilaiyīl introduced the fish-trap (sited) in the seaboard of al-Ḥidd to the west (side), (with the authority of ?) our Shaikh, the honourable and revered Shaikh ‘Īsā b. ‘Alī Āl Khalīfah[57]—the matter being that it lies alongside the fish-trap belonging to Mubārak b. Faḍālah, and in the course of the sea-tide (to and from) it—the said Mubārak preferred a plaint of the detriment it causes his own fish-trap.

The honourable Saiyid, ‘Abdullāh b. Ibrāhīm mediated between them to (arrange) a settlement, and required the aforesaid (‘Abdullāh b. Yilaiyīl) to remove it. So he conceded that to him, and assented to abandoning it on condition that Mubārak pay him what he had expended upon it. It (the fish-trap site) should then remain as it had been, (open) ground, before that, neither he nor anyone else to build (upon) it, so that (the decision) [58] should not be concealed (from people).

Recorded in writing on 15 Rabī‘ (I ?) 1339 (November–December 1920).

Muḥammad b. Biyāt bears testimony to what has been recorded.

Signature of ‘Abdullāh Yilaiyīl (thumb-print)

This was effected in the presence of ‘Abd al-Laṭīf b. Maḥmūd (seal).

Muḥammad b. Maḥmūd bears testimony to what has been recorded (seal).

Khalīfah b. Ḥabtūr bears testimony to this (seal).

The total sum of money which Mubārak paid ‘Abdullāh b. Yilaiyīl was 690 rupees.’

Commentary

Naṣṣūr, said al-Shurūqī, is the diminutive of Naṣr, and this Naṣr b. Madhkūr was the *ḥākim al-Baḥrain*, a sort of governor on behalf of the Shah (*wālī min qibal al-Shāh*). G. Rentz and W. E. Mulligan, in the *Encyclopaedia of Islam*, second edition, s.v. ‘ al-Baḥrayn ’, say of the Persians that their ‘ instruments of policy were often chiefs of the Huwala [59] or other Arabs settled on the

[57] Shaikh ‘Īsā b. ‘Alī retired in 1923, but lived till 1932 (Belgrave, op. cit., 10 and 76).

[58] The Arabic was explained to mean—so that the *kalām* ‘ decision ’ should not be concealed—but the phrase is obscure.

[59] Arabic هَولي, pl. هُوله (so ‘Abd al-‘Azīz), but in al-Ḥanafī, op. cit., 402, هُوَلَه. It is supposedly connected with the root *ḥ w l*, linked with the idea that they are immigrant Arabs. For these events cf., *inter alia*, Gholam-Reza Tadjbakhche, *La question des Îles Bahrein*, Paris, 1960, 43.

Persian coast such as Djabbāra of Ṭāhirī and Nāṣir and Naṣr Āl Madhkūr of Būshahr in the 12th/18th century'. Naṣr Āl Madhkūr was expelled from al-Baḥrain by the Āl Khalīfah in 1197/1783; the expression 'the year of Naṣṣūr' probably refers to that date. So probably the Faḍālah family came into possession of the fish-trap with the conquest of the island by the Āl Khalīfah, supposing always that they did not already hold it at that time.

The second document shows how the shares inherited under Islamic law, by the women in the family, were sold back to their brother.

The third document has to be related to plate I, of al-Ḥidd village on al-Muḥarraq island where, as marked, are sited three traps in succession, running from east to west, called respectively Ibzūr, al-Wusṭā 'the Middle (trap)', and al-Baḥrīyah 'the Sea (trap)'. From the position of the traps it can be seen how the tide runs. Coming up from the south, the tide turns eastwards then southwards to meet the long arm (*al-madd*) of the three traps probably, as it recedes. It will be noticed that the *madd* of the trap called al-Baḥrīyah is only about one-third the length of that of Ibzūr. The tide slipping off the shallow sea-bed at the southern extremity of al-Ḥidd is strained through fish-traps sited in a different direction from those to the north, and presumably rushes into the deeper channel scoured round the south of the peninsula, to the traps facing a direction diametrically opposite from Ibzūr.

Now when al-Shurūqī was about 10 years old, another *ḥaḍrah* was set facing the northern approach to his family's fish-trap. This gave rise to a quarrel between the two parties, and 'Abdullāh b. Ibrāhīm, of a local family of al-Muḥarraq Saiyids, made peace between them (*aṣlaḥ baina-hum*).

The builder of the new *ḥaḍrah*, 'Abdullah b. Yilaiyīl (Julaiyīl) was by origin an Omani who had become a *nawkhudhā*, and, having made some money, he wished to build a *ḥaḍrah*, presumably in order to invest it. About a third of the inhabitants of al-Ḥidd are of Omani descent, but 'Abdullāh b. Yilaiyīl had emigrated there from Oman, and was therefore, relatively a new-comer. Probably he purchased from the Āl Khalīfah Shuyūkh—in the document the ruler Shaikh 'Īsā b. 'Alī—a licence (*ijāzah*) to set up the *ḥaḍrah*, or the Shaikh could make a gift of land for this purpose. In the case at issue there was clearly detriment (*ḍarar*) to the very much older *ḥaḍrah* already there, so the matter was settled by mediation and the new agreement endorsed by Shaikh 'Īsā himself. The compensation paid to 'Abdullāh b. Yilaiyīl, about £52, seems quite a considerable sum for that time.

In document 3 the signatories include the *qāḍī* 'Abd al-Laṭīf Maḥmūd and Ibn Ḥabtūr, merely an ordinary man, a tribesman (*wāḥid 'ādī, qabīlī*)—I have no information on the other witnesses. The witnesses to document 1 are all Arabs, the Banī Ṭawq being a tribe or section of the Faḍālah. The religious element, however, is represented in the *muṭawwa'* (plur. *maṭāwi'*) [60] who is the

[60] Al-Ḥanafī, op. cit., 353 discusses the *muṭawwa'* who is a teacher of boys in a Qur'ān school also. *Muṭawwa'* is further the name of a kind of fish. Al-Ḥanafī refers to Muṭawwa' al-'Amā'ir, the latter a Bedouin tribe. In al-Baḥrain one says *limṭawwa'* (pl. *il-miṭāw'ih*) according to Shaikh 'Abd al-'Azīz.

I

imām al-masjid; some tribes even have a special *muṭawwa*ʻ to themselves. The resentment (though coupled with acceptance) of the professional men of religion which one finds everywhere is expressed in popular lore by the following phrase which must be a sneer at religious hypocrisy, ما ظَنّتِي وَلْدِ ٱلْهَوَا يِدْخَل النار إلّا لِمْطَوَّع طايِحٍ فى قَعَرْها 'I do not think the catamite [61] would enter Hell but the *muṭawwa*ʻ falls into its very depths'.

Al-Shurūqī said al-Silaṭah should properly be spelled with *ṣād*, not *sīn*, the singular being Ṣulaiṭī/Ṣilaiṭī. A Ṣilaiṭī, he said, is a person able to set bones who practises popular medicine. These people would be *mutaṣalliṭīn ʻala 'l-ʻūq* 'able to deal with the *ʻūq*'. He described the *ʻūq* [62] as a sort of illness preventing one from work. The Banī Murrah [63] are skilled in this, and amongst the Murrah one finds the Banī ʻArrāf, mostly in the south of Saʻūdī Arabia. The verse of ʻUrwah b. Ḥizām [64] in which he speaks of repairing to the ʻArrāf (skilled healer) of al-Yamāmah and the ʻArrāf of Ḥajr for the cure of a disease is celebrated. The Murrah used to neglect the observance of the *ʻiddah* in a woman divorced or widowed, and she would marry the next day. This not infrequently raised doubts as to a child's paternity—in such case the child would be taken to the Banī ʻArrāf who are able to distinguish who the true father would be.[65] Al-Shurūqī added that they and perhaps others also could tell to which tribe a man belongs by scrutiny of his features. I believe this would also be possible in south-west Arabia.

The owners of the *ḥaḍrah*, then, do not possess the ground lying outside the limits of the *ḥaḍrah*, and over which the fish must pass at high tide in order to enter the *ḥaḍrah*. They possess only the actual part of the sea-floor upon which the fish-trap itself stands. This muddy, sandy area extending up to the sea-shore is known as *saṭwah*, *saṭū* meaning sea-mud. The *ḥaḍrahs* are built upon the *saṭwat al-bilād*, i.e. the piece of land (*buqʻah*) belonging to the village. The owner of the fish-trap has the right to stop anyone using the *saṭwah* in any way which might cause detriment to the functioning of the *ḥaḍrah* as a trap for fish. The owner can even sell the land upon which his *ḥaḍrah* stands, but the parcel of land or rather sea-bed sold may only be used for siting a fish-trap. The *ḥaḍrah* owner cannot, for instance, fill the land upon which the *ḥaḍrah* stands with rubble and the like so that it can be utilized as a building site—which many owners nowadays would like to do. Nowadays anyway, this *saṭwah* land is considered as belonging to the Government (*Ḥukūmah*), and, of course, with

61 cf. *Banāt al-hawā* 'prostitutes' (*zāniyah*). *Wald al-hawā* is the corresponding term for the male.

62 This is not a word which figures in Lane's *Lexicon*, but the sense can be derived from a classical root.

63 The Murrah are also encountered north of Ḥaḍramawt, but see ʻUmar Riḍā Kaḥḥālah, *Muʻjam Qabā'il al-ʻArab*, Damascus, 1949.

64 Quoted by ʻAbd al-Salām Hārūn in al-Jāḥiẓ, *Ḥayawān*, I, 63. Cf. Masʻūdī, *Murūj al-dhahab*, Paris, 1861–77, III, 352. Ibn Saʻd, *Ṭabaqāt*, ed. F. Schwally, Leiden, 1905–9, I, I, 98, refers to the ʻ*arrāf* of Hudhail who was shown children.

65 This would be the function of the ancient Arabian *qā'if*.

the exploitation of the off-shore oil, the rulers of all Arab states in the Gulf naturally claim as much of the sea-bed as they can, but it might be that before the sea-bed had any value to persons other than the villagers, custom probably regarded the sea areas in their immediate vicinity as belonging to them ; this is to be investigated. It may be observed, however, that at Abu Dhabi the fishing grounds are rented by groups, and they pay an ' anticipated gift ' (*shufīyah/shifīyah*) to the Shaikh—*'ushūr*, the Shaikh resolving all their quarrels about the fishing grounds.[66]

There are, however, boundaries of protection to the fish-trap (*ḥudūd ḥimāyah li 'l-ḥaḍrah*), and as will be seen below one defines the actual area of the *ḥaḍrah* itself by such formulas as *ḥudūd ḥaḍrat-ah min al-barr ilā* ' the boundaries of his fish-trap are from the shore to . . . '. While the principle is that any new *ḥaḍrah* must not in any way cut off the movement of the fish to an already existing trap, it appears that there are no fixed distances round the existing *ḥaḍrah* regarded as appertaining to it, for all depends upon the location of the *ḥaḍrah*—which from plate I can be seen to be entirely dependent on the terrain and flow of the tide. This is quite different from the *ḥarīm al-bīr* which is reckoned by some legal authorities as a fixed area extending to as much as 500 *dhirā'* (about 250 yards) on all sides of a well.

On the west of al-Baḥrain the mouth of the *ḥaḍrah* always faces the south. I noticed from the air that some of the traps on this side seem to be right at the end of the shallow marine shelf around al-Baḥrain, next to deep water. In fact, they say, some *ḥaḍrahs* have to be approached by boat.

' Poaching fish in the area of a fish-trap is regarded as a serious offence ' according to James Belgrave.[67] This applies both to stealing the fish from inside the trap and preventing the fish from access to it. An offender is treated as an ordinary thief.

A deed of sale

In a deed of sale for a *ḥaḍrah* called Kuraikarah, sited on the flat bed of the sea area pertaining to al-Ḥūrah village of al-Manāmah island, its limits are defined as, ' Bounded on the north by the *ḥaẓrah* of al-Qirainīyah the property of Muḥammad b. 'Amr, on the east by the *ḥaẓrah* of 'Alī al-Dūkhī, on the south by the *ḥaẓrah* of Umm Sirwāl, and on the west by the *ḥaẓrah* of the (stone) breakwater of the beach '.

الحظره المسماة كُرَيْكَرَه الكائنة بسَطْوَة بحر الحُورَة من المَنَامه . . .
حدودها : تحدّها من الشمال حظرة القرَيْنيَّة ملك محمد بن عَمْرو ، ومن الشرق
حظرة على الدُوخي ، ومن الجنوب حظرة امّ سِرْوَال ، ومن الغرب حظرة قَيْد البرّ .

[66] These, of course, are distinct from the *saṭwah* of a village. In al-Baḥrain a fishing-ground is *maḥdaq*, pronounced *mḥadaq* (pl. *maḥādiq*), *ḥadāq* being the act of fishing. Fishing-grounds are called *aqwā'* (sing. *qū'*) if they are coral reefs or stony rocky areas (*'irshān* (pl. *'arīsh*)) ; al-Quṭāmī speaks of *qū'ah*, the hard rock sea-floor. T. M. Johnstone has also given me *mismacha/mismakah* ' a fishing-ground for *simach* '.

[67] James Belgrave, *Welcome to Bahrain*, Stourbridge, 1957, 67.

I

The topographic terms are instructive. *Kuraikarah* is a *hair* (pl. *āt*)[68] *fī 'l-baḥr*, explained to me as rocky ground under the sea, a pearl-bed. The *qaid*, in this case running parallel to the shore (fig. 8), I have rendered as

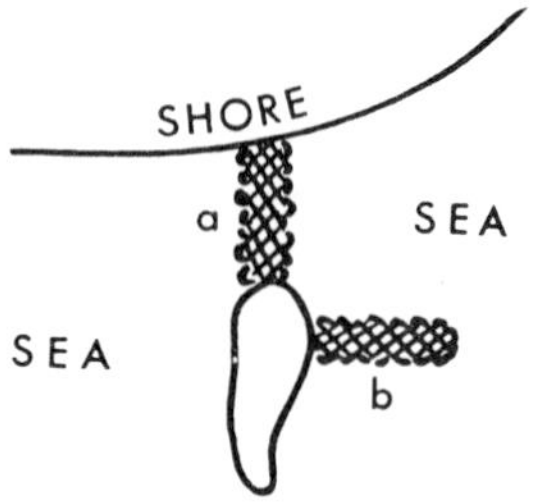

Fig. 8. Stone *qaid* running parallel to the shore.

Fig. 9. a. *Qaid al-barr* to or towards the land.
b. *Qaid al-baḥr* running towards the deep sea (*al-ghazir*/*ghizir*).

'breakwater'—it is a term used in Ḥaḍramawt also in irrigation systems, perhaps for a dike to prevent flood-water eroding an earth bank. *Qaid* was explained as *nabāwah*,[69] which latter is a *rabwah basīṭah min aḥjār ṣighār*, a simple eminence (in the sea) of small stones—a *ḥidd* is the corresponding thing of sand. Shaikh 'Abd al-'Azīz draws a distinction between *qaid al-barr* (fig. 9) which runs from a piece of land towards another piece of land, whereas *qaid al-baḥr*/*qīd il-baḥar* (fig. 9), stretches out from the land in the direction of the deep sea (*al-ghazir*). *Qaid al-qaṣṣār* is an arm or breakwater, extending from the shore to a rock in the sea. A *bulṭ*/*bilṭ* (pl. *bulūṭ*) is an artificially constructed *qaid* or arm of stones, either to extend the existing land by creating an obstacle which will gather sand (fig. 10), or to break the force of the waves beating on the shore and prevent the sea from damaging the land (fig. 11). In the former

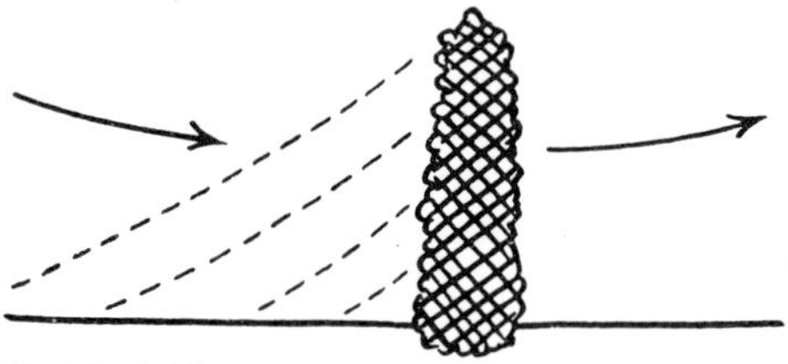

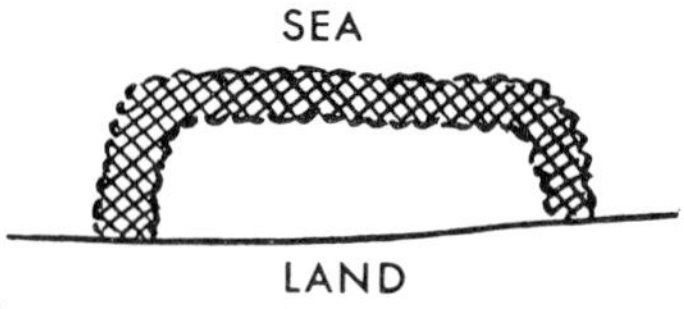

Fig. 10. A *bilṭ* constructed to 'grow land' on to the shore. The *bilṭ* is formed of layers of stones. At low tide the waves move towards the right (as indicated by the arrows), the water passing through the *bilṭ* and laying down sea-sand and debris of all sorts on the left side of the *bilṭ*.

Fig. 11. A type of *bilṭ* constructed to break the force of the waves, and prevent the sea from damaging the land.

[68] This is a well-known word. Its origin is discussed in a rather superficial note by al-Ḥanafī, op. cit., 403. Rāshid b. Fāḍil Āl Bin 'Alī, *Majārī 'l-hidāyah*, Manāmah, 1341/1922–3, 1, calls it *makān maghāṣ al-lu'lu'*, a pearl-bed.

[69] A *najwah* (pl. *najawāt*) is a *nabāwah fī 'l-baḥr*, an eminence in the sea with shells below the surface, but it projects when the water is shallow. Lorimer, op. cit., 2220, calls the *najwah* a submarine mound surrounded by deep water. He says the *hair* is often known from of old, but the *najwah* is a recent discovery and still bears the name of the founder. I have heard rather similar statements.

case the *bulṭ* will run down the shore to the sea, and sand and other matter will be piled up by the sea on one side of it. The *bulṭ* is built of layers of stones by the owner of the land.

Kuraikarah was sold to ʿAbdullāh b. K͟hamīs al-S͟hurūqī, the vendors being no less than five persons, among whom the property was shared as follows :

2 brothers each holding 14 shares	28 shares
2 sisters each holding 7 shares	14 shares
A certain S͟harīfah bint Ibrāhīm	6 shares
Total	48 shares

The purchase price was 350 rupees, about £26. The transaction took place in 1362/1943, and the deed bears the seal of al-S͟haik͟h Salmān b. Ḥamad Āl K͟halīfah, *Ḥākim al-Baḥrain,* and signatures of the *Mudīr* (presumably of the Land Department, *Dā'irat al-Ṭābū* (Turkish *ṭāpū* ' title-deed, etc.') from which this document was issued), and Sir Charles Belgrave. The *Ṭābū* published the names of the vendors as owners, then after the expiry of an appropriate period it published this deed which abrogates all previous deeds. Witnesses' signatures include the vendor who is the *wakīl,* ʿAbd al-Raḥmān b. ʿAbdullāh b. ʿAlī Sālim of al-Manāmah, ʿAbd al-ʿAzīz b. Mahdī al-Bannā', a builder of al-Manāmah, and Aḥmad b. Ḥāsim Taitūn, a S͟hīʿī also of al-Manāmah.

The waqf of a fish-trap

A man may bequeath part of a *ḥaḍrah* to the *awqāf* the income of which will go to pay Qur'ān reciters to recite the Qur'ān in Ramaḍān—my informant on this point was a Sunnī. The other part of the *ḥaḍrah* will go to his children among whom it is distributed in shares, doubtless according to the Islamic law of inheritance. The property is then designated as *mawqūf ʿala 'l-dars wa 'l-s͟hurakā'* ' put in trust for recital of the Qur'ān and the partners '.

The S͟hīʿah possess a considerable number of *ḥaḍrahs* which have been bequeathed to the *awqāf,* in all about 200. About 10% of the S͟hīʿah *ḥaḍrahs* held as *awqāf* are owned in partnership, being owned in part by the S͟hīʿah, in part by the heirs of the original owner known in modern *waqf* documents, as stated, by the term *s͟hurakā'*. The Baḥārnah of al-Baḥrain even have *waqfs* of *ḥaḍrahs* in Qaṭīf in Saʿūdī Arabia, a district which once formed part of al-Baḥrain in its former wider geographical sense including territories of the mainland, and there are still S͟hīʿah there. Belgrave [70] reports that certain *Ma'tams* (literally ' places of assembly for mourning or rejoicing '), S͟hīʿah religious institutions, are wealthy enough to hire reciters from Iraq, Persia, and Qaṭīf, paying them £400 or £500 for the 10 days of Muḥarram. This money is obtained, he implies, from the income of the *awqāf.*

Nowadays *waqf* documents are printed forms replete with alternative

[70] op. cit., 191.

clauses to suit various contingencies,[71] those inapplicable to the case in point to be deleted, but they do reproduce certain colloquial terms and phrases that were probably standard form in the MS documents preceding the printed contract.

The renting of a *ḥaḏrah* is put up to tender by the Shī'ah Waqf Department whose office is at Bāb al-Baḥrain, and it is announced over the Baḥrain radio that such and such *ḥaḏrahs* are open to tender. In former times this used to be made known by the *wazīr* in the villages, he being a sort of village headman, but nowadays it is done by the *mutawallī* in charge of the *waqf*. The latter, however, is not to be paid the *ḍamān*, i.e. the purchase-money paid by the farmer of the *ḥaḏrah*, but it must go direct to the office of the *awqāf* only. It may be inferred that this measure was taken to eliminate abuses and corruption.[72] My informants added that, previously, the announcement that a *ḥaḏrah* was available for letting or farming was probably made in the mosque on a Friday, but the *wazīr* also had a *majlis* at which the matter might be made known, and presumably the contract transacted. If a *ḥaḏrah* has the misfortune to remain unrented for a whole year one says that it *ḥālat*, i.e. has completed a year (*ḥawl*), as Shaikh 'Abd al-'Azīz expresses it, ' without luck '.

The action of renting or farming the *ḥaḏrah* is termed *khufar wa ḍamān*. The renter or farmer (*ḍāmin*) *yuḍammin al-ḥaḏrah*, i.e. hires it as a piece of land from the owner (*mālik*) or *waqf* authority in consideration of a stipulated sum. The *ḍāmin* normally constructs the trap, very rarely does the *mālik* do so. Dr. T. M. Johnstone describes the *ḍamān* as a contract between the owner (*mālik/mālish*) and the *baiyār* or fisherman.

A *ḥaḏrah* contract which I examined at the Shī'ah Waqf Office contained the statement *'aqd iyjār al-ḥaḏrah fī saṭwat al-Ḥidd al-musammāh al-Bida'*, *min quffāl al-ghawṣ sanat 1383 ilā quffāl al-ghawṣ sanat 1384* ' the contract for renting the fish-trap in the sea-bed (adjacent and belonging to) al-Ḥidd, known as al-Bida' from the close of the pearl-diving of the year 1383 (8 October 1963) to the close of the pearling season in the year 1384 (October 1964) '.[73] There were two witnesses and signatures to this quite modern agreement. It may be remarked that the renter-farmer was a tribesman (*qabīlī*), *ergo* a Sunnī.

The standard contract, both in the case of palms (*nakhīl*) and fish-traps (*ḥuḏūr*) commences from *awwal al-mūsim*, and goes on to define the period of *iyjār aw ḍamān* ' renting or farming '. It stipulates that a renter-farmer (*ḍāmin*) who hires the fish-trap from the Waqf Department cannot ask for a *simḥān*— defined to me as a *tanzīl qism min al-ḍamān*, rebate of part of the rent—because of the small quantity of fish taken by the trap.

[71] The contract issued by the *Dā'irat al-Awqāf al-Ja'farīyah* is entitled *'Aqd iyjār—ḍamān*, and the standard form I have relates to agricultural properties, most of the regulations being of a Western pattern.

[72] cf. Belgrave, op. cit., 56.

[73] These times are defined each year in a document published by the Baḥrain Courts with the title *I'lān, mūsim al-ghawṣ li-sanat . . .*, of which I have a copy for 1383/1963–4.

The rent of the fish-trap (*khufar al-ḥaḍrah*) is paid in Baḥrainī dīnārs in three instalments, at the beginning of the season, at mid-season, and at the close of the season. While this is customary for the most part, instalments need not always follow this pattern exactly if the agreement be made differently. The owner (*mālik*) may also receive *idām* of fish in season. A *mukhaffir* is the person who hires or rents, and one says, *Khaffart min ʿinda-k* ' I hired from you ' and *Khaffaraw ḥaḍarāt-hum ʿalā fulān* ' They hired their fish-traps to So-and-so '.

The renter-farmer farms the fish-trap and supplies the fisherman (*baḥḥār*) with the wherewithal to obtain the palm-*jarīd* with which it is made (*yukhaffir-hā wa-yuʿṭī-h li 'l-jarīd*), taking a share of the profit (*murābaḥah*) [74] from him, i.e. 1%, or 2%, a tenth, or two-tenths, according to what is specified in their agreement. The fisherman brings the fish to the *ḍāmin* and sells it to him—apparently this is in addition to the *murābaḥah* which the renter-farmer receives from him. Should the fisherman fall ill and be unable to pay his debts to the *ḍāmin*, he pays in instalments (*yufaṣṣil* [75] *al-fulūs*) over a number of years, just as is the case with the pearling crew who have received advances from the *nawkhudhā* which they cannot repay.

I have noted that the *ḥaḍrah* is sometimes financed by the *jazzāf*/*yazzāf*, acting like the *ṭawwāsh* who advances money, etc., to the pearling crew. The *jazzāf* is a vendor of fish—if prosperous he may be considered a fish-merchant, but notwithstanding his profession is not one of which he or anyone in al-Baḥrain would be proud. The *jazzāf* will finance the *ḍāmin* who has no money, to enable him to hire and construct the *ḥaḍrah*. The *ḍāmin*, who in this case is also the *bawwār*/*baiyār* [76] undertakes to deliver all the catch to the *jazzāf*, and indeed he may only leave apart for the consumption of himself and his family the fish that fetches the lowest price in the market. This process continues until the debt is paid back, when the *ḍāmin* is said to have *istafḍal*, i.e. become able to keep all the catch for himself. Nevertheless the *ḍāmin*, even then, to keep on good terms with the *jazzāf* will often continue to sell his fish to him at a lower price. The debt acknowledged by the *ḍāmin* to the *jazzāf* has added to it a certain percentage which is in effect a species of interest. It is also the custom for the *jazzāf* to supply the *bawwār* with certain provisions during the period of their relationship.

The *bawwār*/*baiyār* is, properly, the man who *ībārī 'l-ḥaḍrah* (from the verb *barā-hā*), i.e. goes at low tide to the fish-trap (*lammā tathbur al-māyah*, or, *līn tathbur al-māyah*) in order to collect the fish in it. His catch on such an occasion is known as *bārah*. The *bawwār* could be a tribesman (*qabīlī*), or indeed of any class, but he would only be a very poor person. He could be the owner of the *ḥaḍrah*, or the *ḍāmin*, or he might be a man hired by either of them.

[74] *Murābaḥah* is well defined in Lane, *Lexicon*.

[75] For the pearlers a *faṣl* is an instalment paid after each *quffāl* (close of season) to discharge a debt or loan.

[76] cf. T. M. Johnstone, *Eastern Arabian dialect studies*, London, 1967, 101. He has suggested that this word might be derived from the Mahri root *b ' r* ' to fish '.

For his pains the *bawwār* would receive remuneration in one of the following forms :

(*a*) part of the catch (*bārah*), up to the maximum of half of it (*nuṣṣ il-bārah*) ;
(*b*) a daily wage ;
(*c*) a daily wage plus *iydām* (class. *idām*), the latter usually being the fish fetching the lowest prices in the market.

It may be remarked in passing that in the morning people not unnaturally hesitate to purchase the *bārat al-lail* or night catch which would not be so fresh as the morning catch (*al-bārat al-ṣibḥīyah*).

Dr. Johnstone has noted that the *baiyār* is a (prawn) fisherman who takes two-thirds of the product of the fishing, out of which he pays expenses, a third for himself, and a third for the canoe (*hūrī*). If he has three fishermen (*baḥāḥīr*) he takes two-thirds of the catch, the other third going to the owner of the fish-trap, and he will have to pay them from his own two-thirds.

Yet another term in use is *marshī*,[77] defined by Dr. Johnstone as a fisherman who does not receive a share in prawn-fishing, but a fixed annual wage, in 1959 about 1,000 rupees. Shaikh ʻAbd al-ʻAzīz adds that the *marshī* is the *bawwār* ' persuaded ' to fish for *ribyān* ' prawns '. He receives (1966) an annual wage of up to Baḥrain *dīnārs* 150, apparently twice what he received a few years ago, and he gets also his family's daily *iydām* of fish.

One can take fish from a fisherman operating a *ḥaḍrah* (though in view of the statement above, perhaps not when he is contracted to sell to a *ḍāmin*). This is fresh fish, known as *ruṭbah*, and he would say, *Kull yawm ana umashshī lak ruṭbah raṭlain, thalāthah* ' Every day I shall bring you fresh fish, two or three pounds '. The agreement would be for a fixed price to be paid per pound of fish, and these *ruṭab* (pl.), *zain shain al-thaman wāḥid* ' good or bad the price is the same '.

All kinds of fish find their way into the *ḥaḍrah*, though naturally the catch differs according to the season of the year. I. MacIvor [78] listed the commoner fishes caught in it, the correct Arabic of which Shaikh ʻAbd al-ʻAzīz has supplied for al-Baḥrain, these being—prawns (*ribyān/rubyān*), crabs (*qubqub*), sardines (*ʻūm*), the *jwāf/yiwāf*,[79] the *ṣāfī*,[80] the *firyāl*, the *qurqufān*, the *waḥar*,[80] the *sibaiṭī* [81] (or *burṭām/birṭām*), the *sūlī*.[82] The *mutābilu* and *malasānee* which MacIvor gives do not appear to be identifiable.

[77] Shaikh ʻAbd al-ʻAzīz states that *marshī* is *ism mafʻūl* from *rāshā* (v.n. *mrāshāh*) ' to persuade ', used in al-Baḥrain specially for children.

[78] *Administration Report of the Persian Gulf Political Residency and Muscat Political Agency* for 1880–1, 56. Cf. the list of fish in Lorimer, op. cit., 2308 f., ' Fisheries of the Persian Gulf '.

[79] Al-Jāḥiẓ, *al-Bukhalā'*, 102, etc. Cf. al-Damīrī, *Ḥayāt al-ḥayawān*, trans. Jayakar, London, 1906– , 501, *juwāf*; Ibn al-Athīr, *al-Nihāyah*, Cairo, 1311/1893–4, I, 219, *juwāfah*, a poor-quality fish.

[80] These fish-names are also to be found in al-Ḥanafī, op. cit.

[81] Al-Ḥanafī, op. cit., *sibaiṭī*; Johnstone, op. cit., 204, *ṣbeeṭī*; Shaikh ʻAbd al-ʻAzīz, *sbīṭī*. It is described as ' a large porgy '.

[82] For MacIvor's ' ghulee '.

When sea-fishing as opposed to trapping, it is only on the rocky sea-floor (*qūʿ*) that one is likely to catch the best fish such as *shiʿrī*, *sūlī*, *hāmūr*, *burṭām*, or *sibaiṭī*, etc. The less-favoured fish such as *lukhum* (ray or roach) and *waḥar*, are caught where the sea-floor is sandy. Where there is mud one takes shark or cat-fish, and where *mashʿūrah* seaweed grows only very small fish like *yimyām* and *yimā*, or very poor quality fish like *lazzāq*.

The plummet is used to discover what type of sea-floor lies below the surface ; it has a coating of fat to bring up sample particles from the sea-bottom. The sea-bottom may be rocky (*qūʿ*), sandy (*biyāḍ*), muddy (*ṭīnah*), or seaweed of the *mashʿūrah* type. From the rocky sea-floor the plummet (*bild*) would bring up red colour (*ḥamar*) or red sand, and one would say, *Hādhā ḥamar*, or *Hādhā ḥamar al-qūʿ*, *hādhā qūʿ*. *Dachch* is sandy bottom with no coral, etc., no use for fishing. One asks the question, *Niqal al-bild* ' Has the plummet brought up anything ? '. To this the answer would be, *Niqal* or *Yanqul*.

According to al-Shurūqī, if the weather is hot the fish do not come very much to the fish-traps.

It may be worth recording that in January 1964 there were what I believe to have been exceptional storms accompanied by severe cold which brought quantities of small fish on to the beaches and even fish of saleable size. Numbers of sea-snakes had been cast up, and about the R.A.F. station on al-Muḥarraq there were two or three turtles on the stretch of beach which were cast ashore in a stupefied condition. The waves had also thrown up much seaweed. The Baḥrainī Arabs who were out collecting fish from the shore or floating dead [83] in the water said they had been killed by the cold. They seemed to have no prejudice against eating fish killed by the storm or cold.

The farming of Ḥaḍrat Bint Kamāl

The information and statistics given here on the fish-trap Ḥaḍrat Bint Kamāl were kindly made available to me by Mr. ʿAlī ʿAbdullāh al-Salmān of the Shīʿah Waqf Department (*Dāʾirat al-Awqāf al-Jaʿfarīyah*) to whose assistance I am much indebted.[84] This fish-trap has been established *waqf* property for about 40 years, having been bequeathed in part in the year 1346/1927–8.

ʿAbdullāh al-Shurūqī stated that each fish-trap has its own recognized name (*kull ḥaḍrah lahā ism maʿlūm*) and the rent of a trap would cost, say, 500, 1,000, 2,000 rupees or more. The Bint Kamāl fish-trap brings in a good income because it is a shrimp-trap, and shrimps fetch a high price. *Al-ḥaḍrah li-ṣaid rubyān akthar ijārah* ' The trap for catching shrimps costs more to hire '.

[83] The Zaidī sect of the Shīʿah does not permit that fish found floating on the sea be sold.

[84] Belgrave, op. cit., 56 : ' It was at this time (1928) that the Government took an important step by placing the administration of Shia Waqf property, which had been dealt with by the Kadhis, in the hands of an elected council of town and village Shias. This was a popular move as in the past much of the proceeds from the property, gardens, houses, fish-traps and shops, which had been bequeathed by Shias in the past for religious purposes, had not been spent on the objects for which it was dedicated '.

There are traps specially for shrimps in some parts of al-Baḥrain, notably at Sitrah. They cost more to rent than others, perhaps as much as £1,000 for a single trap for one season. Lorimer [85] says that prawns (*ribyān*) are caught in March–April and October–November with fine triangular hand-nets. I can recall, however, eating them in December and January, and Shaikh ʻAbd al-ʻAzīz confirms that they are fished all the year round, though the winter prawns are the best and this season is often more abundant.

The following statistics will provide some indication of the annual value of a *ḥaḍrah*.[86] The rent (*ḍamān*), as will be seen, fluctuates between a little over £500 sterling to £1,000. The Shīʻah authorities averred that the tenderer would make at least twice the amount of the *ḍamān*. The *ḍamān* of the Ḥaḍrat Bint Kamāl:

Year			Rupees
1370	commencing	13 October 1950	14,000
1371		2 October 1951	12,300
1372		21 September 1952	11,650
1373		10 September 1953	12,000
1374		30 August 1954	14,550
1375		20 August 1955	10,500
1376		8 August 1956	11,600
1377		29 July 1957	No figure [87]
1378		18 July 1958	10,000
1379		7 July 1959	Income to partners
1380		25 June 1960	7,500
1381		14 June 1961	Income to partners
1382		4 June 1962	8,000

There must without question be some strong reason for the fall in the annual value of the fish-trap to the Shīʻah trustees from a maximum figure of 14,000 to 4,000 rupees. This merits careful investigation. An informant blamed the fishing by foreign companies for prawns in the Persian Gulf, which he said had considerably affected the number of prawns that come to the shores of al-Baḥrain. The quantity had so much diminished, he added, that whereas as much as £3,000 would formerly be paid for the hire of a *ḥaḍrah*, the rent was now very much less. The latter statement is borne out by the statistics quoted above, but no positive proof as to the reason for the fall in the quantity of prawns at al-Baḥrain seems to be forthcoming yet. In 1966 Shaikh ʻAbd al-ʻAzīz stated that the *ḍamān* could lie anywhere between Baḥrain dīnārs 50–200 per annum.

[85] cf. *Gazetteer*, Appendix E.

[86] Although the Muslim year is used, this in itself would make no appreciable difference to the circumstances.

[87] There is no figure for 1377, commencing 29 July 1957, because, instead of dividing the annual income, the partners (*al-shurakā'*) took the entire income for that year, but received nothing in the year following.

Taxation

There are no taxes on fish in al-Baḥrain, but Baḥrainīs informed me that in Sa'ūdī Arabia, always referred to as ' al-Mamlakah ', a fifth (*khums*) is taken on fish. The tribes of the Aden littoral in many places used also to take a fifth on products of the sea including fish, and there are still various ways in which taxes are collected on the fishing industry in the coastal sultanates.

Appendix

The star-calendar as used in al-Baḥrain

Like all maritime and agricultural communities of Arabia, the fisher-folk of the Persian Gulf reckon the seasons in accordance with a calendar consisting of 28 stars—just as they do along the southern coast of the peninsula, though, as indeed I have already remarked elsewhere,[88] the actual star-names employed by the peoples of the Persian Gulf and Gulf of Aden would apply to different periods of the year. So in order to understand what a fisherman means when he states that a certain species of fish shoals along the shore at a certain star, or some wind or current is to be expected, it is essential to be acquainted with the star-calendar he uses. Thus one can establish with considerable precision when, for instance, prawns are likely to be plentiful in the fish-traps. While in al-Baḥrain I did not construct an almanac which would supply such data, but with recourse to the star-calendar *infra* it should be a fairly simple matter to do so by interrogating fishermen.

This star-calendar is derived from the Baḥrainī almanac for the year 1382/1962–3, with additions from the Qaṭarī almanac [89] for the following year. These are the names of the stars for the dawn ascension (*ṭāli' al-fajr*), and some of the winds recorded by Dickson [90] are associated with them, or with the seasons of the year.

A mode of reckoning entirely new to me, and requiring investigation, is the calculation of the days of the year commencing from Suhail. In the 1962 Baḥrainī almanac the first day of Suhail falls on 25 August, coinciding with the zodiacal sign Sunbulah, whereas in the 1963 Qaṭarī almanac it falls on 18 August, coinciding with the 25th of the zodiacal sign al-Asad.[91] This day is Firāq Suhail. The preceding day in each case is the 365th day of the old year, but a leap year should of course make it the 366th day. This system of reckoning

[88] cf. my ' Star-calendars and an almanac from south-west Arabia ', *Anthropos*, XLIX, 3–4, 1954, 436. The Shibām star-calendar commences in January with al-Han'ah.

[89] Al-Saiyid 'Abd al-Raḥmān al-Hāshimī, *al-Taqwīm al-Baḥrainī li-'ām 1382 hijrīyah* (sic) *1962–1963 m.*, al-Ḥūrah, al-Baḥrain, n.d.; 'Abdullāh b. Ibrāhīm al-Anṣārī, *al-Taqwīm al-Qaṭarī bi 'l-tawqīt al-'Arabī wa 'l-zawālī li-'ām 1383 hijrī, al-muwāfiq 1963–1964 mīlādī*, al-Dawḥah, Qaṭar, n.d.

[90] op. cit., 249. Dickson does not seem to have understood the system very exactly, and his valuable data are somewhat confused.

[91] In al-Baḥrain I was told that the Maṭla' Suhail (rising of Suhail) falls on 26 August, and that it rises with al-Ṭarf, i.e. Ṭarf al-Asad, which information corresponds with the table *infra*. The Baḥrainī almanac says that after Suhail, from 28 August, the heat begins to diminish a little.

appears to me to correspond closely to the numbering of the days of the year from *Nairūz* which opens the *Azyab* monsoon in southern Arabia.[92] An informant in al-Mukallā told me recently that nowadays *Nairūz* there falls upon 6 August, equivalent to the 11th of the star Bala‘.

The Qaṭarī almanac divides the year into six seasons of two months each :[93]

Wasmī,	commencing 1st of al-‘Awwā
Shitā’,	9th (approximately) of al-Iklīl
Rabī‘,	5th (approximately) of al-Dhābiḥ
Ṣaif,	1st of al-Mu’akhkhar
Ḥamīm,	9th (approximately) of al-Thuraiyā
Kharīf,	5th (approximately) of al-Nathrah

January	3	الشوله	July	4	الهقعه = الجوزة [94] الأولى
	16	النعايم		17	الهنعه = الجوزاء الثانيه
	29	البلده		30	الذراع = المرزم
February	11	الذابح	August	12	النثره = الكُلَيْبَيْن
	24	بلع		25	الطَرْف = سهيل
March	9	سعد السعود	September	7	الجبهه = الخامسة
	22	الأخبيه		19	الزبرة = السادسة
April	4	المقدّم	October	3	الصرفه = السابعة
	17	المؤخر		16	العوّى = الوكيذب
	30	الرشا		29	السماك = الوحيمر
May	13	الشرطين	November	11	الغفر
	26	البطين		24	الزبانا
June	8	الثريا	December	7	الاكليل
	21	الدبران		20	القلب

The Qaṭarī almanac states that sea-folk (*ahl al-baḥr*) call al-Dhirā‘, Mirzam,[95] and al-Haq‘ah, al-Jawzā’ al-’Ūlā, and al-‘Awwā is called Thuraiyā al-Wasm

[92] cf. my *Portuguese*, 174. At Khawr Fakkān I came across a type of calculation used by fishermen which I had not time to examine properly. They say, *Al-darr dashsh fi ’l-Asad* ‘ The *darr* enters in al-Asad ’. This last I assume is the zodiacal sign (*burj*), according to the Qaṭarī almanac, 25 July to 24 August. Suhail in Qaṭar falling on 18 August within that period, I surmise that the *darr* calculation commences from Suhail. I was told that the sardine arrives *fī thalāthīn min al-darr*. Mr J. Wilkinson of Shell has confirmed to me that in Buraimi the *darr* commences from Suhail, adding that the *darr* system which he has studied, is not to be found in Oman proper, but it is used in the Ẓāhirah. My investigations in Socotra seemed to indicate that it was not known there, but I found the Shaḥrā of Ḥallānīyah island of the Kuria Muria group which I visited through the kindness of the High Commissioner, Sir Richard Turnbull, in 1966 use a *darr* system, 12 *darrs* to the year, each of about 30 days. Mr G. Tibbetts has pointed out to me that Ibn Mājid (G. Ferrand, *Instructions nautiques*, Paris, 1921–8, I, 143b–145b) has composed a poem on the *darr* system. It appears to consist of 10-day periods commencing on the 150th day of Nairūz in the zodiacal sign al-Thawr, i.e. about September. This would appear not to be the Nairūz discussed in my *Portuguese*, but further investigation would be lengthy.

[93] Al-Baḥrain has only four seasons—Ṣaif, 24 June ; Kharīf, 25 September ; Shitā’, 23 December ; Rabī‘, 22 March.

[94] This incorrect spelling is in the original.

[95] Al-Dhirā‘ is also called Dhirā‘ al-Asad, falling in the period al-Asad. As I did not register local pronunciation for these names, vocalization is according to the lexicons.

by the cultivators. Both almanacs give some indications of rains and prevailing winds. The Qaṭarī almanac states that al-Qalb always has 14 days, and in leap years an extra day is added to Sa'd al-Su'ūd.[96]

[96] Since this article was written, Dr. T. M. Johnstone's *Eastern Arabian dialect studies* has appeared, containing in easy systematic transliteration, some of the vocabulary of this study. I have not aimed at consistency between the standard and colloquial forms of Arabic, but have registered it as spoken to me, by persons educated and uneducated. Differences, as one expects, will be apparent in our vowelling according as an individual speaker uses standard or dialectal Arabic, even when a word is not so classical as to figure in the lexicons. If we differ occasionally over consonants—Arab writers themselves often differ in this respect also. While I accept that persons speaking dialectally might employ other vowels than I have recorded I have retained the spellings as I heard them in 1963. I am much indebted to Dr. Johnstone for looking through my article and commenting upon it.

ADDENDA

p. 488, l. 22: the date of Ibn al-Mujawir's composition of the *Tārīkh al-Mustabṣir* can now be regarded as being the early 7th/13th century.

II

Customary law among the fishermen of al-Shiḥr

The fishermen's laws,[1] here translated and commented upon, form one of a collection of such documents which I made prior to 1965 as part of an extensive study of the fishing communities on the southern coasts of the former Aden Protectorates. Fisher law, like maritime customary law,[2] is well established and cases relating to the sea I have come across in *fatāwā* collections may show that its basic principles have long been understood. A later document, more comprehensive than this by far, seems almost like an attempt by the Quʿayṭī Government to codify the customary law of the sea which of course is in no way founded upon the *sharīʿah,* though it does not seem to conflict with it. Sayyid Muḥammad Saʿīd Midayḥij, who explained the obscurities of this little *Qānūn al-baḥr,* translated below, stated that the customary law of every fisher village, such as Quṣayʿar, Raydat ʿAbd al-Wadūd, and al-Ḥāmī, differs in one respect or another, but it is broadly similar everywhere. The jurisdiction of al-Shiḥr fisher law, administered by local headmen, extends from al-Maʿīnah to east of al-Mudawwarah and while outsiders are permitted to fish in this area they are subject to the custom of al-Shiḥr.[3]

This present document is limited to dealing with the law of contract between the fisherman/seaman (*al-baḥḥār*) and the 'owner', here known as *al-nākhūdhah,* to casting from sambooks, off-shore trawling with the seine-net (*jarīf*), and some miscellaneous items. The 'owner' of the *jarīf* is usually a rich man or one who can lay his hands on money easily. He pays the seamen (*baḥḥārah/baḥarah*) some money in advance on the understanding

Customary law among the fishermen of al-Shiḥr

قانون البحر

واجبات الناخوذه للبَحَّره

١ متي اراد النوخذه بحره للشغل عنده

يتقاول مع البَحَّار بنجم الجبهه او الزبره ويتحوّل البحار عند الناخوذه بنصف نجم الفرغ .

٢ فى حالة مرض البحار يكون سهمه يمشى مثل غيره ولو مات فى اول السنه سهمه يمشى الى غلاق السنه .

٣ العادة ولو ما ذكرت فى الشروط

على الناخوذه يدفع قهوه كل سنه للبحار روبيه واحده [١] ، وفى كل عيد ، الحَجّه وشوّال ، أربع آنه [٤] ، وللربان حق السنبوق روبيه [٤] ، وحق الهورى روبيه [١] .

٤ في حالة رفض الناخوذه للبحره بدون سبب

اذا رفض الناخوذه البحار بدون عذر فالدين التقدوم الذى مع البحار معاد يلزم دفعه للناخوذه فالحق يسقط .

واجبات البحّار

٥ عند ما يرفض البحار الشغل

ما يحصل البحار الفسخ إلّا بنصف القلب . وعلى شرط يُشعِر النوخذه من نجم الجبهة او الزبره .

٦ عند ما يكون البحار مديون للنوخذه

عند ما يُشعِر البحار النوخذه المستاجر عنده فى النجوم المذكورة يدفع نصف الدَّين الذى عليه . والنصف الآخر يدفعه لغاية نصف القلب . وإن تأخر شيء من الدين الى بعد نجم القلب ولو بعض من الفلوس فالبحار ما يحصل الفسخ من النوخوذه الأول بل يشتغل عنده الى نهاية السنة .

٧ فى حالة تلاعب البحار وتكاسله من الشغل وهو مستاجر اذا تحقق الخِلاف عند البحار فى عدم ركوبه للبحر بدون عذر او بعذر غير مقبول يلزم على البحار مقابل العطال يدفع كل يوم للنوخذه خمس روبيه ٥ بصفة غرامه عليه ، واذا كان الرهان يكون عليه عشرين [٢٠] روبيه .

٨ طُروح القِشار

الى السنابيق متوازنة فوق القشار وخالف واحد من السنابيق يخسر عشر روبيه [١٠] لصاحب السنبوق الذى صار عليه التعطيل وتوقيفه يومَيْن .

٩ الغديف ليس حرّ

الى قَدْ العَيْد قليل للبحار يغدف الى الظهر ويعود ثانى يوم ،ولا خرج العصر يخسر ١٥ روبيه للحكومة ، ويتعطل ثلاث من الغديف . وان كان الحوت واجد مُباح الغديف الى بعد العصر ويرفع من البحر .

١٠ شروط الحوى

فى العيد شروطهم الذى يحصله يحويه ، وبعد الحوى يعطى اصحابه خبر ، ويخرجون وان حد قريب وقت الحوى يحوى معه . والى قد الجرف ناصِبه والخطاء على الأولات الذى يصل عنده يحويه . ولا لاصحابه باقى اهل الجرف شئ فيه ، بل للذى يحوى . والى خرج العيد قبله لا يمكن احد يخرج له إلّا من الذى جريفه الأَوّلى القَبْلى . إلّا ان ما له قصد به ، يخرجون له .

١١ تَبْروح العَيْد

عند الجُرُف خارج البلدان كل من جاء يُبرِّح . وداخل البلد كُلّا له محل يبرح فيه ، إلا ان جاء غريب ينفّس له المقدّم . ومولا الجريف الى دخل داخل الدَّوْر لازم مقدم الرُّبُع ينفّس له .

١٢ وضع الصيد الحُسوس فى القشار

الى وضعه واحد فى القشار وكلا منه الصيد يشرد من الاصطياد . وأدب عامل هذا خمس ايام يتحيّر سنبوقه وعشرين روبيه خساره للحكومه .

١٣ مرض الربان

الى مرض الربان يعطى صاحب السنبوق المقدم خبر ، والمقدم يأمر أهل الربع كل يوم واحد يخرج عوضًا عنه الحق يَنْسَم .

١٤ مرض ربان الجرف

يكون مثل ما سبق يعوض بغيره من اهل الجرف .

موافق ٦ مارس سنة ١٩٤٦

سالم عُبَيْد بِاَلْكَوْر   عمر عبد الله حُنَيْن   محفوظ عَوَض با حَباره

قانون الجرف

الشروط التى للناخوذا

١٥ يستاجر الناخوذا صاحب الجَريف ربان للجريف وبحره ابتداءً من نجم جبهه وزبره ، ومدة الاجار من هذه النجمين ، والفسخ الى طلوع الجريف ، وهو آخر يوم نجم السماك .

وما كان من قرضه بطرف الربان والبَحَرَه يكون دفعها بنجم السماك . وان خرج نجم السماك والقرضه باقيه ، فما دفعها الربان او البحار ، فهو يكون ملزوم للنوخوذا ، ومَساد له شىء فسخ بعد ما يخرج نجم السماك .

١٦ من عطّل منهما ربان او بحار اى عطل جريف الناخوذا بغير عذر شرّعى عليه خساره للناخوذا ، وهى على الربان عشر روبيه خساره . وعلى البحار خمس روبّيه خساره .

الشروط التى على الناخوذا

١٧ اذا عطّل الناخوذا او فسخ فى الجريف ، وله قرضات بطرف البحره او الربان فلا له إلّا ، اذا خدم الربان او البحار ، يدفع للناخوذا ثلث فى القرضه التى بطرفه ، وما بقى يدفع ربع من رزقه ، وثلاثة ارباع له .

١٨ اذا فسخ الناخوذا فى الربان او احد من البحره وعنده قرضه فليس يستحقّ بشىء منها الناخوذا .

١٩ على الناخوذا اذا جرى أمر الله على الربان او احد من البحره فقسمه يكون لازم لحتى يطلع الجريف .

٢٠ على الناخوذا قهوة للربان عشرين روبيه [٢٠] ، وللبحار ثمان روبيه [٨] من غير شرطه .

الشروط التى فيما بين النواخيذ اصحاب الجرف

٢١ اذا مثلًا اثنين جرف يتطاردون احدهن من قِبْله والآخر من شرق ، يتطاردون على المحوى فكل من تقدم منهم ، وطرح كَدّه فى بطن المحوى ملك المحوى . واذا ما طرح فى بطن المحوى وطرح إلّا فى الشّاعِبَة او فى الرابع فليس له شىء فى المحوى . وان طرحوا اثنينهم فى الشاعبة والآخر فى الرابع فليس لهم شىء فى المحوى كلهم . وان ركب جريف من البحر او اجاء من محل آخر وطرح فى بطن المحوى ملكه عليهم .

٢٢ كل جريف ملزوم عليه ان يخلى فرق فيما بينه وبين الجريف الآخر محوى من الغَلِّه او الخُور بقدر اذرع ٣٠٠ .

٢٣ لكن لحق جريف الاخر غارق فليس له معبر لحتى يخرجه . وان وجده قاصر من الحبال لازم عليه يدركه بالحبال ان هو من البحر او من البر .

شرط لاصحاب الجرف على اهل السنابيق

٢٤ من بعد ما يغور الجريف اى يعطف بحبل القِدّامى فليس لصاحب السنبوق غديف فى الرَّحب وان غدف فى رحب الجريف فهو صار مخالف ومتعدى .

ابهام محفوظ بن سعيد بن عوض بَالرَّعيه   ابهام محمد بن احمد بن عَجْلان

مَقَادِمة البَحَرَه آل الجرف   ٣/٦/٤٦

that it will be paid back from what they are owed at the close of the trawling season (*mūsim al-jarīf*). The advance cash paid the seaman/fisherman is spent on food and clothing for his family and on other things at feast times (*mawāsim al-aʿyād*). The fishermen are bound to the *nākhūdhah* by debts which must be repaid if the fisherman is to leave his employ.

The *jarīf* and the crew that work it, at one time forty men, have been described in a 1950 fisheries report.[4] 'Sardine seine nets used on the Quʿayṭī coast, consist of a simple bag braided from strong hand-spun cotton. The mesh decreases towards the cod-end where it is too small for the fish to gill themselves. One examined was approximately thirty feet wide at the mouth, six feet wide at the cod-end and fifty yards long. To either side of the mouth are attached wings of palm-rope, braided in meshes, each several yards long. These are buoyed above with wood floats[5] and ballasted below with stone sinkers. Such wings are upwards of one hundred yards long, and may be much longer. To either wing is attached several hundred yards of hauling warp. The net is shot from a sambook or a large *hūrī* [dug-out canoe].' The sambook takes up its station seaward of the sardines which shoal so thickly at certain times of the year that they can be seen from the air as black patches in the sea, and the shore crew pull on the eastern wing to bring it inshore. The western wing, which is a little longer than the eastern, is also drawn into shore so that the sardines or other fish are enclosed by a crescent. The wing meshes are too open to stop the fish escaping to the open sea; as the fish are reluctant to face the fishermen, the fishermen take up positions in the water behind these wings shaking and splashing them so that the mouth of the bag offers the only quiet path of escape. Clauses 21-23 define what is the procedure when two beach seines are trawling close together.

It must be explained that the *jarīf*-seine means, by extension, the fish it contains or the crew working it. Al-Majraf Quarter in al-Shiḥr can only be so named after the *jarīf*-seine. Al-Qishār/Qashār of clause 8 is a rocky area of the sea-bed with holes and little caves in which fish abound.

The Law of the Seamen

A. DUTIES OF THE NĀWKHŪDHAH TOWARDS THE SEAMEN

1. *When the Nawkhudhah wants seamen to work for him*

He contracts with the seaman from al-Jabhah (22 February), or al-Zabrah (7-19 March), and the seaman starts work[6] with the Nākhūdhah at half-way through the star al-Fargh (24 September).

2. *In the case of illness of a Seaman after he is hired*

When a seaman is sick his share goes on like the others; and even should he

die at the beginning of the year his share will go on to the end of the year.
3. *The Custom(ary Payment), even though not mentioned in the terms (of employment)*
It is the Nākhūdhah's obligation to pay the seaman a gratuity (lit., coffee) each year, of one rupee and four annas at each feast, al-Ḥajjah and Shawwāl.[7] The sambook Captain (*Rubbān*) shall receive four rupees, and the canoe Captain two rupees.
4. *In the case of the Nākhūdhah dismissing the Seamen without (due) cause*[8]
If the Nākhūdhah dismisses a seaman without a reason, then the debt, *i.e.*, the advance given the seaman, is no longer required to be repaid to the Nākhūdhah, and the obligation is cancelled.

B. DUTIES OF THE SEAMEN

5. *When the Seaman refuses work*
A seaman will not be able to break his contract[9] until half-way through al-Qalb (12 June), and on condition of giving the Nawkhudhah notice from the star al-Jabhah (22 February), or al-Zabrah (7-19 March).
6. *When the Seaman is in debt to the Nawkhudhah*
When the seaman gives notice to the Nawkhudhah by whom he is hired at the afore-said stars, he must repay half of the debt he owes, and the other half he must repay half-way through al-Qalb (12 June) at the latest. If (repayment) of any of the debt be delayed until after the star al-Qalb, even though it be but a few coppers, the seaman will not be able to break his contract with the first Nawkhudhah, but will work with him till the end of the year.
7. *In the case of the Seaman, when hired, idling or being lazy in his work*
If disobedience (to these conditions) be proven of the seaman in his not going to sea, without (due) excuse, the seaman, to make recompense for not working,[10] must pay the Nawkudhah five rupees each day by way of a fine on him; and if it be a Captain he must pay twenty rupees.
8. *Casting at al-Qishār*
When the sambooks lie opposite each other above al-Qishār, and one of the sambooks disobeys (the rules),[11] it is fined ten rupees in favour of the owner of the sambook which has suffered loss,[12] and is banned (from the sea) for two days.
9. *Casting is not unrestricted*
When the sardines are few the seaman may cast until noon and return a second day, (but) if he puts to sea in the afternoon[13] he is fined fifteen rupees (payable) to the Government, and is prevented from casting for three days. If fish is abundant casting is permitted until after the afternoon,[14] then he must leave the sea.
10. *The conditions of trawling*
Where sardines are concerned their conditions are that he who finds fish trawls it (and tells his fellows after trawling and they put to sea); and if anyone be near-by when he goes to trawl, he trawls with him. If the nets are already set up and he passes beyond the first (nets), whatever (shoals) he

comes upon he may trawl. His fellows the other members of the (other) *jarīf* crews have no share in this[15]—on the contrary it goes to the man who trawls. If anyone goes out to the sardines before him (*i.e.*, the man who has discovered the shoal), no-one else may go out to them but he whose *jarīf*-net is the first to the front, except if he (the man whose right it is to have them) does not want them (the sardines), they may go out to them.

11. *Exposing the sardines to dry in the sun*

When the sardines (*juruf*)[16] are outside the town area[17] any comer may dry (his sardines) in the sun, but inside the town (al-Shiḥr) each person has his place in which to expose them to dry, though if a stranger[18] come, the headman will make room for him. When the owner of a *jarīf*-net enters within the (beach area lying between) the (ends of) the encircling wall, the headman (of the Quarter)[19] must make a space for him (the stranger).[20]

12. *Throwing rotten fish into the sea*[21]

When anyone throws it into al-Qishār and the fish eat of it they leave the fishing. For a person who does this the punishment is five days' ban[22] on his sambook and twenty rupees fine to the Government.

13. *Sickness of the Captain (Rubbān).*

When the Captain is ill, the owner of the sambook informs the headman,[19] and the headman issues an order to the people of the Quarter (*rubuʿ*) for one (of them) to go forth in his stead until he recovers.[23]

14. *Sickness of the Captain of the Jarīf-Crews*

It will be the same as the afore-going, someone else of the *jarīf*-crew taking his place.

6 March, 1946

Thumb-marks of Sālim ʿUbayd Bil-Kawr, ʿUmar ʿAbdullāh Ḥunayn, Maḥfūẓ ʿAwaḍ Bā Ḥabārah.

C. THE LAW OF THE JARĪF-CREWS

15. *Stipulations in favour of the Nākhūdhā*

The Nākhūdhā, owner of the *jarīf*, hires a Captain (*Rubbān*) for the *jarīf*-crew and seamen, to commence from the star al-Jabhah (22 February) and al-Zabrah (7 March), and the hired man's term (starts) from these two stars. Termination (of the contract) may be (arranged) up till the (time of) the withdrawal of the *jarīf*-trawl,[24] which is on the last day of the star al-Simāk (17 April). At the star al-Simāk any loan outstanding with the Captain and seamen will be repaid, but if the star al-Simāk passes by, and the debt still remains unpaid by the Captain or seaman, he will continue under contract to the Nawkhūdhā, for there will be no termination (of his present contract) after the passing of the star al-Simāk.

16. Whosoever of them, be he Captain or seaman, renders the Nākhūdhā's *jarīf* idle in any way without any customary sanction[25] will be liable for a fine (to be paid) to the Nākhūdhā which, in the case of the Captain will be ten rupees, and in the case of the seaman five rupees fine.

D. STIPULATIONS AGAINST THE NĀKHŪDHĀ

17. If the Nākhūdhā does not employ, or dispenses with the *jarīf* (crew?) and has loans out with the seamen or Captain, he is entitled to nothing except if the Captain or the seaman has worked, he shall repay the Nākhūdhā one third of the loan made to him, and, of the remainder he shall repay one quarter in return for his maintenance, but the other three quarters shall be his.
18. If the Nākhūdhā dispenses[26] with the Captain or any of the seamen, and the latter owes (the Nākhūdhā) a debt, the Nākhūdhā is not entitled to (the return) of any of it.
19. If God's command befall the Captain or any of the seamen the Nākhūdhā (continues) responsible, for his share is a due (charge) until the *jarīf* is withdrawn.
20. The Nākhūdhā must pay the Captain a gratuity (lit., coffee) of twenty rupees, and the seamen one of eight rupees without (express) stipulation.[27]

E. STIPULATIONS AMONG THE NĀKHŪDHĀS AND JARIF-MEN

21. When, for example, two *jarīf*-crews are racing one another, one coming from the west and the other from the east, racing for an area to be trawled *maḥwa*)—whichever one of them arrives there first and weighs its anchor in the middle[28] of the area to be trawled, thereby obtains possession of the (said) area to be trawled. When, however, it does not weigh (anchor) in the middle of the area to be trawled, but only (on the east or west side of it) where the flank rope and net, or the tug-rope (will be),[29] it has no right at all to the area to be trawled. If two (crews) weigh (anchor), one of them at the flank-rope, and the other at the tug-rope, neither one of them thereby establishes any right at all to the area to be trawled; and if a *jarīf*-crew sails in from the sea (side), or arrives from any other place, and weighs anchor in the middle of the area to be trawled, it (thereby) establishes possession of it, and they do not.
22. Each *jarīf*-seine is responsible for leaving a space of three hundred *dhirāʿ* [150 yards] between itself and the next *jarīf*-seine in a trawling area on al-Ghallah or al-Khawr.[30]
23. Anyone coming upon another person's *jarīf*-seine foundering may not pass on until he has brought it to shore. If he finds it short of ropes he must come to its aid with ropes whether he is coming from the sea or from the shore.[31]

F. STIPULATIONS IN FAVOUR OF THE JARĪF-MEN AGAINST THE SAMBOOK CREWS

24. After the *jarīf*-seine is submerged, *i.e.*, is lashed by the eastern rope,[32] the sambook man may not cast (his nets) in the area (to be contained by the east and west ropes of the *jarīf*-seine).[33] If he does cast in the area (to be

occupied) by the *jarīf*-seine he is acting in a wrongful and contentious manner.

Thumb-prints of Maḥfūz b. Saʿīd b. ʿAwaḍ Ba-ʾl-Raʿiyyah and Muḥammad Aḥmad ʿAjlan, headmen of the seamen and *jarīf*-crews, 3/6/46.

The coastal star calendar

RABĪʿ
January 1, al-Hanʿah; 14, al-Dhirāʿ; 27, al-Nathrah.
February 9, al-Ṭaraf; 22, al-Jabhah.
March 7, al-Zabrah; 20, al- Ṣarfah.

ṢAYF
April 2, al-ʿAwwā; 15, al-Simāk; 28, al-Ghufar.
May 11, al-Zubān; 24, al-Iklīl.
June 6, al-Qalb; 19, al-Shawl.

KHARĪF
July 2, al-Naʿāyim; 15, al-Baldah; 28, Bulaʿ.
August 10, Suhayl; 23, al-Dhābiḥ.
September 5, al-Khibāʾ al-Awwal/al-Kharāʾ Lawwal; 18, al-Khibāʾ al-Thānī/al-Kharāʾ al-Thānī.

SHITĀʾ
October 1, al-Dalū; 14, al-Ḥūt; 27, al-Naṭḥ.
November 9, al-Buṭayn; 22, al-Thurayyā.
December 5, al-Barakān; 18, al-Haqʿah.

Notes

1. I am indebted to Mr L.J. Hobson, formerly of the Aden Political Service, for giving me this document.
2. Compare my 'Maritime customary law off the Arabian coasts', *Sociétés et compagnies de commerce en Orient et dans l'Océan Indien,* Actes du huitième colloque international d'histoire maritime (Beyrouth, 5-10 September 1966), Paris, 1970, 195-207.
3. On al-Shiḥr and customary law there, *see* my preliminary remarks in 'Wards and quarters of towns in South-West Arabia', *Storia della città,* Roma, 1978, 43-8, unfortunately full of printing errors.
4. W.A. King-Webster, 'Aden Fishery Department report', no. 4, covering a voyage from 10 October to 4 December 1950.

5. In fact these floats are the broad butts of palm-*jarīds*.
6. *Taḥawwal* was paraphrased as *yabtadi l-ʿamal* to begin the work. I should be inclined to link it with *ḥawl*, a year.
7. Al-Ḥajjah is al-ʿĪd al-Kabīr, and Shawwāl is ʿĪd al-Fiṭr. At both feasts one customarily gives one's servants presents.
8. *Rafaḍ* was explained as *istaghnā ʿan al-baḥḥār*. Compare Stace, *An English-Arabic vocabulary* (London, 1893), 178, 'to turn off' with the synonym *ṭarad*. *See also* M.A. Ghanem, *Aden Arabic grammar* (Aden, 1960), 89, *raft*, expulsion, dismissal; and Lane, *Lexicon*, *rafaḍa*, drive away.
9. One says *fasakh al-ʿahd* in the sense of breaking a contract. C. de Landberg, *Glossaire datînois* (Leiden, 1920-42), renvoyer, disloquer, détacher, in Oman, ausziehen. Compare H. Jacob, *Perfumes of Araby* (London, 1915), 235, *fasaḥ* means one's congé, but no dismissal could be complete without money (*fusâḥât* [*sic*]).
10. One says *ʿaṭṭal al-sambook*, left the sambook without work, idle; but compare M.A. Ghanem, op. cit., 149, *ʿaṭāl*, damage.
11. Amplified as *khālaf al-niẓām*. He might do so, for example, by going in some other direction after fish, looking for a better place.
12. By loss, perhaps, damage is meant, through a collision possibly.
13. *I.e.*, from about 3:00 p.m. onwards.
14. For *naṣab*, *see Ḥaḍramoût* in C. de Landberg, *Etudes sur les dialècts de l'Arabie méridionale* (Leiden, 1901), 724.
15. While the general import of this clause cannot be in doubt, the ambiguity of the pronouns makes it difficult to render with certainty.
16. Sometimes *juruf* can mean sardines.
17. Outside the walls all *jarīfs* may expose their sardines to dry without hindrance, but in al-Shiḥr because of the confined space there has to be an agreed procedure.
18. At the sardine season (*mūsim al-ʿayd*) people from Shiḥayr, Quṣayʿar, etc., come to fish at al-Shiḥr.
19. In al-Shiḥr the *muqaddam* of the Ḥāfah (Quarter) will be a seaman (*baḥḥār*), i.e., a fisherman. The Rubuʿ is a part of the Ḥāfah-Quarter proper, and al-Ḥawṭah Quarter, for example, is divided into four quarters with a headman for each.
20. The stranger will be placed on the outside parts of the space between the two ends of the wall, and the native of al-Shiḥr nearer the centre.
21. *Ḥusūs* is *ʿayd*-sardine or fish which has lain a day or two and become unfresh.
22. *Taḥayyar* is explained as *tawaqqaf*.
23. *Nasam, yinsam*, explained as *yakhush hawā*, take air, *istashfā*, to be cured. *Nisimit* means, I have recovered, *Ḥaḍramoût*, 722, *yansam*, se reposer and *nasam*, soulagement.
24. *I.e.*, at the end of the sardine fishing season.
25. *Sharʿ* means here *ʿādah*, custom. One says *ʿalayk sharʿak*, and *al-shurūʿ al-ṭawīlah* means *al-aṣl al-ṭayyib*. A customary excuse in this case would be that one is required to attend a wedding, etc.
26. *Fasaḥ*, explained as meaning *lā yurīd*, he does not want, at al-Shiḥr. *Compare* Stace, op. cit., 122, *fasaḥ*, to permit.
27. The wife and children of the seamen get a set of clothing (*kaswah*) at both feasts (*al-ʿīdayn*), as well as the seamen.
28. I am indebted for additional explanation here to Sayyid Aḥmad al-ʿAṭṭās who, after consulting the *muqaddams* of al-Shiḥr on certain points, replied to me by letter. *Baṭn al-maḥwā* means 'the middle of the area in which the *jarīf*-net trawls (*yaḥwī*) and where it places its anchor.'
29. The *shāʿibah* means the ropes and nets which the *jarīf* casts on both flanks, and is on the west and east sides (ʿAṭṭās). The *rābiʿ* is the rope of the *jarīf*-trawl extending from the *jarīf*-trawl to the shore so that the seamen who are on the shore may draw it in (ʿAṭṭās).
30. Al-Khawr (ʿAṭṭās writes al-Khūr) Means an ʿayqah, a low-lying inlet covered with salt water. ʿAlawī b. Ṭāhir, *al-Shāmil* (Singapore, printed 1941 but never published),

85, calls the ground between Sharj and al-Mukallā al-ʿAyqah, but it is not entirely covered in water. For both *khawr* and *ʿayqah*, compare *Gl. Dat.*, pp.658 and 2348, *Inst. Naut.*, I fol.86b. But it is a technical term among the seamen of al-Shiḥr meaning west, *i.e.*, the direction of Ḥāfat al-Khawr, the Khawr Quarter, lying west of al-Shiḥr wall (ʿAṭṭās). Al-Ghallah is a landing-place (*makān khayṣah*) from which to the west lies al-Qishār (al-Qishār min-hā wa-qiblah), but from here to the east is called Ghallah (ʿAṭṭās). I understand from Sayyid Aḥmad's note that this clause simply means one must leave 150 yards between one's *jarīf*-trawl and the next on both the *east and west* sides.

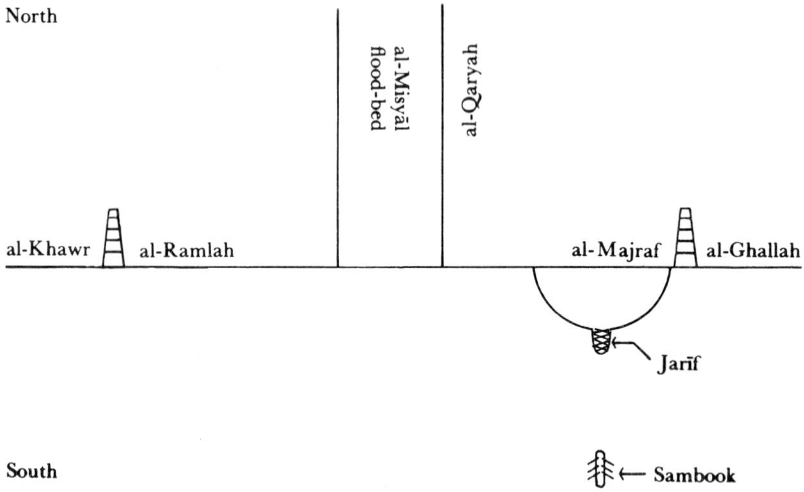

The shore of al-Shiḥr showing the quarters al-Ramlah and al-Majraf within the walls and the *Jarīf* seine pulling into shore with a shoal of sardines.

31. *I.e.*, whether he is putting out to sea to fish or returning from fishing.
32. Al-Qidmāy or al-Qudūm is a technical word meaning from the wall of al-Shiḥr and eastwards (*min sūr al-Shiḥr wa-sharq*). So *ḥabl al-Qidmāy* simply means the eastern rope of of the *jarīf*-seine (ʿAṭṭās).
33. *Al-raḥib* or *raḥib al-jarīf*—the *raḥib* is the place lying in the middle between the ropes and the nets of the *jarīf*-trawl on its eastern and western flanks (ʿAṭṭās).

III

Star-Calendars and an Almanac from South-West Arabia

Contents:

I. Introduction
II. Correlations of the Star-Calendar with the Solar Year
1. Generalities
2. Zodiacal Signs
3. Rain-Stars
4. Cold Spells
5. Heat Spells
6. Popular Saws on the Mansions of the Moon
7. The Months of the Syrian Christian Year
III. A Yemenite Almanac
1. Preliminary Remarks
2. The Yemenite Almanac of Muḥammad Ḥaidarah
3. Acknowledgements
IV. Bibliography

I. Introduction

The star-calendar of the South Arabian cultivators and mariners has been carefully and thoroughly examined by CARLO VON LANDBERG (7) who dealt with it from the historical and philological aspects, correlating material from Arabic literary sources with his own field-work. Embedded in a lexicon, known to specialists alone, his study of the subject is not generally available at a time such as the present when so much agricultural development has begun to take place in Southern Arabia. Moreover LANDBERG's calendar is based mainly upon information collected by him in the Wādī ʿAmaḳīn, whereas there seem to be numerous methods of computation of the agricultural year which do not entirely coincide with the calendar known to him.

The portion of this article concerned with the star-calendar will contain certain material nearly identical with that of LANDBERG, but set forth in tabular form for use in the field. For further reference and more profound research, recourse must be made to his Glossaire Daṯînois, and to the bibliography (p. 459). This article is, in some sense, a supplement to LANDBERG embodying new field-work and drawing on new written sources of information. It is not however suggested that it is in any way a synthesis of the works cited in the bibliography though nearly all of them have been consulted. The bibliography itself enables one to form some conception of the area over which the star-calendar is used, apart from Southern Arabia which alone concerns us here, and indeed it is probably used over wider territories still.

EDWARD LANE's Lexicon contains much of the earlier Arab stellar lore, but the latter is most readily available to us in the works of AL-BĪRŪNĪ (18), (21). ḤAMDĀNĪ (ob. 945 A. D.) shows by inference that the star-calendar was used by cultivators in the 10th century [1], and there has in fact been a continuous tradition of astronomical writing up to the present time in Southern Arabia [2]. As a science, astronomy had an importance for men of religion too, that is to say a practical importance, and the Syrian AL-SHAIZARĪ for instance, lays down a rule that the muezzin must know the "Mansions of the Moon and form of the stars of each Mansion" [3], so as to perform the call to the dawn prayer at the appropriate time.

No attempt is made to identify the stars, but they have been very exactly described by AL-BĪRŪNĪ, and LANDBERG has made identifications which may assist those interested in that aspect of the study.

II. Correlations of the Star-Calendar with the Solar Year

1. Generalities

Table I is a correlation between the European solar calendar and the star-calendar of the Arabs of Ḥaḍramawt, compiled by a certain AL-SHIBĀMĪ [4] who, as his name implies, must have been a citizen of Shibām, the commercial centre of the country. A list of the 28 stars of the year is given at the foot, each star-period running for 13 days, except *al-Haḳ'ah* which lasts 14 days. The first star of each season, of which there are four in Ḥaḍramawt, is known as the Father (*al-Abū*) [5]. The table was copied as printed here, but all other sources make the year commence with the period called *Ṣaif* by AL-SHIBĀMĪ, i. e. the beginning of April.

---

[1] *Ṣifat Djazīrat al-'Arab*, edit. D. H. MÜLLER (Leiden 1884), p. 154.

[2] Cf. Materials for South Arabian History II (Bull. School Oriental and African Studies). (London 1950), 13, 2, p. 593 for Ḥaḍrami writers. The comparatively recent author AL-SHAWKĀNĪ says in his *Al-Badr al-Ṭāli'* (Cairo, 1348 H.), II, 222 that he had written a book *Djawāb al-Sā'il 'an Tafṣīl al-Ḳamar Manāzil* (Answer to the Inquirer on the Division of the Moon into Mansions), a work doubtless available in the Yemen.

[3] *Nihāyat al-Rutbah fī Ṭalab al-Ḥisbah* (Cairo 1365 H. 1946 A. D.), p. 111. He refers to IBN ḲUTAIBAH, "*K. al-Anwā'*" as the standard work on the Mansions (cf. BROCKELMANN, GAL I, 122). A Ms. of this work is available in Oxford (Bodl. I, 1033) which will probably contain ancient popular rhymes and lore of the type to which reference is made infra. AL-SHAIZARĪ died in 1193 A. D.

[4] BĀ ṢABRAIN AL-SHIBĀMĪ, for his biography, see Materials for South Arabian History, op. cit., p. 593. This table, entitled *Ḥisāb al-Shibāmī*, is said to have been printed (in Arabic).

[5] *Abū* is used so with the definite article also when describing the head of a trade-guild in Ḥaḍramawt.

Table I

Table of the Arabic Stars according to the European Months, known as *Ḥisāb al-Shibāmī* [5a].

Al-Rabīʿ						*Al-Ṣaif*						*Al-Kharīf*						*Al-Shitā*					
January		February		March		April		May		June		July		August		September		October		November		December	
1	[1]1	1	6	1	8	2	[8]1	1	4	1	9	2	[15]1	1	5	1	10	1	[22]1	1	6	1	10
2	2	2	7	2	9	3	2	2	5	2	10	3	2	2	6	2	11	2	2	2	7	2	11
3	3	3	8	3	10	4	3	3	6	3	11	4	3	3	7	3	12	3	3	3	8	3	12
4	4	4	9	4	11	5	4	4	7	4	12	5	4	4	8	4	13	4	4	4	9	4	13
5	5	5	10	5	12	6	5	5	8	5	13	6	5	5	9	5	[20]1	5	5	5	10	5	[27]1
6	6	6	11	6	13	7	6	6	9	6	[13]1	7	6	6	10	6	2	6	6	6	11	6	2
7	7	7	12	7	[6]1	8	7	7	10	7	2	8	7	7	11	7	3	7	7	7	12	7	3
8	8	8	13	8	2	9	8	8	11	8	3	9	8	8	12	8	4	8	8	8	13	8	4
9	9	9	[4]1	9	3	10	9	9	12	9	4	10	9	9	13	9	5	9	9	9	[25]1	9	5
10	10	10	2	10	4	11	10	10	13	10	5	11	10	10	[18]1	10	6	10	10	10	2	10	6
11	11	11	3	11	5	12	11	11	[11]1	11	6	12	11	11	2	11	7	11	11	11	3	11	7
12	12	12	4	12	6	13	12	12	2	12	7	13	12	12	3	12	8	12	12	12	4	12	8
13	13	13	5	13	7	14	13	13	3	13	8	14	13	13	4	13	9	13	13	13	5	13	9
14	[2]1	14	6	14	8	15	[9]1	14	4	14	9	15	[16]1	14	5	14	10	14	[23]1	14	6	14	10
15	2	15	7	15	9	16	2	15	5	15	10	16	2	15	6	15	11	15	2	15	7	15	11
16	3	16	8	16	10	17	3	16	6	16	11	17	3	16	7	16	12	16	3	16	8	16	12
17	4	17	9	17	11	18	4	17	7	17	12	18	4	17	8	17	13	17	4	17	9	17	13
18	5	18	10	18	12	19	5	18	8	18	13	19	5	18	9	18	[21]1	18	5	18	10	18	[28]1
19	6	19	11	19	13	20	6	19	9	19	[14]1	20	6	19	10	19	2	19	6	19	11	19	2
20	7	20	12	20	[7]1	21	7	20	10	20	2	21	7	20	11	20	3	20	7	20	12	20	3
21	8	21	13	21	2	22	8	21	11	21	3	22	8	21	12	21	4	21	8	21	13	21	4
22	9	22	[5]1	22	3	23	9	22	12	22	4	23	9	22	13	22	5	22	9	22	[26]1	22	5
23	10	23	2	23	4	24	10	23	13	23	5	24	10	23	[19]1	23	6	23	10	23	2	23	6
24	11	24	3	24	5	25	11	24	[12]1	24	6	25	11	24	2	24	7	24	11	24	3	24	7
25	12	25	4	25	6	26	12	25	2	25	7	26	12	25	3	25	8	25	12	25	4	25	8
26	13	26	5	26	7	27	13	26	3	26	8	27	13	26	4	26	9	26	13	26	5	26	9
27	[3]1	27	6	27	8	28	[10]1	27	4	27	9	28	[17]1	27	5	27	10	27	[24]1	27	6	27	10
28	2	28	7	28	9	29	2	28	5	28	10	29	2	28	6	28	11	28	2	28	7	28	11
29	3			29	10	30	3	29	6	29	11	30	3	29	7	29	12	29	3	29	8	29	12
30	4			30	11			30	7	30	12	31	4	30	8	30	13	30	4	30	9	30	13
31	5			31	12			31	8	1	13			31	9			31	5			31	14
				1	13																		

[1] *al-Hanʿah*
[2] *al-Dhirāʿ*
[3] *al-Nathrah*
[4] *al-Ṭarf*
[5] *al-Djabhah*
[6] *al-Zabrah*
[7] *al-Ṣarfah*
[8] *al-ʿAwwā*
[9] *al-Simāk*
[10] *al-Ghufr*
[11] *al-Zubān*
[12] *al-Iklīl*
[13] *al-Ḳalb*
[14] *al-Shawl*
[15] *al-Naʿāʾim*
[16] *al-Baldah*
[17] *al-Mirzam*
[18] *al-Suhail*
[19] *Bā ʿUraiḳ*
[20] *al-Khibā*
[21] *al-Fargh*
[22] *al-Dalū*
[23] *al-Ḥūt*
[24] *al-Naṭḥ*
[25] *Buṭain*
[26] *al-Thuraiyā*
[27] *al-Barakān*
[28] *al-Haḳʿah*

[5a] The transliteration of the Arabic words in this article has been made according to the system of the Encyclopaedia of Islam. N. B. The small number to the left of the first day of each star refers to the name in the lower portion of the table.

Table II

Amongst the Wāḥidī tribes to the west of Ḥaḍramawt the following Star-Calendar is in use:

Rabī'	*Ṣaif*	*Kharīf*	*Shitā*
1) *Djabhah*	8) *Kalīl*	15) *Suhail, al-Khirih*	22) *Thuraiyā*
2) *Zabrah*	9) *Ḳalb*	16) *Sa'd*	23) *Barākān*
3) *Ṣarfah*	10) *Shawlah*	17) *Nāḥiz*	24) *Hiḳā'*
4) *'Uwwā*	11) *Na'ā'im*	18) *'Araḳ*	25) *Hinā'*
5) *Simāk*	12) *Baldah*	19) *Khāmis*	26) *Dhirā'*
6) *Ghawfar*	13) *Ḳuwaidim*	20) *Sādis*	27) *Nitharah*
7) *Zibān*	14) *Mirzam*	21) *Sābi'*	28) *Ṭaraf*

These names are, with slight dialectical variants, the same as the list reported by LANDBERG from the Wādī 'Amaḳīn. For *Nāḥiz* LANDBERG has *Ḳatrah,* a star also appearing in the Yāfi'ī calendar.

According to my chief informant in Ḥaḍramawt, 'ABDULLĀH RAḤAIYAM, the following seasons were distinguished in Tarīm:

Season	No. of days	Beginning of Season		
Rabī'	89	6th Jan.,	or	6th of *Han'ah*
Ṣaif	94	22nd March,	or	3rd of *Ṣarfah*
Kharīf	93	7th July,	or	6th of *Na'ā'im*
Shitā	98	25th Sept.	or	8th of *Farigh*

These dates differ slightly from those of the Almanac which is translated below (p. 445). RAḤAIYAM's calendar also differed in that he gave the star *Han'ah* the odd fourteen days. He seemed to regard the extra day as intercalary and called it *kabīsah* (pl. *kabā'is*), while a second intercalary day was called *kabīsat al-kabīsah.* He said that these intercalary days could be added to *Haḳ'ah* and *Han'ah.* In LANE's Lexicon the star *Djabhah* is given the fourteen day period.

In Ḥaḍramawt one expresses the date as, "The 1st day in *Baldah* (equivalent to) the 26th of *Sha'bān,* year 1366 (*awwal yawm fī Baldah, sitt wa-'ishrīn Sha'bān, sanat 1366*)".

In comparing the various star-calendars of antiquity and the present day, I have discovered that the 28 stars listed by the authorities correspond, more or less, in nomenclature, though there are variant forms and alternatives for some star-names. There is however an enormous variation in the particular Mansion of the Moon which is said to appear at a given period; while this may have a simple explanation to those acquainted with astronomy, it is at first somewhat astonishing to the uninstructed layman, especially as the variations occur in territories at no great distance from each other. The following comparisons are made in the case of *Djabhah,* for South Arabia; let it be said that further afield, in the Sudan, this star-name appears in the latter half of August [6].

[6] See (15). Sudanese stellar lore evidently derives nearly from that of Arabia.

Table III

Month	Day	Sun-Stations	Agricultural Stars	Evening Ascension	Morning Ascension
January	10	*Sa'd al-Dhābiḥ*	*Rābi' Awwal* (Evening)	*al-Nathrah*	*al-Na'ā'im*
	23	*Sa'd Bula'*	*Rābi' Ākhir* (Evening)	*al-Ṭarf*	*al-Baldah*
February	5	*Sa'd al-Su'ūd*	*Khāmis al-Ṣawāb*	*al-Djabhah*	*al-Dhābiḥ*
	18	*Sa'd al-Akhbiyah*	*Sādis al-Ṣawāb*	*al-Zabrah*	*Sa'd Bula'*
March	3	*al-Fargh al-Muḳaddam*	*Sābi' al-Ṣawāb*	*al-Ṣarfah*	*Sa'd al-Su'ūd*
	16	*al-Fargh al-Mu'akhkhar*	*Ẓāfir Awwal*	*al-'Awwā*	*Sa'd al-Akhbiyah*
	29	*Baṭn al-Ḥūt*	*Ẓāfir Thānī*	*al-Simāk*	*al-Muḳaddam*
April	11	*al-Sharaṭain*	*Simāk (al-Far')*	*al-Ghufr*	*al-Mu'akhkhar*
	24	*al-Buṭain*	*Ghurūb Kāmah*	*al-Zubānā*	*al-Rishā (Baṭn al-Ḥūt)*
May	7	*al-Thuraiyā*	*Ghurūb al-Thawr*	*Iklīl*	*al-Sharaṭain*
	20	*al-Dabarān*	*Ṭulū' Kāmah*	*al-Ḳalb*	*al-Buṭain*
June	2	*al-Haḳ'ah*	*Ṭulū' al-Thawr*	*al-Shawlah*	*al-Thuraiyā*
	15	*al-Han'ah*	*Ṭulū' al-Ẓulm al-Awwal*	*al-Na'ā'im*	*al-Dabarān*
	28	*al-Dhirā'*	*Ṭulū' al-Ẓulm al-Thānī*	*al-Baldah*	*al-Haḳ'ah*
July	12	*al-Nathrah*	*'Alib*	*Sa'd al-Dhābiḥ*	*al-Han'ah*
	25	*al-Ṭarf*	*Suhail*	*Sa'd Bula'*	*al-Dhirā'*
August	7	*al-Djabhah*	*Rawābi' Awwalah*	*Sa'd al-Su'ūd*	*al-Nathrah*
	20	*al-Zabrah*	*Rawābi' Ākhirah*	*Sa'd al-Akhbiyah*	*al-Ṭarf*
September	2	*al-Ṣarfah*	*Khāmis 'Alān* *	*al-Fargh al-Muḳaddam*	*al-Djabhah*
	15	*al-'Awwā*	*Sādis 'Alān*	*al-Fargh al-Mu'akhkhar*	*al-Zabrah*
	28	*al-Simāk*	*Sābi' 'Alān*	*Baṭn al-Ḥūt*	*al-Ṣarfah*
October	11	*al-Ghufr*	*Awwal Fāri'*	*al-Sharaṭain*	*al-'Awwā*
	24	*al-Zubānā*	*Rabī' Kāmah* (Evening)	*al-Buṭain*	*al-Simāk*
November	6	*al-Iklīl*	*al-Thawr* (Evening)	*al-Thuraiyā*	*al-Ghufr*
	19	*al-Ḳalb*	*al-Nadjmain* (Evening)	*al-Dabarān*	*al-Zubānā*
December	5	*al-Shawlah*	*Ẓulm Awwal* (Evening)	*al-Haḳ'ah*	*al-Iklīl*
	15	*al-Na'ā'im*	*Ẓulm Thānī* (Evening)	*al-Han'ah*	*al-Ḳalb*
	28	*al-Baldah*	*Suhail* (Evening)	*al-Dhirā'*	*al-Shawlah*

* Vocalisation doubtful.

AL-SHIBĀMĪ	(Ḥaḍramawt)	22nd February.
GLASER	(Yemen)	5th February.
Almanac	(Taʿizz)	22nd January.
LANDBERG	(W. ʿAmaḳīn)	21st January.
Wāḥidīs	(Aṣbaʿūn)	about beginning of January.

The 21st and 22nd of January are really the same date, because an Arab day of 24 hours (*waḍaḥ* in Mukallā) commences at 6 p. m. on what we should regard as the previous evening, so that 6 a. m. on the 22nd of January is according to the Arabs, 12 o'clock and 12 noon is their 6 o'clock. AL-BĪRŪNĪ's list of Mansions would make *Djabhah* appear about the beginning of January, like some of the South Arabian calendars.

The most abnormal star-calendar seems to be that used amongst the Yāfiʿīs, and I was given a list of stars and seasons, but as unfortunately I cannot trust the accuracy of my informant it is not printed here. Star-calendars are said to vary from district to district in Yāfiʿī territory which is very mountainous.

Table III is extracted from GLASER (5), but corrected, for his article is full of errors and misspellings of a misleading nature. The Agricultural Stars (GLASER's *Maʿālim al-Zirāʿah*, the Almanac's *Nudjūm al-Zirāʿah*) also appear in the Almanac, but GLASER's datings fall two days behind those of the Almanac, the corresponding dates in the Christian era have been taken from the Almanac. For instance GLASER's commencing date, the 9th of October, is here the 11th of October. Variants in nomenclature exist but are not given, as they are unimportant. The Almanac inserts against the 19th of October, the "Rise of *al-Thuraiyā* in the Evening", against April 25th it has the "Setting (*Ghurūb*) of *al-Thuraiyā* for 50 days", and against June 14th, "Appearance of *al-Thuraiyā*".

2. Zodiacal Signs

In Wāḥidī territory I was given a list of the three Zodiacal signs which occur in each of the four seasons, called *burūdj*, as in Classical Arabic. These do not correspond to the signs as given in the Almanac [7].

Rabīʿ	*Ṣaif*	*Kharīf*	*Shitā*
Saraṭān	*Ḥamal*	*Sunbulah*	*Ḳawzah*
ʿAḳrab	*Asad*	*Djadī*	*Mīzān*
Ḥūt	*Ḳaws*	*Thawr*	*Dalū*

3. Rain-Stars

Each season, I was told in Dawʿan, has a period or periods when rain usually falls, known as *ḥamīm* or *ḥamīmah* [8], with the exception of *Shitā*;

[7] I cannot unfortunately exclude the possibility of simple error here, though my informants were generally reliable.

[8] LANE, Lexicon, gives *ḥamīm* as a period in the star *al-Dabarān* (*Aldebaran*) when there is little or no rain; it also means intense heat, and then the rains following the heat.

though the best star in which one might hope for rain is *al-Naṭḥ*. When this information was given me I was quite unaware of a passage in Ḥamdānī [9] listing the stars under which rain falls, and corresponding in general to the following :

Rabī'	The *ḥamīmah* comes at *Ṣarfah, Zabrah, Djabhah.*
Ṣaif	The *ḥamīmah* comes at *Shawl* and *Ḳalb, Simāk* and *'Awwā.*
Kharīf	The *ḥamīmah* falls in *Suhail, Mirzam, Na'ā'im* [10].

The exact dating of these stars can be discovered by reference to Table I.

Ṭabarī (edit. De Goeje, etc.) III., IV. 2424 states that the Prophet forbade asking rain of the stars (*al-istisḳā' bi-'l-Kawākib*) which was a practice of the Pre-Islamic Arabs (cf. *Kanz al-'Ummāl* of 'Alī b. 'Abd al-Malik al-Muttaḳī [Ḥaidarābād, 1312-4 H.], VIII, 177).

4. Cold Spells

In Wāḥidī territory I was told that the three coldest days in the year (we had such a cold spell about mid-December) were known as *al-ḍarbah* (the blow), but in Daw'an, *al-ḍarbah* was said to be a cold period which might last from three to thirteen days, occuring in *al-Barakān, al-Dhirā'*, or *al-Nathrah.* It was said to come usually about three times a year, but it might occur in *al-Haḳ'ah* or *al-Han'ah.* The period known as *al-'Adjūz* is marked in the Almanac (infra).

5. Heat Spells

According to Raḥaiyam of Tarīm, there is a week in summer (*Ṣaif*) called *al-Muthammanāt,* the last 4 days of *Shawl* and first 4 of *Na'ā'im* (28th June to 5th July) which is intensely hot. Another period *al-Arba'īnīyah,* of 40 hot days is distinguished, but he did not give me the dates. This latter term is also used in Syria, Persia, etc. [11]. Meteorological information exists for the eastern portion of the Wādī Ḥaḍramawt, so this could be checked with mean temperature records.

6. Popular Saws on the Mansions of the Moon

Lane (Lexicon) says that the ancient Arabs had 28 proverbial sayings relating to the risings of the 28 Mansions of the Moon, such as, "When *al-Sharaṭān* rises, the season becomes temperate". A number of such sayings are also quoted by Al-Bīrūnī [12]. In contemporary Ḥaḍramawt similar pro-

[9] *Ṣifat Djazīrat al-'Arab,* op. cit., p. 154.

[10] For Arabic jingles on the rains at *Suhail,* see my Prose and Poetry from Ḥaḍramawt (London 1951), p. 164, No. 51.

[11] Cf. Western Arabia and the Red Sea (14), p. 608.

[12] Chronology of Ancient Nations, transl. C. E. Sachau (London 1879), p. 336. Cf. *Kanz al-'Ummāl,* op. cit., IV, p. 178 ; Marzūḳī K., *al-Azminah wa-'l-Amkinah* (Ḥaidarābād 1918), II, p. 161.

verbial sayings are attached to all the stars of the agricultural year, unquestionably very ancient. At *al-Ḳalb* for instance, one says, "The palm-tree becomes red with ripe dates (*al-Ḳalb, tiḳlab al-nakhlah fīh bi-ruṭab*)". Again, one says at *al-Djabhah*, "Man's brow pours with sweat then (*al-Djabhah, yaṣubb bi-'l-'araḳ fīhā min djabhat al-insān*)". Another type of saying is the *radjaz* or working-song, a couplet, such as that reported by SNOUCK HURGRONJE [13] :

If the *'Ulyā*-wind [14] blows on the 7th of *al-Shawl* [15],
The pigeon no longer comes to the field-bank.

There are numerous others. Another important date is what one might call the Autumn term, at which palm-plantations and the like are pledged, falling about the 7th of *al-Thuraiyā* in Ḥaḍramawt [16]. In Wāḥidī territory at *al-Baldah* there is a feast, not an *'īd* in the Islamic sense, but probably a kind of harvest festival when people meet together and shake hands. It is then that grain is brought by the peasant to the tribesman, this present being known as *'ādah* (custom), probably a sort of tax.

Of the many superstitious beliefs attaching to the stars I should like in particular to mention that concerning the *Bint al-Nathrah*, Daughter of *al-Nathrah*, brought to my attention recently by MUḤAMMAD LUQMAN. The Yemenites believe that a girl born at this Mansion (*manzalah*) lacks a hymen. They therefore register that she was born under this star with the *Ḳāḍī*, lest, says my Aden informant, there later fall suspicion of something else. The practice continues at Aden to this day, and the *Ḳāḍī* even makes his own private investigations to ascertain that the girl was actually born under this star, before he will issue a certificate as to the date of birth. The certificate can then be produced to the parties with whom the marriage is being contracted so that all is in order. Medical superstitions are too common to mention in detail, and appear throughout the almanac. In Ḥaḍramawt a good star for blood-letting is *Ṣarfah* (commencing 20th March).

7. The Months of the Syrian Christian Year

This form of solar calendar is employed in the Yemen, Lahej, and among the Ṣubaiḥī tribes of the Western Aden Protectorate which are subject to Lahej. I have not heard that it is used in any other province of South-West Arabia. The Christian months as given by the Almanac of MUḤAMMAD ḤAIDARAH differ however, in certain month-names from the Ṣubaiḥi calendar. The two calendars are as follows :

---

[13] *Sa'd èl-Suwênî* (19), p. 411. See also Prose and Poetry from Ḥaḍramawt, op. cit., Pref. p. 39.

[14] According to SNOUCK, the North Wind (see footnote no. 89, infra).

[15] SNOUCK has badly transcribed the Arabic which he avers has no metre, but this is not so. If transcribed as it would be pronounced, the metre is apparent,

Lā hab-ba-til-/ 'Ul-yā bi-sā-/bi' bish-Shawl/
Mā 'ā-d yib-/ raḥ fil-Ḥa-wī-/lī-'al-'awl/.

[16] For fuller details, see Materials for South Arabian History, op. cit., p. 592.

Almanac	Ṣubaiḥī Star-months (SHUHŪR AL-NUDJŪM)
Kānūn I (the 19th is Jan. 1st)	*Kānūn al-Awwal*
Kānūn II	*Kānūn al-Ākhir* or *al-ʿAḳab*
Shubāṭ	*Shubāṭ*
Mārt (*Adhār*)	*Adhār*
Nīsān	*Nīsān* (*Naisān*)
Māyis (*Aiyār*)	*Mabkar*
Ḥazīrān	*Ḥuzairān*
Tammūz	*Tammūz*
Aghustūs (*Āb*)	*Āb*
Ailūl	*Ailūl*
Tishrīn I	*Tishrīn al-Awwal*
Tishrīn II	*Tishrīn al-Ākhir* or *al-ʿAḳab*

Like some other South Arabian agricultural calendars, the Ṣubaiḥī year is divided into three seasons which they name *Shitā* (commencing in *Nīsān*), *Ṣaif* (commencing in *Tammūz*), and *Shimāl* (commencing about the end of *Ailūl*, the name certainly meaning the South-East monsoon). *Shimāl* continues for about six months. It is significant that the year begins in April, as in the case of the star-calendars of other tribes.

The calendar-months of the Almanac are those of the Ottoman financial year [17] which at one time corresponded in number to the Moslem year, but through lack of adjustment, the financial year 1362 fell in the Moslem year 1365. The history of the Ottoman financial year is however irrelevant here and, in any case, complicated [18]. On March 1st 1917, the Ottomans changed from the Julian to the Gregorian reckoning, and this adjustment was presumably made in the Yemen (where the Turkish authorities collected the taxes on behalf of the Imām) at the same time. The Moslem year cannot of course be used by the fiscal authority as, being lunar, it does not correspond to the harvest-seasons. In this connection, RYCKMANS (17) holds that in ancient Arabia a stellar calendar was used for economic, and a lunar calendar for religious purposes.

I did not discover if the Ṣubaiḥīs use the Julian, Gregorian or any other reckoning for their star-months as they call the Syrian Christian calendar, and in any case their reckoning may not be scientifically very exact. Lahej [19] however, like the Yemen, uses the Gregorian reckoning, and if the Syrian calendar or Ottoman financial year was introduced during the Turkish occupation in the first World War the adjustment was made then.

For several reasons set forth elsewhere I am persuaded that the

[17] Professor P. WITTEK kindly enlightened me on this point, supplying the reference, J. MAYR, Probleme der islamischen Zeitrechnung (Mitteilungen zur osmanischen Geschichte). (Hannover 1926), II, III-IV, pp. 269-304.

[18] It is discussed in J. MAYR, op. cit.

[19] The Admiralty's Western Arabia and the Red Sea (14), generally highly accurate, seems to be mistaken here, suggesting that the Julian reckoning is still used in Lahej, but from other information it supplies, it is clear that the Gregorian reckoning is used.

use of the Syrian calendar in Ṣubaiḥī country is older than either of the two Ottoman occupations of the Yemen. As I have shown [20], the district was in close contact with Christian tribes about the beginning of the Moslem era, if not actually itself Christian. The Syrian Christian months are used by Ḥamdānī [21] in the tenth century of the Christian era to indicate the precise seasons at which the harvest falls in the Yemen. It may of course be objected that, as a man of learning, he could be expected to have some acquaintance with the Syrian calendar. On philological grounds however there are circumstances which indicate a non-Ottoman origin for the Ṣubaiḥī use of the Syrian calendar. It is the Syrian names that are used for every month, and not *Mārt, Māyis,* and *Aghustūs*; but there is one exception, *Mabkar* is used for *Adhār*. This is in itself significant for the term *Bikār* (from the same root) is cited by Glaser [22] as the name of the season mid-June to mid-August, a season which would commence in the Ṣubaiḥī month *Mabkar* [23]. This name is then purely South Arabian. My informants seemed to be very imperfectly acquainted with the Moslem months and their sequence, and they knew them by names differing from the standard nomenclature of the Moslem year [24].

III. A Yemenite Almanac

1. Preliminary Remarks

The almanac (*taḳwīm*), portions of which are translated or summarised here, was compiled by a certain Muḥammad Ḥaidarah [25] and covers the Moslem year 1365 H. which commenced on Dec. 5th, 1945 (i. e. lst *Muḥarram*). In order, however, to show the sequence of events from January to December, I have made the translation commence as if the compiler had dealt with a single solar year according to our reckoning.

According to the compiler's preface, he composed this almanac in the South Yemenite city of Ta'izz for the crown-prince Saif al-Islām Aḥmad, now Imām of Ṣan'ā', and he avers that it is used in the Government offices of the Yemen. It is based, he says on the almanac of al-Imām al-Mahdī [26],

---

[20] Cf. Notes on Ṣubaiḥī Territory (Le Muséon). (Louvain 1953). 46, p. 128.

[21] *Ṣifat Djazīrat al-'Arab,* op. cit., p. 199. Nashwān b. Sa'īd is noted in Brockelmann. S. GAL I 528 as having written a poem on the Greek months.

[22] Die Sternkunde (5), p. 89. A list of seasons given by him notes *Ḳiyāẓ* which finishes about the end of April, *Dithā* ending about mid-June, *Bikār* ending about mid-August, and *Ṣurāb* which ends about the last days of December.

[23] *Bikār* and *Mabkar* may refer to the rains which fall about this period, for certain words from the same root in Class. Arabic have this sense, and the information in the Almanac would assist such an interpretation. Glaser, op. cit., gives a reference to the word *Bikār* in the pre-Islamic inscription numbered Glaser 158.

[24] Their names for the months of the Moslem year seem to correspond to those given by Landberg, Glossaire Daṯînois (7), p. 1449 which in themselves are curious.

[25] See (12).

[26] I am unable to say whether a specific Imām is intended by this title or whether it is a general title applied to the late Imām Yaḥyā.

and the reckoning (*ḥisāb*) of ʿALĪ B. ʿALĪ AL-YAMĀNĪ, formerly Shaikh al-Islām in Ṣanʿāʾ. One need not be surprised at astronomical knowledge in a religious dignitary, for as been noted, supra, the star-calendar has its religious as well as its practical uses. The author himself is described as "the *shaikh*, the scholar, the astronomer (*falakī*) at Taʿizz".

The almanac displays that curious blend of science and superstition common to all almanacs. The preface contains forecasts of portentous events, such as the appearance of a Mahdī who will be victorious over his foes; little captions at the sides give snippets of information such as "Manākhah is the Paradise of the Yemen, 500 feet above sea-level, with sweet water and fruits, and always cool", and "Djabal al-Manār is the highest mountain in the Yemen, 5000 feet high, and 10 miles E. of Ibb" [27].

MUḤAMMAD ḤAIDARAH claims that his almanac is not only indispensable to the official, but also to the peasant who must know the rising and setting of the agricultural stars (*nudjūm al-zirāʿah*), but he adds, significantly, "Now, concerning the seasons of sowing (*badhr*) and harvesting (*ḥiṣād*), I am unable to include them all, on account of the great differences between each district (*nāḥiyah*) and wādī". The last statement I have seen myself, to be true, and this fact was known to the early Arab geographers, including, of course, ḤAMDĀNĪ.

The almanac contains beside what is given here in translation, the Moslem date, the Ottoman Financial Year which is used for purposes of taxation, the European date, a column with the Mansion of the Moon for each day of the year, and a corresponding column showing the agricultural stars of which there are of course 28, but as the Moslem year is less than the solar year by about 11 days, the latter are not covered by the almanac. The sign of the Zodiac in which the Sun lies is also given. At the beginning, a list of the official festivals (*al-mawāsim al-rasmīyah*) of the Yemen is inserted. These festivals are:

New Moon of *Muḥarram*, Feast at the opening of the *Hidjrah* Year.
10th *Muḥarram*, Fast of *ʿĀshūrāʾ* Day.
12th *Rabīʿ* I, Birth of the Apostle of God.
17th *Rabīʿ* I, Birthday of H. M., Our Lord (*djalālat Mawlānā*), the Commander of the Faithful *al-Mutawakkil ʿala ʾllāh* [28].
1st *Djumādā*, Birthday of his Noble Highness, Our Lord (*Ṣāḥib al-sumūw al-mufakhkham, Mawlānā*), the Honourable Crown-Prince (*Wali ʾl-ʿahd al-akram*).
1st *Radjab*, Feast of the "Friday of *Radjab*" of the people of Ṣanʿāʾ, and the Mashriḳ [29].
27th *Radjab*, Feast of the Honourable Night-Journey.

---

[27] It would be interesting to know where the compiler has obtained this information, unless it be from Turkish sources.

[28] The late Imām Yaḥyā.

[29] The Eastern Yemen.

15th *Sha'bān,* The Noble Night of Mid-*Sha'bān* (*Sha'bānīyah*). General Examination of Schools, preparatory to the holiday of the month of *Ramaḍān.*

30th *Ramaḍān,* End of the Fast, *Al-Waḳfah al-Ṣaghīrah* [30].

1st *Shawwāl,* Blessed *'Īd al-Fiṭr,* lasting 5 days.

1st *Dhu 'l-Ḥidjdjah,* First of the Ten of Moses [31], lasting 8 days.

9th *Dhu 'l-Ḥidjdjah, 'Arafah* Day, *Al-Waḳfah al-Kabīrah* [30].

10th *Dhu 'l-Ḥidjdjah,* Blessed Feast of the Sacrifices, lasting 8 days.

18th *Dhu 'l-Ḥidjdjah,* Going forth from the towns for *al-Nushūr,* and the mentioning on the *Yawm Ghadīr Khumm,* of the *Imāmate* of the Successor (*al-Waṣīy*) [32].

Two of these are specifically Yemenite festivals, most of the rest ordinary Moslem holidays. During *Ramaḍān,* Government offices are closed, and official duties performed in the evening only. Friday is a holiday.

The "Friday of *Radjab*" is said to be the date when the Yemen accepted Islam. The *'Īd al-Nushūr* on the 18th of *Dhu 'l-Ḥidjdjah* is a curious feast; its name would imply that it had something to do with a Spring festival [33]. According to Col. MUḤAMMAD HASSAN [34] the Ṣan'ānis go out from Ṣan'ā' to an eminence some three miles away. Tents are set up, and some units of the army also are brought out for the occasion. At 9 a. m., in the presence of the Imām and his ministers, poets and orators recite poems and addresses on the subject of the renewal of allegiance to the Imām. The Imām himself takes his rifle and fires at a special target [35], then his sons do likewise, and others follow suit. The ceremony concludes with an equestrian display, terminating at 12 noon. To this doubtless ancient ceremony, it is probable that the Zaidī rulers of the country have attached the ceremony of the Day of the Spring of Khumm, the latter a little place between Mecca and Medina known to the mediaeval Arab geographers such as YĀḲŪT. It was here, according to my Aden correspondent, on the 18th of the month of *Dhu 'l-Ḥidjdjah* that, on his last pilgrimage, the Prophet made a declaration in which he said, "Of whomsoever I am Lord (*Mawlā*), 'Alī is his Lord" [36]. The Zaidīyah and Bohras took this tradition and use

---

[30] This name is taken from the halt at *'Arafāt* (*al-wuḳūf li-'Arafāt*), and applied first to the *'Īd al-Aḍḥā,* and then by analogy to the Feast of the Breaking of the Fast at the close of *Ramaḍān.* See G. E. VON GRUNEBAUM, Muhammadan Festivals, 1951.

[31] Presumably the Ten Commandments.

[32] 'Alī the Prophet's son-in-law is intended.

[33] According to LANE it means the becoming green of land after summer rain, or the first herbage coming forth after the *Rabī'* rain.

[34] *Ḳalb al-Yaman* (Baghdad 1947), p. 161. Cf. E. ROSSI, L'Arabo parlato a Ṣan'â', p. 93.

[35] This is quite a common part of ceremony in S. Arabia. When we visited the 'Awdhalī Sultan in 1940 he fired a shot over our heads at a mark on the other side of the wādī.

[36] This tradition and the incident are quoted in the *Madjmū' al-A'yād* of MAIMŪN B. AL-ḲĀSIM AL-NUSAIRĪ ed. R. STROTHMANN, Der Islam (Berlin 1943), 27, p. 45, and AL-BĪRŪNĪ, The Chronology of Ancient Nations (8), p. 333, etc.

it in support of their contention (which they maintain in common with all Shī'ah that 'Alī, Muḥammad's son-in-law, was his lawful successor). They give orations at a place where there is water in which they revile Muḥammad's successors Abū Bakr and 'Umar for taking, as they assert, the place of 'Alī.

2. The Yemenite Almanac of Muḥammad Ḥaidarah

January

1
2 Mustard (*tartar*) is planted [37].
3 The tree leaves fall.
4 The cold in Aden increases.
5 The time is now suitable to plant madder (*fuwwah*) if planted.
6 Much dew.
7
8 The ground is trenched.
9 Abundance of oranges (*burtuḳāl,* but this word also means snuff).
10 Much hoar-frost (*ṣaḳī'*) and intense cold.
11 Drinking of water at sleep is forbidden.
12 Beef is avoided.
13 The face of the earth becomes green.
14 The *ḳiyāḍ* [38] of white *'alas* (rye) [39].
15 Rain is expected. Grafting of vines [40].

---

[37] According to the Military Report on the Aden Protectorate (Simla 1915), *tartar* means mustard-seed.

[38] *Ḳiyāḍ*, evidently employed here simply in the sense of harvest, was defined by my Aden correspondent as the season in which fruits and chick-peas (*'atar*) are planted, after the grains are harvested. A variant spelling is *ḳiyāẓ*, similar to the Class. Arabic *ḳaiẓ*, summer season. The peasants of the highland plain of the Yemen, according to E. Rossi, L'Arabo parlato a Ṣan'â' (16), p. 151, have four seasons, *ḳiyāḍ* raccolto del tardo inverno, *dithā'*, raccolto di primavera-estate, *'allām*, stagione delle piogge, *ṣurāb*, raccolto di autunno. These names resemble terms used in the pre-Islamic inscriptions. Landberg, Gloss. Daṯ, op. cit., gives *maḳīẓah* from the same root, meaning Spring. The seasons for *dhurah* in Tarīm are, *mūsim* (*dhurah*) planted after the floods *ṣaif*, planted at *Zabrah*, cropping in three months, and *shitā* planted at *Baldah*, cropping after six months.

[39] *'Alas* was said by my Aden source to mean wheat (*burr*, *ḳamḥ*), but a very white variety of it. Rossi, op. cit., renders it simply as grano. P. W. R. Petrie & K. S. Seal, A Medical Survey of the Western Aden Protectorate 1939-40 (Colonial Office, Dec. 1943), Middle East No. 66, p. 106 is the authority for the rendering as rye. 'Abd al-Wāsi' b. Yaḥyā, *Tārīkh al-Yaman* (Cairo 1346), p. 283, gives another name *nusūl*, by which it is known in the Yemen.

[40] The Yemen has many types of vine, distinguished, like palmtrees in Ḥaḍramawt, Iraq, etc., by name. 'Abd al-Wāsi', op. cit., p. 284 (cf. 1947 edition, p. 30) gives a list of these names, distinguishing 24 varieties of which 18 are grown in the vicinity of Ṣan'ā'. Some of these may be named after localities, such as *Tabūkī* from Tabūk in N. Arabia, *Shāmī* (perhaps from Syria or N. Arabia). Curiously enough many of these names are known to Ḥamdānī, *Ṣifat Djazīrat al-'Arab*, op. cit., p. 196, including the variety still known in the Yemen as *Rūmī*, which I suppose must therefore be rendered as Greek. For a list see also Muḥammad Ḥassan, *Ḳalb al-Yaman*, op. cit., p. 83.

16 Planting of sour pomegranates (a small sour green type).
17 The grain harvest (*ḥiṣād*) begins in Tihāmah.
18 *Fattah* (flat bread chupattis shredded into pieces) with honey must be eaten.
19
20 Hot foods are eaten.
21 The inside of the earth becomes hot.
22 Trees are planted.
23 The first of *Rabī'*. The sun is in *Dalw* (Aquarius).
24 Sugar-cane cut [41].
25 Roses collected till *Nīsān* (April 14th).
26 Cutting out of clothes is praised.
27 Small palms transplanted.
28 Planting of figs ends.
29 Young vine-shoots (*gharīsah*) transplanted.
30
31 Contracting of marriage is praised.

February

1
2 In Aden the cold increases greatly.
3 The eating of nuts (*djawz* [42]) and almonds (*lawz*) is praised.
4
5 Tamarind (*ḥumr*) collected.
6 Birds court.
7 Cotton and melons (*baṭṭīkh*) are planted.
8 Camel-stallions on heat.
9 Milk and cream abundant.
10 Substances move in the body.
11 Vines are planted.
12 From now, fruit-bearing trees are planted.
13 Birth of Muḥammad.
14 The harvest (*ḳiyāḍ*) of wheat (*burr*), harvested (*yuḥṣad*) till the end of *Shubāṭ* (March 13th).
15 Sugar-cane planted.
16 Male goats (*tais*) and rams (*kabsh*) come on heat.
17 Sesame (*djuldjulān*) is sown in Tihāmah at the running springs (*ghail*).
18 Birth of His Majesty, the Imām of the Age, the Commander of the Faithful in *Rabī'* I, 1286 H. (1869-70 A. D.) at al-Shahārah, and acknowledged Caliph in 1322 H. (1904-5 A. D.) in the Mosque (*Djāmi'*) of Ṣan'ā' [43].

---

[41] According to 'Abd al-Wāsi', op. cit., p. 284, brown sugar is manufactured in the Yemen where it is called *'Aṭawī*. In Ṣan'ā' sweetmeats used at feasts and marriages, are manufactured.

[42] Rossi, op. cit., gives this word the simple meaning of nuts, but properly it means walnuts, which according to Western Arabia and the Red Sea, op. cit., are grown in the Yemen.

[43] The notice refers to the late Imām Yaḥyā Ḥamīd al-Dīn, murdered on the high road in 1948 along with a number of his close relatives.

19
20
21 Sowing of grain in al-Rabādī and Ḥubaish [44].
22 The last of the blowing of the South Wind (*Azyab* [45]). *Al-Ḥūt* (Pisces).
23 Sexual desire (*bāh*) begins to be noticed.
24 Sugar-cane is planted.
25
26 Sap runs in the wood to the top of it.
27 Night and day are equal.
28 Cabbage (*lahānah* [46]), i. e. *kurunb* (an Egyptian word) is planted.

March

1 Grape-vines (*ʿinab*) are irrigated.
2 Day begins to grow longer.
3 End of tree-planting.
4 The wolf comes out, and lice (*barāghīth* [47]) stir.
5 Pruning (*taḳlīm*) of grape-vine-stocks.
6 Contracting of marriage is avoided.
7 Eating of roast meat (*ḥanīdh*) approved.
8 For eight days copulation is avoided.
9 The pollenating winds (*lawāḳiḥ*) blow, the heat appears, wild beasts and the goat (*māʿiz*) bear, and birds sing.
10
11 The first of the violent wind (*al-ḥusūm*), and the days of intense cold (*al-ʿadjūz*) [48].
12 Roses and flowers planted.
13 Grapes are abundant in some districts.
14 *Adhār*. The opening of financial year 62 [49].
15 Indian banana (*al-mawz al-Hindī*) planted.
16 Black cummin (*ḥabbah sawdāʾ* [50]) is sown in hot districts and taken after half a year. The days of violent wind (*al-ḥusūm*) come to an end.

---

[44] Al-Rabādī (perhaps Ribādī or Rubādī), is said to be the name of a group of villages.

[45] Rossi, op. cit., p. 244, gives "vento da sud, *Azyab* o *ʿAdanī*". Mukallā dhow-men described it as a wind blowing westwards. It is found in Hamdānī, op. cit., p. 154 and Abū Makhramah, see O. Löfgren, Arabische Texte zur Kenntnis der Stadt Aden im Mittelalter (Uppsala 1936-50), p. 37. It seems to be known in Aethiopic.

[46] Cf. Rossi, op. cit., p. 164, cavolo.

[47] *Barghūth* is said by my Aden correspondent to have this sense in Taʿizz. In Class. Arabic it means gnats, etc.

[48] According to my Aden correspondent, the intensely cold days of *al-ʿAdjūz* come at the very end of the cool weather, and are so named because the old woman (*ʿadjūz*) and old man feel the cold then acutely. Lane, Lexicon, gives this period as coming under the star *al-Ṣarfah*.

[49] The opening of the Ottoman financial year 1362, corresponding to 11th *Rabīʿ* II, 1365 H.

[50] Also called *kuḥṭah* (ʿAbd al-Wāsiʿ, p. 286) a very well known condiment and medicament in Arabic literature. R. Dozy, Supplément aux Dictionnaires Arabes,

17

18

19 Young locusts (*dabīb*) begin to appear.

20 Contrary tempestuous winds.

21 Grain is winnowed at Ibb and Djiblah and its district.

22

23

24 The sun in its height at the Zodiacal sign (*burdj*) *al-Ḥamal* (Aries), remaining in it a month.

25

26 The *khaṭāṭīf* (birds of prey) (called *sawaiyad*) appear [51].

27 Goat-meat (*djidā'*) eaten.

28 Red millet (*dhurah*) sown, and harvested after seven months.

29

30

31 Snakes open their eyes, and the pollenating winds (*lawāḳiḥ*) blow.

April

1

2 Beginning of the fall of the rains in the East and South-East of the Yemen until August.

3

4 The light parts of the waters of the seas evaporate, and their waters thicken [52].

5

6 Sowing of grain in al-Suḥūl and Djiblah etc.

7 Turbulence of the sea.

8 Copulation is praised.

9 Times of bleeding and cupping.

10 Trees in leaf.

11 *Mashwī* (sheep stuffed with rice and eggs) eaten.

12 Contracting of marriage praised.

13 Presentation of respects to Kings approved.

14 Grain sown in the hot and temperate districts, and plucked after four months, and, in Tihāmah, "vigna sinensis" (*didjr*) etc., is plucked after 75 days.

15

16

I, 242, quotes the following relevant notice from the author Ibn al-Baiṭār (ob. 646 H. 1248 A. D.), "des grains noirs qui viennent du Yémen et dont on se sert pour guérir les maladies des yeux".

[51] *Khaṭāṭīf* seems to be a plural of *khuṭṭāf*, a bird of prey, of the desert, according to Lane's Lexicon.

[52] The month of April has a very high humidity, on an average only two degrees below that of July; perhaps this is what the writer of the almanac intends.

17 Planting of the broad bean (*fūl*) is suitable in cold places, but it is unsuitable elsewhere; it is harvested after six months.
18
19 The beginning of coming down of the rains on the hill-slopes. Palm-trees are fertilised.
20
21 The insects of the ground appear[53].
22 End of the ascendance of the sun.
23 *Ṣaif*. The sun is in *Thawr* (Taurus).
24 The heavy rains on the hill-slopes begin, and the heat intensifies.
25
26 The summer (*Ṣaif*) of Aden continues on till September.
27 The blowing of the North Winds (*Shimāl*).
28 Planting of *dhurah* begins in Ḥudjarīyah at the end of the month[54].
29 Strawberry (*tūt*) and lemon drink used.
30 Winnowing.

May

1
2 The heat in Aden intensifies.
3 Birds bear chicks.
4
5 The South Wind (*Djanūb*) blows, and the wādīs expand and over-flow.
6
7 Light clothing is worn.
8 Lemon drink is used.
9 Eating of salty things is avoided.
10
11 Eating of *fattah* with meat is found agreeable.
12 The blood stirs (i. e. sexual potency is said to increase).
13 Fruit bunches, and almonds become ripe.
14 Sowing of barley (*sha'īr*) in some of the country.
15
16
17 The *Kāwī* wind[55] blows in Aden.
18 Eating the flesh of male goats (*tais*) is forbidden.
19

[53] In Dhala (Ḍāli') about this date, I noticed an insect with transparent wings which seemed to come out of a hole in the ground. On discovering this, our Awlaḳī soldiers began to eat the insect which they knew under the name of *shaḍawī* or *shaḍī*. Perhaps allusion to this is intended, or the statement may be general. For similar practices cf. DJĀḤIẒ, *Ḥayawān* (Cairo) VI, 398.

[54] This entry falls about the end of *Djumādā* I, and it is no doubt this Moslem month which the writer means.

[55] LANDBERG, op. cit., p. 2598, gives the sense of "contrarier" for *kāwī*. The Arabs say that this *Kāwī* wind helps to ripen the date-crop (*Kharīf*).

20 Sowing of *gharb* (giant millet, "sorghum vulgare" [56]) and pennisetum (*dukhn*) in al-Ḥudjarīyah.
21 Copulation is avoided.
22
23
24 The first of the breezes (*bawāriḥ* [57]). *Al-Djawzā'* (Gemini).
25
26 Contracting of marriage is avoided.
27
28
29 The *Ṣabā* wind stirs, and sea-voyaging is good.
30 The cool wind stops in Aden.
31 The Friday of *Radjab* [58] in Ṣan'ā' and the East (Mashriḳ).

June

1
2 Planting of pennisetum (*dukhn*) in Tihāmah and the hot districts.
3 The falling of the rains begins in the high ground, and the flowing of floods (*sail*) in Tihāmah [59].
4
5 The crops become green.

---

[56] Military Report, op. cit., gives *gharib*, coarse red jowari. PETRIE, loc. laud., gives a variant, *ghorba*. In Wāḥidī territory *gharbah* was defined to me as a kind of *dhurah* resembling the *Ḥabashī* type. In Lahej according to the *Hadīyat al-Zaman* (see footnote no. 59, infra) the best kinds of *dhurah* (*dhourra*) are *Bukr*, *Baiḍā'*, *Ṣaif*, and *Shām*. It is also known there as *Hind*. *Bukr*, called *Haimar* in Dathīnah, is red, *Ṣaif* is white.

[57] In Ḥaḍramawt, *barāḥ* a breeze, this pl. is probably connected with it.

[58] The first Friday in *Radjab* is a feast-day. See MUḤAMMAD ḤASSAN, op. cit., p. 162, and supra. By the Mashriḳ, the eastern part of the Yemen is meant.

[59] The mountain rains cause floods which come down the wādī beds to the Tihāmah plains where bunds are ready to catch the water and lead it into the fields. A list of the names for the irrigation bunds and fields is given by AḤMAD FAḌL B. 'ALĪ MUḤSIN AL-'ABDALĪ, *Hadīyat al-Zaman* (Cairo 1351 H.), p. 32, which is worthy of note because these terms occur in placenames. "The bunded channels leading the water off the main wādī (*a'bār*, sing. *'ubar*) split off into lesser channels (*sawāḳī*), called *ashrudj*, singular *shardj*. Every *ḥaḳl*, (which is any land which is bunded [*tasawwam*] with a wall of earth to retain the water) is called a *faladj* or *dahil*, and what is smaller than that is called a *djirbah*, while what is smaller than a *djirbah* is called a *fakhkhah*. A field at the upper part of the channel (*'ubar*), or the wādī, is (*marda'*), and fields at the lower part of the channel or the wādī are called *mantā'*. Each *djirbah* or *dahil* has a name (of its own) by which it is distinguished from others." — N. B. the vowelling of *fakhkhah*, *marda'*, and *mantā'* is doubtful as the text is unvowelled and these words are not known to the lexicons. Another land measure is the *ḍamad* equivalent to the Ḥaḍramī *hidjdj*, the amount of land which a pair of oxen can plough in one day. Other terms of topographical interest in use in Ḥaḍramawt are *muṣfā*, a wide channel carrying water from a well to the crop *mi'dāh*, a kind of conduit from the well to the field of clay or stone, and *kadhīr* the clay mound between each *maṭīrah* or irrigation-bed. In Yemen the *maṭīrah* probably corresponds to the *libnah* ('ABD AL-WĀSI', op. cit., p. 289) which is 10 cubits (*dhirā'*) square.

6 Disease and plague are lifted by the permission of God.
7 The hot winds (*samā'im*[60]) commence, and the grapes turn black.
8
9 The West Wind (*Dabūr*)[61] blows.
10 Rice is planted.
11 Bodies become strong (or lusty)[62].
12 The ground becomes split[63].
13 Diseases (*'āhāt*) and bad smells are lifted.
14 Sowing of *gharb*[64] in hot districts.
15 Pennisetum (bullrush millet, *dukhn*) planted in the mountains.
16 The heat intensifies to a peak in Aden.
17 The longest day in Ṣan'ā', 13 hours, 20 minutes.
18 The fall of the rain-period (*nuḳṭah*[65]) on Tuesday night[66].
19 The *Kāwī* wind intensifies.
20 Entering the Bath is found agreeable.
21 The times for using acid (*ḥawāmiḍ*) (drinks or fruits ?).
22 The rain-period (*nuḳṭah*) of Cancer (*Saraṭān*).
23 The period of the blowing of the hot wind (*samūm*) for 70 days. *Al-Saraṭān* (Cancer).
24 Winnowing of wheat (*burr*) in Ibb from the 15th of June to the 11th July[67].
25 Collection (*ḳiṭāf*) of honey[68].

---

[60] Landberg, op. cit., *samūm*, vent brûlant.

[61] The name *Dabūr* was known to Ḥamdānī, op. cit., p. 154. The following terms, used by seamen, were given me by a Mukallā dhow-master, *Dābir*, West (*dabbar*, to go westwards), *Mākhir*, East (*makhkhar*, to go east-wards).

[62] The word is *ḳuwwah*, "strength", usually, "sexual powers".

[63] By the splitting of the ground, the muddy surface of the wādīs and their fields may be intended. This ground is run smooth by the floods, then dries in the heat; its surface cracks and peels. After this entry, in brackets, appears the word *al-taḥārīḳ* which I have not been able to have explained satisfactorily. Perhaps the senses given in the lexicons may apply, i. e. "the burning up of herbage by the heat".

[64] See footnote no. 54, supra.

[65] The term *nuḳṭah* (lit. "point"), is applied to rain which falls on a certain known day, and then ceases abruptly.

[66] This notice really relates to the 17th of June, because, according to Arab reckoning, the 18th of June would commence at 6 p. m. on the 17th of our June.

[67] The text says, "between the two *Ẓulms*", for which see table III, col. 2.

[68] For bees and honey, see H. Scott, In the High Yemen, London 1942, pp. 58-9, & W. H. Ingrams, Bee-Keeping in the Wadi Du'an (Man). (London 1937.) 37, No. 33. The following notes are supplementary. Summer honey is called *Shawwah* or *Ṣaifī* in Daw'an (Du'an); it is said to be hot, the bees feeding on *sumur* (acacia) flowers. The autumn honey is called *Kharfī* because it comes in *Kharf* (i. e. *Kharīf*) at the star *Buṭain* (which would fall at the end of *Kharīf* according to Landberg's *'Amaḳīn* calendar, i. e. 7th-21st October); it is cool, the bees having fed on *'ilb*-flowers (Zizyphus Spina-Christi). In Daw'an there are usually two honey-seasons (*wādjibāt*) each year, but there are sometimes three seasons, one falling in the season *Rabī'*, besides *Kharf* and *Ṣaif*. During *Rabī'* the bees feed on *'ilb*, and the honey is called after the season, *Marba'ī*, like the two other collections. It is interesting to find that Pliny Secundus reports two types of incense, Carthiathum and Dathiathum, which

26 Evening of the blessed Night-Journey [69].

27 The waters sink into the earth, and the breeze (*hawā'*) lightens.

28 Sowing of Arab wheat (*burr*, a small-eared Yemenite kind smaller than that found elsewhere), and barley.

29 Sowing of *khashkhāsh* (poppies, usually pronounced *khashkhash*) [70], and *mūmah* [71]. It is plucked after five months.

30 Close of the season of rain in the hill-slopes (probably the flat part at the foot of the mountains is meant).

July

1 The lengthening of the day comes to an end.

2 The heat intensifies in the Yemen.

3 Drinking of purgatives avoided, except of necessity.

4

5 Perfuming oneself with scents is found agreeable [72].

6

7 The hot winds (*samā'im*) blow for 51 days.

8 Planting of bullrush millet (*dukhn*) in the hot and the temperate districts.

9 *Kinib*-millet [73] and *ṭuhuf* (*ṭahaf*, "myrica gale") are planted, and harvested after 70 days.

10

11 The last of the breezes (*bawāriḥ*).

12

13 The Night of Mid-*Sha'bān* (i. e. 14th *Sha'bān*) [74].

14

15 Sowing of white *'alas* (a sort of wheat) [75], *ḥilbah* (fenugreek, Trigonella Foenumgraecum) [76], barley, and lentils (*'adas*), and plucked after three months.

---

would correspond to *Kharfī* or *Kharīfī*, and to a word *Dithā'ī*, from the season *Dithā* for which see footnote no. 38, supra. The text with the Ḥimyarite equivalents may be consulted in C. Conti Rossini, Chrestomathia Arabica Meridionalis, Roma (1931), p. 22. Al-Baḥr al-Zakhkhār (Cairo 1949) III, 408 gives the Yemenite equivalents *Rabī'ī* and *Ṣaifī*.

[69] Koran, Sūrah, XVII.

[70] Rossi, op. cit., p. 168, *khishkhāsh*, semi di papavero, sedativi.

[71] *Mūmah*, the cotton seed (*bazrah kuṭnah*), made into medicaments placed on boils to bring out the solidified pus.

[72] Yemenites are greatly addicted to perfumes. One perfumes one's clothes by sitting over a censer of burning frankincense as in ancient Baghdad. They often sprinkle perfumes over departing guests. There are many other such practices.

[73] For *kinib*, see C. v. Landberg, Arabica V (Leiden 1898), p. 213; Western Arabia, op. cit., p. 595. *Ṭuhuf* is known to Hamdānī, op. cit., p. 199. See Löfgren, Arabische Texte, op. cit., glossary.

[74] A Moslem feast; it was also the date of the pilgrimage to the tomb of the Prophet Hūd from ancient times.

[75] See footnote no. 39, supra. See also Landberg, Gloss. Daṯ., op. cit.

[76] Petrie & Seal, op. cit., p. 109. The seeds are ground and water is added to the powder. This water is then poured off, and the remaining paste beaten with fresh water. It froths, thickens, and rises; at this stage various condiments, vegetables, or even fish are added.

16
17 Dates collected.
18
19 Rise of the Yemenite Sirius (*al-Shi'rā al-Yamānī*).
20 Locusts die.
21 Male goat meat (*tais*) is harmful at mid-day.
22
23
24 The first of *Kharīf* of the Yemen. The sun in *Asad* (Leo).
25 Planting of lucerne (*ḳaḍab*) is not suitable during the whole of the season of *Kharīf* only; rain and cold are harmful to it.
26
27
28 The night begins to increase.
29 Care must be taken not to overload the stomach.
30 One bathes in cold water.
31 The first of the seven days of *al-Bāḥūr* [77].

August

1
2 The lice (*barghūth*) go [78].
3 Cutting out of clothes is praised.
4 Entering the Bath is found agreeable.
5 The rain-flood (*sail*) is expected in Tihāmah.
6 Affections of the eye (*ramad*) increase in the heat.
7 Carrots (*djazar*) and water-melon (*baṭṭīkh*) planted.
8 Copulation is forbidden on account of the intense heat.
9 Wheat (*burr*) reddens.
10 The waters sink (into the earth).
11 Grapes and figs become sweet.
12
13 All fruits ripen.
14 The first of the Bear Constellation (*Banāt Na'sh*) appears.
15 Plucking of fruits begins.
16 Planting of *gharb* (giant millet, "sorghum vulgare") in part of Tihāmah.
17 Planting of *dhurah* in Zabīd and Rima'.
18 The latter part of the night becomes cold.
19 The rain-floods (*sail*) irrigate Tihāmah.
20 The eating of fruits is agreeable.
21 Milk is scanty in the udder.
22 Contrary winds.
23 The Blessed *Lailat al-Ḳadar* [79]. *Al-Sunbulah* (Virgo).

---

[77] *Al-Bāḥūr* is the name applied to the hottest days of July (Class. Arabic).
[78] See footnote no. 47, supra.
[79] The Night of Power, Koran, Sūrah, XCVII.

24 Leaves of the trees change.
25 Quinces (*safardjal*) and pomegranates (*rummān*) abound.
26 Contracting of marriage praised.
27 Melons (*ḥabḥab*, a large green melon with pink flesh), garlic (*thūm*), and onions (*baṣal*) planted.
28 Curdled milk (*al-laban al-rāyib al-ḥaḳīn*) is employed (for its cooling properties).
29 *Fidjl* (radish, "Raphanus sativus") (i. e. the Yemenite *baḳal*) is planted.
30
31 Cessation of the rains in the mountains.

September

1 The second (star) of the Bear Constellation.
2 The rain-floods (*sail*) cease in Tihāmah.
3 The third (star) of the Bear Constellation.
4 Planting of *bīnī* and *ḥidjnah* in Tihāmah, red and white *dhurah* (*al-thālithī*) [80].
5
6
7 The heat abates, and fish are born.
8
9
10 Ripe dates (*ruṭab*) and grapes abound.
11 The fourth (star) of the Bear Constellation.
12 The *Kāwī* (wind) lightens in Aden.
13
14
15 Fruit-trees planted in some of the sunny districts.
16 Time of planting the vine and pruning it.
17 Cucumbers (*khiyār*) are collected in the provinces (*nāḥiyah* [81]) and administrative district of Ta'izz.
18 Sour pomegranates, figs, and the prickly pear (*al-balas al-Turkī*, or *al-shawkī*) [82] are planted.
19
20 Grapes are plucked. The sap rises to the top of the trees.
21 Green grain (*djaḥīsh al-ḥubūb*) in the *liwā'* (a larger administrative district than the *nāḥiyah*) of Ta'izz.

---

[80] *Bīnī* and *ḥidjnah* must, from the text, be respectively, species of red and white *dhurah*. They also seem to be called *al-Thālithī* which my Aden informant avers, refers to the star, by name *al-Thālith* at which these grains are planted, i. e., the third (*thālith*) star of the Bear Constellation.

[81] For administrative units of the Yemen see Nello Lambardi, Divisioni amministrative del Yemen. Con notizie economiche e demografiche (Oriente Moderno). (Roma 1947), 27, pp. 142-62.

[82] Rossi, op. cit., p. 164, *balas Turkī*, fico d'India, the prickly pear, so called in Ta'izz, in Ḥaḍramawt, *tīn barshūm*.

22 The winds blow.
23 The crows come to the fields. *Al-Mīzān* (Libra).
24 The fifth (star) of the Bear Constellation.
25 Planting of *al-Khāmisī* [83] in Tihāmah, grain.
26 Quinces and pomegranates come to an end.
27
28 Grain-harvest in the districts of the Mashriḳ (i. e. Eastern Yemen).
29 The end of the *Kāwī* wind in Aden.
30 The heat in Aden ends.

October

1
2 Sixth (star) of the Bear Constellation.
3 Planting of *Sābiʿī* [84] in Tihāmah, grain and "vigna sinensis" (*didjr*).
4
5 Chick-peas (*ḥumuṣ*) [85] planted along with *dhurah* (*ṣawmī*) [86] and plucked after three months.
6 Aden winter, from October till March.
7 Abundance of lemons (*līmūn*).
8 The harvest (*ḳiyāḍ*) [87] of broad beans (*fūl*) and curcuma (*ʿuṣfur*, syn. *kurkum*) in the Mashriḳ.
9 The harvest (*ḳiyāḍ*) of barley (*shaʿīr*) in al-Djanadīyah and the surrounding districts.
10 Lemon drink is made.
11 Moist foods are employed.
12 Drinking of water at night is disapproved.
13 Eating of *tharīd* (*fattah* [88] with meat) is found agreeable.
14 The *Ṣabā* wind rises [89].

---

[83] An allusion to the grain planted at the fifth star of the Great Bear constellation. E. Glaser, Sternkunde, op. cit., says that *Khāmis*, *Sādis*, and *Sābiʿ*, are the chief stars of the Bear. Cf. table II, col. 3.

[84] Grain planted at the seventh star of the Bear.

[85] Another term for *ḥumuṣ* is *ṣunburah* or *ṣumburah*.

[86] According to the Military Report, supra, *ṣawmī* is red *jowari* (millet), better than *gharib* (see footnote no. 56, supra).

[87] For *ḳiyāḍ*, see footnote no. 38, supra.

[88] *fattah*, flat chupatti bread shredded into pieces and moistened; here it is made into a dish known as *tharīd*, a dish which the Prophet greatly liked.

[89] The *Ṣabā* wind according to Hamdānī is also known as *al-Ḳabūl* (op. cit., p. 154), and it blows in the reverse direction from the *Dabūr* (see footnote no. 61, supra). According to the Mukallā dhow-master, the main winds are *Azyab* which blows to the west, *Shamāl* which blows from the land to the sea and takes one to India (known in the Wāḥidī area as *al-Ḳaws*), *al-Barrī* which takes you from the land to sea, *al-Rudūd* (or *al-Ḥādjah*, or *al-Ḥāyah* in local pronunciation) which carries you to the coast. A calm is *al-Ḥawāl* or *al-Shawāl*. Dawʿanis gave me the following names for winds in their Wādī, *al-ʿIlyā* (in *Ṣaif*) a wind blowing inland from the direction of the sea, *al-Shimāl* (in *Shitā*) a wind from the North, *al-Sharḳī* a wind from the East, *al-Ḳiblī* a wind from the West; they knew also *al-Ḥabbānī* (lit. from Ḥabbān) blowing from

15 In al-Ḥudjarīyah the green ears (of *dhurah, djaḥīsh*) begin.
16
17 The movement of copulation abates.
18
19
20 The harvest (*ḥiṣād*) of grains in part of al-Ḥudjarīyah.
21 Looking at the clouds is praised.
22 The drinking of soft drinks (*al-muraṭṭabāt*, such as pomegranate, almond, *tūt* [strawberry] [90] lemons and limes) is found agreeable.
23 The first of *Shitā'*. The sun in *'Aḳrab* (Scorpio).
24 Blood-letting (*ḥadjāmah*) and bleeding (*faṣd*) disapproved.
25 The waters sink (into the earth), and the sea is disturbed.
26 The *Azyab* wind [91] blows from the N. E. to the S. W.
27
28
29 Planting of wheat (*burr*) in the Ta'izz district (al-Ta'izzīyah) where there are perennial streams (*ghail*) until mid-*Tishrīn* II (November 28th).
30
31 When wood is cut, it does not get wormed [92].

November

1 Presentation of respects to Kings approved.
2 The harvest (*ḥiṣād*) of grains, *gharb* (giant millet, "sorghum vulgare"), begins in al-Ḥudjarīyah.
3 The cool breeze (*hawā'*) begins.
4
5 Mosquitos (*nāmis*) and gnats (*ba'ūḍ*) abound.
6 The cold at the latter part of the night is harmful.
7 Violets (*binafsadj*) planted.
8 The contracting of marriage is found agreeable.

---

the South-West at the time of floods (*sail*). I have little doubt but that the names given by Ḥamdānī, in his section on winds, might be discovered to be still in use.

[90] The *tūt* is probably the strawberry, though lexicons give mulberry. My Aden source says that it is made into a drink in the Yemen which does the blood good and temperates the heat. Almonds are ground and made into a sort of drink but I do not consider it very palatable.

[91] For *Azyab*, see footnotes nos. 45 and 89, supra. According to another source this wind lies between the *Shamāl* and *Dabūr* winds. The opposing wind to the *Shamāl* is the *Djanūb*, blowing from the Yemen.

[92] In Daw'an, *'ilb* is cut in winter and seasoned for one year. Then nothing will eat it. If it is cut in *Kharīf* the white wood will be eaten by the white ant, but not the red parts. Ibn al-Mudjāwir, *Tārīkh al-Mustabṣir*, ed. O. Löfgren (Leiden 1951), p. 81, says that wood cut at the period known as *Layālī al-Bīḍ*, or during the waning of the moon will get wormed. At 'Azzān an insect vaguely resembling a bee and called *khanḍarūr* flew into our room and the Arabs said that it eats wood; perhaps it is Vespa orientalis, see H. Scott, In the High Yemen, op. cit., p. 66.

9 The eating of rice of *al-zurbiyān* (a dish containing the ingredients, meat, ghee (*saman*), saffron, cardamom (*hail*), cloves (*ḳurunful*), an Indian or perhaps even Persian dish) is found agreeable.
10 The contracting of marriage is disapproved.
11 Wool is worn.
12 Kites (*ḥid'ah*), vultures (*rakham*)[93], and (the birds of prey called) *sawaiyad*, go to the low-lying land (*ghawr*, which is also a name applied to the western part of Tihāmah).
13 Ants hide themselves in the hollow of the earth.
14 In Ibb; "The bull (*thawr*) does not enter in *al-Thawr* (the star)[94]."
15 The barley harvest (*ḥiṣād al-sha'īr*), lasting half a month in the mountains.
16
17 The South winds (*riyāḥ al-djanūb*) blow.
18 Insects hide themselves.
19 Moistness of the atmosphere commences.
20 The sea is disturbed.
21 Mists (*ghaim*) are frequent.
22 In al-Ḥudjarīyah the harvest (*ṣurāb*, Rossi, raccolto di autunno) of *dhurah* begins. When wood is cut it does not worm.
23 *Al-Ḳaws* (Sagittarius).
24-30th No notices as they fall in the next Moslem year.

December

1-4 Notices also lacking.
5 Drinking of cold water at night is forbidden.
6 Toasted and dry foods (*ḳawālī* [sing. *ḳalīyah*] *wa-nawāshif*) are eaten (by toasted foods locusts or maize-cobs [*hind*] parched in girdle pans or ovens [*afrān*], or green ears [*djaḥīsh*] etc. are meant, and by dry foods, almonds, broad beans [*fūl*] etc.).
7 Harvest (*ḳiyāḍ*) of wheat (*burr*)[95] after the *dhurah*.
8 Flies and mosquitos (*nāmis*) begin to die.
9 The sniffing of scents is praised.
10 The cold intensifies in Aden.
11 Mid-day (*ẓuhr*) reaches 6 hours, 22 minutes.
12 The waves of the sea become long.
13 *Tāsū'ā'*-day (9th *Muḥarram*).
14 *'Āshūrā'*-day (10th *Muḥarram*); it is customary to fast it[96].

---

[93] Vulture (*rakham*), but this dictionary rendering may not be exactly what one would call a vulture in English.

[94] "The bull does not enter in *Thawr*", is a widely used and ancient cliché, explained to me as meaning that beasts do not go out to the fields at this star.

[95] According to 'Abd al-Wāsi', loc. laud., the best variety of wheat grown in the vicinity of Ṣan'ā', is a whitish-red type called *al-Burr al-Bawnī*, deriving its name from the plain of al-Bawn, about six miles N. of Ṣan'ā'. Another type is called *al-Samrā'*, said to be reddish-black, and is grown in *Shu'ūb* and *al-Ṣāfiyah* near Ṣan'ā'.

[96] The 9th and 10th of *Muḥarram* are important as feast days and fast days, the latter being traditionally the date of the death of the Prophet's grandson Ḥusain

III

15
16 Grain (*ṭa'ām*) is planted in part of Tihāmah.
17 Ants hide themselves in the belly of the earth.
18 The eating of fowl-flesh is disapproved [97].
19 Steam comes out of mouths.
20 The eating of pigeon-flesh is agreeable.
21 The eating of fish is disapproved.
22 *'Īd Ḥannūkah* with the Jews.
23 *Al-Djady* (Capricorn).
24
25 Tree-leaves fall.
26 In Ṣan'ā', the shortest day, 11 hours, 20 minutes.
27
28
29
30 Eating of beef and oranges (*turundj*) forbidden.
31 Blood-letting (*ḥadjāmah*) is forbidden [98].

3. Acknowledgements

I am indebted to Mr. Muḥammad Luqman of Aden for assistance in the difficult task of interpreting the Almanac which I had unfortunately little time to study with informants while in Aden. Where he was unable to offer a reliable explanation he sent my queries to Aden and obtained replies from scholars there; these anonymous answers have been referred to as "my Aden correspondent". I am indebted to Professor Sidney Smith for drawing my attention to an article in the Jl. of Hellenic Studies, and for notes made from his lectures on pre-Islamic South Arabia, and to Professor Wittek for information on the Ottoman financial year.

---

at Kerbela, and of course, a particularly significant date to the Zaidī (Shī'ah) rulers of the Yemen. For notes on customs observed in S. Arabia on these dates, see C. v. Landberg, Etudes sur les Dialectes de l'Arabie Méridionale, Daṯînah (Leiden 1909-13), p. 1547, and Gloss. Daṯ., op. cit., p. 2296.

[97] About this season in the Wāḥidī Sultanate I was unable to obtain fowls or eggs. The Arabs said the eggs were being set; this may perhaps be the underlying reason for the entry here.

[98] Very little has been written about agriculture in the Aden Protectorates. Apart from pre-1939 material, there might be mentioned, B. J. Hartley, Dry Farming Methods in the Aden Protectorate (Middle East Supply Centre, Agricultural Report no. 1). (Cairo 1944), pp. 37-42, Anon, Farmers in Arabia (Sport & Country). (London 1943, July 23), p. 50-1 (ill.), and the annual reports on agriculture of the Aden Government. A number of reports on specific aspects of agriculture have been made to the Colonial Office. Some are listed in Doreen Ingrams, Survey of Social and Economic Conditions in the Aden Protectorate (Asmara 1949), which has a section on agriculture (pp. 109-124).

IV. Bibliography

(1). Cerulli E., Le stazioni lunari nelle nozioni astronomiche dei Somali e dei Danākil (Rivista degli Studi Orientali). Roma 1929-30. 12. pp. 71-8.
(2). — — Nuovi appunti sulle nozioni astronomiche dei Somali (Rivista d .Stud. Or.). Roma 1931-2. 13. pp. 76-8.
(3). Conti Rossini C., Sul calendario astrologico degli Habab (Rassegna di Studi Etiopici). Roma 1946. 6. pp. 83-92.
(4). Ferrand G., Introduction à l'Astronomie nautique arabe. Paris 1928.
(5). Glaser E., Die Sternkunde der südarabischen Kabylen (Sitzber. d. mathem.-natw. Kl. d. Kais. Akad.d. Wiss.). Wien 1885. Bd. 91, Abt. 2. pp. 89-99.
(6). Hommel F., Über den Ursprung und das Alter der arabischen Sternnamen und insbesondere der Mondstationen (Zeitschr. d. deutsch. morgenländ. Gesellschaft). Leipzig 1891. 45. pp. 592-619.
(7). Landberg C. von, Glossaire Daṯînois. Leiden 1920-42. II. pp. 1092 seq. & 1449 (Contains additional bibliography).
(8). — — Etudes sur les Dialectes de l'Arabie Méridionale I, Ḥaḍramoût. Leiden 1901. p. 585.
(9). Lethielleux J., Au Fezzan — le Calendrier agricole (IBLA). Tunis 1948. 11. p. 73.
(10). Littmann Enno, Sternensagen und Astrologisches aus Nordabessinien (Archiv für Religionswissenschaft). Leipzig 1908. 11. pp. 298-319.
(11). Al-Ma'lūf, Amīn Fahd, An Astronomical Glossary (English-Arabic). Cairo 1935. (Not available to me.)
(12). Muḥammad Ḥaidarah, *Taḳwīm Ṭawāli' al-Yaman li-Sanat* 1365 (i. e. 5th Dec. 1945). Aden, Fatāt al-Gazīrah Press. (The Almanac used in this article.)
(13). Musil Alois, The Manners and Customs of the Rwala Bedouins. New York 1928. p. 7 seq.
(14). Naval Intelligence Division. Geographical Handbooks. Western Arabia and the Red Sea, B. R. 527 (Restricted). London 1946. Appendix E., The Arab Cọuntryman's Division of the Year into Seasons, and his Weatherlore. p. 605 seq.
(15). Owen T. R. H., Notes on an Arab Stellar Calendar (Sudan Notes and Records). Khartum 1933. 16. pp. 67-71.
(16). Rossi Ettore, L'Arabo parlato a Ṣan'â'. Roma 1939. pp. 150-1.
(17). Ryckmans G., Rites et Croyances pré-islamiques en Arabie méridionale (Le Muséon). Louvain 1942. 55. p. 175.
(18). Sachau C. E., The Chronology of Ancient Nations. London 1879, ch. XXI. (Al-Bīrūnī's *Athār al-Bāḳiyah.*)
(19). Snouck Hurgronje C., Sa'd èl-Suwênî, ein seltsamer Walî. Zeitschrift für Assyriologie. Strassburg 1911. 26. pp. 221-39.
(20). Weinstock S., Lunar Mansions and Early Calendars (Jl. of Hellenic Studies). London 1949. 69. pp. 48-69. (Contains Arabian material.)
(21). Wright R. Ramsay, The Book of Instruction in the Elements of the Art of Astrology of Al-Bīrūnī. London 1934.
(22). M. B. S. Dictons sur la saison d'été (IBLA). Tunis 1952. 15. pp. 300-302.
(23). Nilsson Martin P., Primitive Time-Reckoning. Lund 1920. p. 251.
(24). Renaud H. P. J., Le Calendrier d'Ibn al-Bannâ' de Marrakesh. Paris 1948.
(25). Sbath P., Le livre de temps (857 H.) (Bull. de l'Inst. d'Egypte). Cairo 1932-3. 15. pp. 235-57.
(26). Tresse R., Usages saisonniers et dictons sur le temps dans la région de Damas (Revue des Etudes Islamiques) 1937.

IV

A Socotran star calendar

Tom Johnstone's departure closed our long association, personal and professional, that commenced in 1957 with an invitation to apply for an Arabic lectureship at the London School of Oriental and African Studies.

During the last year of the British in Aden and the tragic betrayal of the South Arabian Federal rulers, Major Peter Boxhall persuaded the British Middle East Command to sponsor an expedition to Socotra – Tom, my wife and I were among the civilian members attached to it – and it was then that he commenced his field study of the Socotri language. Though the imminence of the disaster in South Arabia was much on my mind we enjoyed our month in this lovely island, entranced by its dream-like quality and its strangeness – as of a lost world – and we spent convivial evenings in its tropical climate. I had to return to Cambridge to keep term, but Tom was able to stay on and take the rare opportunity to visit ʿAbd al-Kūrī and other islets off the Horn of Africa.

With such limited time I did not attempt to learn Socotri, especially since Ḥaḍramī colloquial Arabic is spoken or understood along the coasts. To complement research on the fishing villages and ports of the Ḥaḍramī littoral and their people, I concentrated on contact with the coastal communities and general observation of the island I knew I could never expect to revisit. Part of my work is already published,[1] but a full account of it is to appear in Brian * Doe's forthcoming *Socotra.*[2]

In all enquiry into Arabian agriculture, fisheries, and navigation, it is essential to establish the local star calendar in use. So I set about discovering this, but, as I had found in the Aden Protectorate and Jīzān, there are difficulties in obtaining informants with an accurate knowledge of the calendar. In the previous year at Ḥallāniyyah of the Kuria Muria islands I had tried in the brief few hours spent there to elicit data on star calendars employed locally from the headman (*ʿaqīd al-balad*) Saʿīd Muḥammad b. ʿAbdullāh. We spoke in Arabic, which these islanders do not know well as they are Shaḥrī speakers, and though I report below what he told me, I cannot vouch for its accuracy. It might serve, however, towards further enquiries.

The Socotri calendar overleaf was supplied to me by Khamīs b. Maḥāmīd al-Malḥamī – so called because he solders, (*yulaḥḥim*) – the *nākhudhā* of the dhow hired by the expedition, in Arabic characters. Tom, at my request, put this Arabicized Socotri into the transcription he adopted for Socotri and I think he checked it through with his Socotri speaking informants.

The division of the year into four seasons given me by Khamīs may possibly be an Arab convention only, not properly native to Socotrans; the people of the island perhaps only recognize and differentiate between the Kaws and Azyab winds or monsoons. Kaws is reckoned to last five months, though this must be approximate, and according to one of my informants, the closure of the sea is from Mudawrik[7] (*al-qufāl min Mudawrik*, equivalent, I presume, to Midrak (18–30 April)) and ended by Midrak (21 September). A source less reliable than the *nākhudhā* Khamīs, thought Kaws, and Kharīf[8] also, would begin with the Arabic star al-Qalb when the sea is turbulent – as indeed appears in the name of the Socotri star, *Jiʾish*, itself. On 25 and 26 March when we were on the island the Kaws wind blowing was said to be exceptionally early and was not expected until two months later. Azyab/Izyab begins at Shibēloh/Shiballah/Shibilluh/Shiballuh, variously pronounced, equivalent to al-Dalū, the first star of the season Ṣayrib,[9] on 4 October. This marks also the opening of the sea (*al-futūḥ*) and ends at the star al-ᶜAwwā, Socotri Karbāloh (5–17 April), according to the calendar. This is a little less than the seven months my informants assigned to it – but these dates are of course to be understood as approximate, and others in fact said Azyab ends at Midayrik=Midrak (18–30 April).

Ṣayrib was given as the local name for Shitāʾ, etymologically equivalent to Yemeni *ṣurāb*. About 1 Ṣayrib they plant bulrush millet (*dukhn*), which ripens after three months; it was said to be cultivated at Rās Mūmī, in cold places (*al-maḥallāt al-bāridah*) and in the eastern area of the island, but not at Sūq.

A list of seasons in the Yemen given by Glaser[10] and corresponding to the above, runs: Qiyāẓ/Qiyāḍ, Dithā, Bikār, Ṣurāb; Thomas,[11] for Oman, gives Shitā, al-Qayḍ, Kharīf, Ṣurab; and Wilkinson,[12] (Izki), Shitāʾ Ṣayf, Qayẓ Rubᶜ.

Each star period has a duration of 13 days except Sūd/Sūd al-Kabīr, corresponding to al-Naṭḥ, which has a duration of 14 days. An extra intercalary day in a leap year will probably be added to one of the stars.

The star commencing 12 February is called Miḍbah, stated to be equivalent to Ḥaḍramī *maḍbī*, Socotri *ḍábe*,[13] meaning to roast sheep or goat mutton on stones that have been thrown on to a wood fire. There is an entry in the Taᶜizz almanac at 7 March: 'eating of roast meat (*ḥanīdh*) is approved'.[14] Similar dietary recommendations are generally found in almanacs.

For Karbāloh I have noted also the word Dūtaʾ, which looks like a star name, but I have no further information.

Karbāloh (8 September) has an alternative name, Midayraykah/Midayrik. This appears to be a diminutive of Midrak, as its alternative name Midrak al-

The Star names of the calendar in Arabic[3] and Soqotri

The season of Rabī' (Spring)				
al-Han'ah	In Ḥaḍramawt commences on January 1st.	Ma'ōdīf	4 Jan	The Arabic *ghadaf*, to fish, is said to correspond with it.
al-Dira' (sic)		Fanzak	17 Jan	
al-Natrah (sic)	Al-Nathrah is 'the interstice that is between the two moustaches, against the partition between the two nostrils' in the lion (Lane).	Ḥafānī/Hanśiyyoh	30 Jan	
al-Ṭaraf	The fore or hinder part of a beast (Lane).[4]	Miḍbah	12 Feb	Cf. Ḥaḍramī *maḍbī*, meat grilled on hot stones.
al-Jabhah/Yubhah	The forehead of Leo (Lane).	Qānī d-'ilhoh	25 Feb	The two horns of the cow. (Arabicized Di'labā)
al-Zubarah (sic)	The shoulder-blades or mane of a lion (Lane)[4]	Di'dī	10 Mar	The front legs of the cow.
al-Sarfah	A single star behind the *kharātān* of the Lion, it is on the hinder part of the tail of the Lion . . . and is also called the sheath of the penis of the Lion (Lane).	Ṣafáqhon	23 Mar	Back, rear, hindquarters.
The season of Sayf (Summer)				
al-'Awwā	The haunch of the Lion	Karbāloh	5 Apr	Tail-hair. Azyab ends at this star
al-Simāk	Al-Simākān are 'the two hind legs of Leo' (Lane).	Midrak	18 Apr	
al-Ghufar	Ghufr, the young of the mountain goat.	Shibēloh/Shiballah	1 May	Having conceived, of an animal. A rain season *mūsim maṭar*).
al-Zubān	Zubānayān, the two horns of the scorpion.	Ber Sūd	14 May	
al-Iklīl/Kalīl	The head of Scorpio.	Sūd	27 May	Kaws wind at sea; closure of the sea *mughallaq al-baḥr*) to shipping.
al-Qalb	The heart. Cf. the star Qalb al-Asad.	Ji'ish	9 Jun	Means 'turbulent'. Strong Kaws wind on land.
al-Shūl	Shawlah, 'the part that it raises of the tail of the scorpion' (Lane).	Qadāham[5]	22 June	First beginning of the ripening of dates.

The season of Kharīf (Autumn)				
al-Na'āyim		Fanzak	5 Jul	
al-Baldah	al-Baldah, the breast of the Lion.	Hafānī	18 Jul	
al-Mirzam		Qānī d-'ilhoh	31 Jul	
al-Suhayl		Dā'īdī (=Di'dī Supra?)	13 Aug	Sulaymān al-Mahrī[6] advises: 'Beware of a landfall in Suhayl at Socotra at the end of a monsoon for fear of its current'.
Bā 'Urayq/'Arayq		Ṣafáqhon	26 Aug	
al-Khibā'		Karbāloh/ Midayraykah/ Midrak al-Ṣaghīr	8 Sep	The Kaws wind weakens.
al-Far' (sic)		Midrak/Midrak al-Kabīr	21 Sep	Sometimes there is rain. Canoes (*hūrī*) can go to Qalansiyyah.
The season of Shitā' (Winter)		Ṣayrīb		
al-Dalū	The bucket	Shiblēloh/Shiballah	4 Oct	Al-Futūḥ season commences, and Azyab wind, especially at al-Ḥūt.
al-Ḥūt	The fish	Ber Sūd/Sūd al-Saghīr	17 Oct	Rain from Shibēloh to Sūd, fourteen days.
al-Naṭḥ	The ram's horn.	Sūd/Sūd al-Kabīr	30 Oct	
al-Buṭayn	The belly of the ram (Lane).	Qadāham	13 Nov	
al-Turayyā (sic)		Shīmah	26 Nov	*Shīmah* is white sand – cf. *shīmih* (*Lex. Soq.* 418).
al-Barakān		Māqad	9 Dec	*Māqad*, a stick left smouldering with which to make fire next day.
al-Haqa'ah	Rosette of hairs on the upper part of a horse's chest.	Ma'ōdīf	22 Dec	

Ṣaghīr, Lesser Midrak, would confirm – Midrak proper being called Midrak al-Kabīr, Greater Midrak. A plural, Midawrik/Mudawrik, the Midraks,[15] was also mentioned to me. I do not know if Midrak has a specific sense in Socotri apart from being the proper name of a star, but in Arabic *adraka* can mean 'to ripen' and *mudrak* 'ripe' (of fruit), and this star comes at the end of Kharīf, the season of the date harvest.

Ber/Bar[16] Sūd, and Sūd, often pronounced Sūt, are sometimes distinguished as Sūd al-Ṣaghīr and Sūd al-Kabīr, Lesser and Greater Sūd – I have no information as to what Sūd means.

Shīmah is white sand (CA, dust, earth); perhaps at this star one can see the sandy sea-floor and the sea is clear, not turbid. Sir Thomas Roe[17] logs, on 20 August at Qalansiyyah, '6 fathome water, grosse sand'. On 26 November he remarks that at Tamara or Delisha: 'The Ground is whyte sand but Rockye soe that you must wotch or boye your Cables' – i.e., keep them from rubbing. We found in 1967 that, after a heavy storm, the sea-bed near the shore had altered very considerably. Roe arrived at Socotra on 20–21 August and his vessels suffered badly from the gales of the Kaws monsoon.

Māqad, the smouldering stick, is appropriate to December when, particularly in the mountains, it must be extremely cold and a fire needed for warming.

From Qānī to Karbāloh the star names form the cow (*ilahā*, in Arabicized Socotri) or four parts of it, starting with its two horns, then the front legs (*yaddāt*), then the hindquarters and finally its tail-hair. The corresponding star names of the Arabic calendar refer to parts of the lion – these I have translated for comparison. The names and their history are of great antiquity and much has been written on them.[18] The Socotrans, as apart from the 'foreigners' of the coast, are keepers of cattle of the humpless breed. They also export ghee to Africa, though the actual dealers in it are probably 'the Arabs of Socotra who come to Zanzibar to fetch millet, slaves and ivory'.[18] The Qarā tribesmen of the mountains behind Dhofar of the Arabian mainland who also speak one of the Modern South Arabian languages (Jibbālī) of the south, are cattle-men like the Socotri mountain men (Jabaliyyah). It would be interesting to discover whether the Qarā use a star calendar with star-names relating to the cow, not the lion. There are today no lions on Socotra or other cattle predators of course; whether this was always so is an open question.

In any event the Socotran calendar with its markedly local character must surely be of considerable antiquity. It has so far survived the 'takeover' of Arabic which has been affecting the non-Arabic tongues of the Arabian mainland over the centuries. It is significant in marking a culture distinct from that of the Arab tribes of the mainland.

For comparison I quote – with the reservation already mentioned above – the calendar from the Shaḥrīs of the Kuria Murias. Their year is divided into four quarters and was said to have 18 star periods; if so, one name is missing from my notes, and I am uncertain of the number of days assigned to the individual stars.

al-Shitā	*al-Ṣayf*	*al-Kharīf*	*al-Futūḥ*	al-Ruᶜūt[22]
al-Nasar	Iqtayt	al-Shillī[20]	al-Kaydhib	al-Suhayl/
Nusūr		al-Qalb	al-Ḥaymar	al-Sirmām
Ṭayr		al-Shābik[21]	al-Ṣidr	al-Siminād
		al-Thurayyā		Bālīd
		al-Shaᶜrā		Ḥaṭām[23]

Some of these, e.g. al-Shillī, are also the names of winds. Ḥaṭām is when the sardine (*ᶜawmah*) shoals arrive.

On Socotra the rain season is from Shibēloh to the end of Sūd (1 May–8 June). Sometimes there is rain in Midrak/al-Farᶜ (al-Fargh). My notes also mention a rain period in Shibēloh-Sūd, corresponding to 4 October–12 November, but while this may be so, it looks suspiciously like an error of mine. However, as official meteorological observations were printed for the use of the RAF–some of them restricted at the time, even before World War II–it should not be difficult to verify.

Notes

1 'The Ports of Aden and Shihr', *Studies in Arabian history and civilisation*, London, 1981, various reprints, cap. XII, 213. My wife reported to Save the Children on women and child health, mostly at Hadibu.

2 Tom Johnstone's researches are included among his papers presented to Durham University by his widow. His *Mehri lexicon* (London, 1987), will of course be useful in further researches on Socotra and his Socotran data should extend Wolf Leslau, *Lexique soqotri*, Paris, 1938, based mainly on the texts collected by the Austrian Südarabische Expedition about 85 years ago.

3 The Arabic stars are given in the form written and checked orally with the *nākhudhā*.

4 Landberg, *Glossaire datînois*, Leiden, 1920–42, 1098, says al-Zabrah = class. al-Zubrah = Zubrat al-Asad.

5 Cf. Leslau, *Lexique.*, *máqdaḥ*, 'récolte, aire'.

6 G. Ferrand, *Instructions nautiques . . .*, Paris, 1921–28, II, 58a, *al-maḥdhūr al-awwal min nadkhah Suhaylī Suquṭrī fī ākhir mūsim khawfan min maddihā.*

7 Mudawrik is a plural that would correspond to Arabic Madārik. Cf. some Ḥaḍramī speakers' *yibawkī/yibōkī*, *yikawdhib/yikōdhib*, 'he weeps much', 'he lies much'.

8 Kharīf also means soft ripe dates. Explained also as *kharf*. Cf. Ḥaḍramī and Oman usage, *Gloss. daṯ.* 585.

9 Etymologically the same as the Yemeni *ṣurāb*, 'autumn cropping' for which see my 'The Cultivation of cereals in mediaeval Yemen', *Arabian studies*, I, 1974, 54; 'Star-calendars and an almanac from south-west Arabia', *Anthropos*, Posieux (Fribourg), 1954, 442.

10 Quoted in R. B. Serjeant and Ronald Lewcock, *Ṣanᶜāʾ; an Arabian Islamic city*, London, 1983, 32.

11 Bertram Thomas, *Arabia Felix*, London, 1938, 18.

12 John C. Wilkinson, *Water and tribal settlement in south-east Arabia*, Oxford, 1977, 111.

13 Leslau, *Lexique.*, 359.

14 'Star calendars . . .', *op. cit.*, 447. *Yuḥmad akl al-ḥanīdh*, but for the sense of *yuḥmad*, *maḥmūd*, see Latham, 194, *infra* with which I concur.

15 See n. 7, *supra*.

16 *Ber/bar*, son, as also in Dathīnah.
17 *The Embassy of Sir Thomas Roe to the court of the Great Mogul*, 1615–1619, ed. Sir William Foster, London, 1899, Hakluyt series, no. 1, 31–30.
18 For astronomical calendars in the Yemen in general, see the comprehensive study of David A. King, *Mathematical astronomy in medieval Yemen: a bibliographical survey*, Malibu, 1983. The star names that have not been previously reported do not appear in the valuable study of Paul Kunitzsch, *Über eine anwāʾ-Tradition mit bisher unbekannten Sternnamen*, Beiträge zur Lexikographie des klassischen Arabisch, nr. 4. München, 1983.
19 G. S. P. Freeman-Grenville, *The French at Kilwa island*, Oxford, 114.
20 Al-Shullī (so pronounced) was described to me in al-Shiḥr as a dust rousing wind from the sea which howls!
21 This star-name is found also in Wilkinson, *Water*, 110.
22 Perhaps for al-Ruʿūd, thunderings?
22 Ḥaṭām was said to be equivalent to al-Ḥūt. On the Ḥaḍramī coast it is at this latter star that fishing with the *jarīf* trawl is said to commence. For the *jarīf* see my 'Customary law among the fishermen of al-Shiḥr', *Middle East studies and libraries: a felicitation volume for Professor J. D. Pearson*, ed. B. C. Bloomfield, London, 1980, 197, with also the Ḥaḍramī coastal star calendar.

ADDENDA

p. 94, l. 21: see Brian Doe, *Socotra Island of Tranquility* (London, 1992).

V

ḤAḌRAMAWT TO ZANZIBAR: THE PILOT-POEM OF THE NĀKHUDHĀ SA'ĪD BĀ ṬĀYI' OF AL-ḤĀMĪ

The text of the *qaṣīdah* which follows, containing sailing directions from the Ḥaḍramī port of Sayḥūt to Zanzibar (Fig. 1), was transcribed at al-Mukallā in 1964 along with another poem describing the route from Muscat to Mocha. These texts were provided by my friend Shaykh Muḥammad 'Abd al-Qādir Bā Maṭraf whose grandfather was a nākhudhā of al-Ḥāmī, a little port east of al-Shiḥr. I was later able to consult a better text with 'Awaḍ Mubārak Bagraf in al-Ḥāmī, where I discussed it with several nākhudhās.[1] Finally, in al-Mukallā in Bā Maṭraf's house we had the poems commented upon by a nākhudhā whom I was able to employ. Bā Maṭraf subsequently carried out further enquiries in al-Ḥāmī and has published both poems as a 'Study in the Yemeni Heritage (*dirāsah fī 'l-turāth al-Yamanī*)' in 1977, though so regional a claim is not quite appropriate; Bā Ṭāyi''s poem in fact belongs to the common stock of sea-lore of the Persian Gulf, Oman, and elsewhere. Bā Maṭraf seems to have obtained some particulars unknown to us in 1964 about Bā Ṭāyi' and he refers to a book of his, *Shakhṣiyyāt lā tunsā*, containing biographies of well known seamen and mariners of al-Ḥāmī, but he does not say whether it has been published. However, he is able to give us the date 1217 (1802 A. D.) for the composition of the poem.

Sa'īd Sālim Bā Ṭāyi' was born at al-Ḥāmī at the beginning of 1180 (1766) and went to sea at the age of nine. By the age of twenty he was a competent seaman and he became outstanding in his day among the mariners of southern Arabia. For instance, I was told that Bā Ṭāyi' became blind, but on one occasion when sailing from Basrah he is said to have had the lead cast and to have recognised his position merely by smelling the traces of sea-floor brought up by the lead when raised.

However, I do not consider Bā Ṭāyi' to be a versifier in the same rank as the celebrated pilots some three centuries before him, Aḥmad b. Mājid and Sulaymān al-Mahrī[2]; nor in this particular poem does he exhibit so extensive a knowledge of navigation, though admittedly when I showed these earlier poems to the local nākhudās of al-Ḥāmī, they said that their own knowledge of navigation had, in comparison, declined. The language of all three poets is similar and they all pad out their lines with hackneyed phrases and conventional clichés such as *yā ṣāḥ* and *yā fahīm*, traditional to the mnemonic poem. Bā Ṭāyi' is perhaps a little clumsier in his handling of language, though his text is essentially a guide for simple men in colloquial Arabic, and he may have had more knowledge of literary Arabic than he cared to display.

1 They included Karāmah b. Mubārak b. Hamīsh from Sayḥūt and the nākhudhā 'Abdullāh b. Nāṣir b. al-Asad of Ḥaṣwayn (a Mahrī). They said that they have texts of Ibn Mājid in their houses but they do not use them nowadays. Other persons who helped us were al-Shaykh 'Alī Jābir Ma'mūr mīnā al-Mukallā, the Harbour-master, the Shaykh 'Alī b. Bū Bakr Bā Mu'aybid, and Shaykh 'Umar 'Abdullāh al-Kasādī who drew the compass. The nākhudhā from al-Dīs al-Sharqī, and Sālim 'Awaḍ al-Shuqrī also helped.

2 Their writings contain short pieces on the route to Zanzibar. See Ferrand, 1921–28, I: 67; II: 73–74; Shumovsky 1957: 86v–89v. Al-Quṭāmī, 1964: 65–67, gives a much more detailed route (which seems, however, to have some mis-spellings or misprints) and a useful section (104–108) on Barr al-Sawāḥil.

V

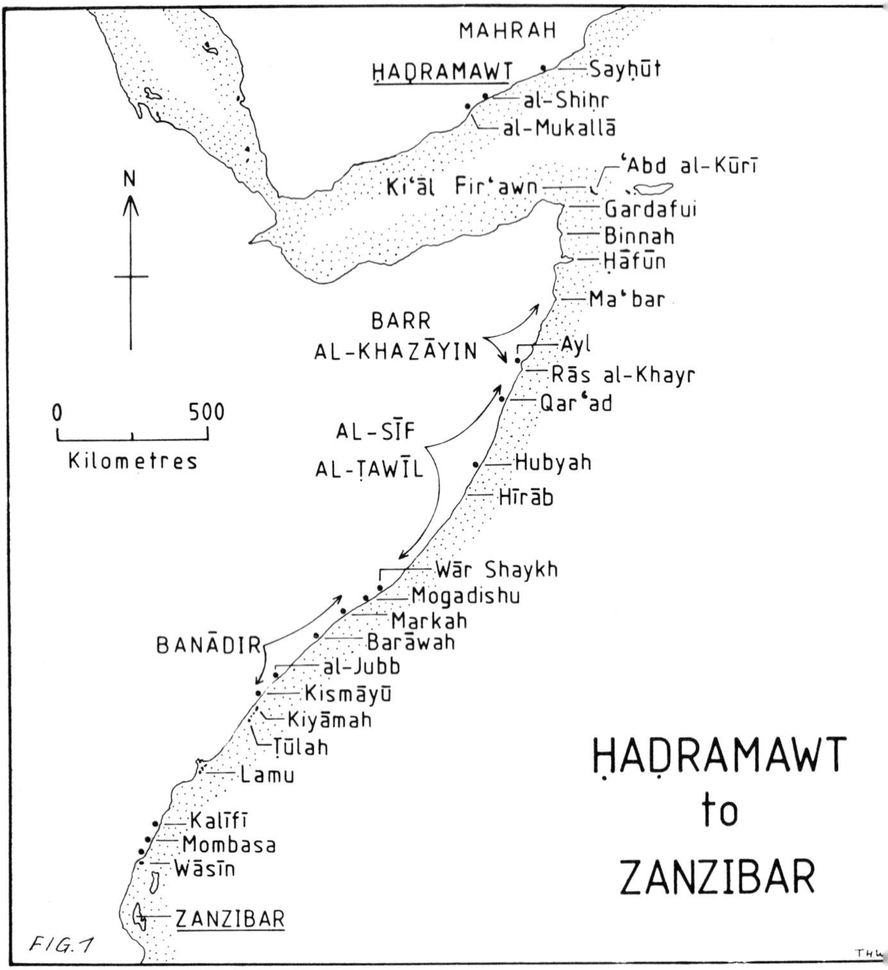
MAHRAH
ḤAḌRAMAWT
Sayḥūt
al-Shiḥr
al-Mukallā
N
Ki'āl Fir'awn
'Abd al-Kūrī
Gardafui
Binnah
Ḥāfūn
Ma'bar
BARR
AL-KHAZĀYIN
Ayl
Rās al-Khayr
Qar'ad
0
500
Kilometres
AL-SĪF
AL-ṬAWĪL
Hubyah
Hīrāb
Wār Shaykh
Mogadishu
Markah
Barāwah
BANĀDIR
al-Jubb
Kismāyū
Kiyāmah
Ṭūlah
Lamu
Kalīfī
Mombasa
Wāsīn
ZANZIBAR
ḤAḌRAMAWT
to
ZANZIBAR

FIG. 1

For a real life account of voyaging down the East African Coast one cannot do better than read the excellent description of the passage by dhow from Kuwait, in Alan Villiers' *Sons of Sindbad* (esp. 63 *seq.* and 126 *seq.* on Lamu). I was taken to see his nākhudhā, al-Najdī in Kuwait some years ago by a daughter of the famous nākhudhā, the late 'Īsā al-Quṭāmī, author of a pilotbook in wide use. Even she however found al-Najdī's colloquial Arabic difficult to understand.

Bā Maṭraf from his personal experience found that most Yemeni (which for him includes Ḥaḍramī) sailors know by heart snatches of Bā Ṭāyi' and his poems are still used by them and by the sailors in the Gulf. Bā Maṭraf's final text has some variants from my 1964 text. These are listed below omitting mere differences of spelling. Sometimes, as in verse 11/4 which is obviously slightly corrupt, I have translated his more recent reading. There is one marked difference in quatrain 28/4 where, in my text Zanzibar is called 'the land of masters and slaves'. Bā Maṭraf's version describes Zanzibar as 'the land of cloves', and he considers this evidence that Zanzibar did know the clove before 1828, contrary to the standard account that Sayyid Sa'īd b. Sulṭān introduced the clove to Zanzibar from Mauritius in 1828 (Ommanney 1957: 23). It is most probable that Bā Maṭraf's version of this line is a substitution for the original allusion to slaves, unpalatable since slaving and slave-owning have gained such ill-repute.

The poem is composed of quatrains in *basīṭ* metre, *mustaf 'ilun fā'ilun mustaf'ilun*, and in the same conventional form with the opening *Ilāhiyyah* verses addressing Allah and the *khatm*.[3] It has seemed to me worth while to print the Arabic since Bā Maṭraf's text is unvocalised whereas the vocalisation here is as I heard it from the lips of the nākhudhā, not a man educated in literary Arabic. Popular spellings are retained though formally incorrect. The grammar and morphology are those of colloquial Arabic. As pronominal suffixes are used rather loosely one cannot always be certain of the sense. In the English rendering, alternative readings are also given where of interest, separated by an oblique stroke. Interpretation or clarification of text is in square brackets.

The Compass (dīrah)[4]

The instructions for taking and altering bearings are easily understood by reference to the diagram of the compass (Fig. 2). The compass is divided into two sections, each called *nuṣṣ al-dīrah*, split by a line from the North Pole (*al-Jāh*) to the South Pole (*al-Quṭb*); the east half is called *al-Maṭla'* (the Ascension) and the west half *al-Mughīb* (Decline, Setting). The compass is divided into 32 rhumbs (*khann*, pl. *khunūn*) each named after a star, with 15 points on each side, plus of course North and South. Both halves use identical names for the fifteen points, and these have to be distinguished as to whether they are in the *Maṭla'* or the *Mughīb* sector. Each star is regarded as having an opposite star in the other half, e.g. *al-Thurayyā* of *al-Maṭla'* has *al-Jawzā'* of *al-Mughīb*, the latter being called *Raqīb al-Thurayyā*, the two points being East by North and West by South respectively. The interval between each rhumb (*khann*) is further divided into quarters so that there are in all 128 compass points. If the helmsman is given the order *inzil* he turns the ship northwards according to the number

3 See commentary to verses 1 and 29.
4 See Ferrand 1921–28, III, 99, with an illustration of the 'rose azimutale des Arabers' (91).

Fig. 2 Al-Dīrah, The Compass

of rhumbs ordered, e.g. *inzil bi-khannayn* means drop two rhumbs from say, *al-Suhayl of al-Mughīb* to *al'Aqrab* of *al-Mughīb*. To turn in a southerly direction he would be ordered *iṭla'*. Parallel orders would be given for the Maṭla' or eastern half of the compass.

Time

The term *zām* (probably from the Arabic *zamān*) denotes a three hour spell, or the distance travelled in this time under average conditions at four knots, i.e. about twelve nautical miles. It is however also used for a six hour watch (cf. commentary to verse 12). *Waḍaḥ/waḍḥ* (pl. *awḍāḥ*), is twelve hours from dawn to dusk or from dusk to dawn. *Yawm*, as used in verse 4/1 is 24 hours.

Bā Maṭraf's readings

1/2 *Wasaṭ baḥr-an wa-barr* (adopted in my translation)
1/3 *al-Nabiyy Ṭāhā*
2/2 *ṣallū*
5/1 *tajī-k 'abdu-hu*; 5/2 *muqaṭṭa'ah hākadhā ba 'ḍrib mithāl*
6/2 *fī ḍaw*
7/4 *wa-'intabih luh*
8/3 *kadha 'l-Rashād Ayl dhī mā-hi 'l-qarāḥ* (adopted)
9/4 *fa-lā takhshā*
10/3 *waḍḥayn lā*
11/4 *niḥnā wiyyāk ma 'ftakk al-lizām* (adopted, my text is unmetrical)
12/2 *yā ṣadr al-samar* (adopted); 12/3 *Murūtī*, without *al-*
14/1 *li-ayn*
15/2 *mustajāb*
16/1 *Kiyāmā* 16/4 *wa-'shjār*
17/4 *tijī Kuwāyūh wa-dhāk*
18/1 *Khawr Yāyā* 18/3 *maqtū'āt jīn*
20/4 *tushāhid-ha 'l-khiyām*
21/1 *mūsim-hun/hin*
22/4 *wa-aqdam amām* (adopted)
24/1 *wa-illā* 24/2 *la-hā 'alāmah* 24/3 *zilla-hā* (adopted)
25/3 *kullan*
26/2 *wa-ba'd-hā* 26/3 *wa-'l-qabd fī 'l-Sindibār illā wa-bān*
28/3 *wa-hū* 28/4 *wa-'dkhul bilad al-qurunful fī salām*

In 9/4, one text I saw had *ydhā* for *idhā*.

قصيدة سعيد سالم با طايع
من سَيْحُوت إلى زِنجبار

يا رَبِّ سَالَكْ تُسَهِّلْ ما عَسَرْ   عَلَى عَبِيدَكْ فِي آلبَرْ وَآلْبَحَرْ
بِجَاه أَحْمَدْ نَبِي خَيْر آلْبَشَرْ   رَحِيم قُدُّوس سَالَكْ يا سَلَامْ
وَبَعْد يا صاح أَسْمَعْ ما نَقُولْ   لا غابَت آلشمس صلّى عَآلرَّسُولْ
نقول يا آللّهْ بِتَوْبَهْ وَلْقَبُولْ   أَغْفِرْ وسَامِحْ وسَهِّلْ فى آلْكلامْ

سَيْحُوت مِنْها نَسَرْنَا بِآلْعَشِى   مِن بَعْد ما قَدْ قَضَيْنا كُلِّ شِى
والقَبْض فى مَطْلَع العَقْرَبْ وَفِي   مِن بَعْد لَيْلَهْ نُوَفِّي كلِّ زامْ
نجم الجِمارَيْن أَقْبَضْ فيه يومْ   وليلَتُهْ با يِبَيِّنْ لَكْ رُسُومْ
أَوْضَاح بَعْد الثلاثهْ فى النجومْ   سُهَيْل مَطْلَعْ قَبَضْ بُهْ وِآسْتَقامْ

تَجِيكْ ذا عَبْد أُلْكُوري جِبالْ   مُقَطَّعَهْ بَاضَرُبْها لَكْ مِثالْ
كَفال فِرْعَوْن فى اليُسْرَى تَزالْ   كذاك جُرْدُفْ مُخاطَبْ فى الكلامْ
وَلْقَبْض أَرْجَعْ كذا فى السِّنْدِبارْ   مُغِيب يا صاح مَعْ ظَوْءِ آلنَّهارْ
بِنَّهْ وبُنْها وَخُتْها يا حَذارْ   أَحْذَرْ تُحَذَّرْ مَعَا وَقْتَ آلْغَمامْ
حَافُون باتِيك عَالِي مُرْتَفِعْ   مِنْ بَعْد ما قد تَوَخَّرْ له دَفَعْ
سِهَيْل تَخْضَرْ مُغِيبَهْ وَآسْتَمِعْ   مَوْلَ آلدَّرَكْ وُآحْتَكِمْ لُهْ لا يَنَامْ

بَرِّ الخَزايِنْ ثلاثهْ فى آلوَضاحْ   أُلْاُولِي بِيَنِّهِنْ مَعْبَرْ وَرَاحْ
وكذا الرّشاد آلذى مَاهِ آلْفَراحْ   إِنْ غاب هاذاك ثانيها تَمامْ

وثالث الوضح راس الخير جاك   والمرتفع برّ با يصبح معاك
نجم الحمارين يسمع لك دعاك   إذا قبضته فلا تلحق ملام

سيف الطويل له علامة ظاهرة   قرعد وهبيه معك مستاخرة
وضحين له يا خليلي شاهرة   وبعد هيرابهم سبعة زوام

وإن قلت زايد بغاها يا نصيح   هذا إذا قدّ معك مسبار ريح
نشوطهم با تجيها مستريح   نحن رآياكم مفتّك الزام

مجراك هذا في النجم الحمر   هو الحمارين يا أهل السمر
وساير البرّ واتبع له وسر   إلى المروتي تبيّن لك علام

شجرة طويلة هنا بعد الكتوف   كتفين هي حمر كلّهن با يشوف
وعاد فارة مريكوزة تشوف   من بعدهن با نفرقهن سهام

والعقرب الآن نجمك إتبعه   لوين ما سار قد نحنا معه
برّ البنادر جميعه يجمعه   إلى براوة ومن بعده سلام

على الحمارين قد نحنا صحاب   إذا دعينا دعانا يستجاب
أقبطه يا صاحبي طيّب فطاب   نصايح الحبّ هذا بالوهام

وذا جزايرْ كِيامهْ يا فهيمْ   طُولَهْ طِوالَهْ تبعها وَٱستقيمْ
أمُّ ٱلحَوادير فى ٱلمجرا تقيمْ   شِجار فى ٱلبَرّ تنظرهن قيامْ

والنجم هذاك لى قلنا عليه   وإن تقلّب ٱلقبض خذ حَنّه إليه
وسايِر البَرّ وتقرّب إليه   وذا كُوايُوه ذاك البرّ هامْ

وخور بايه وشِعْبهْ فى اليمينْ   إحذرْ تنزّلْ معا لِى نازلينْ
أنظرْ إلى السّبُع مقطوعهْ تبينْ   وماندا راسها هذاك قامْ

وكتف شَبيلهْ معكْ فيها ضَربْ   إتبعْ لها وإنت نازلْ مقتربْ
دوّرْ شِراعَك وسايرْ وَٱجتنبْ   لا تتبعِ ٱلحُوشْ تركبْ عَ ٱلعظامْ

وكُن موسّطْ فى ٱلخورْ المليحْ   وخاطرَكْ فيه سالِى مُستريحْ
إنْ كنت كاتبْ وقارى يا فصيحْ   أُنظرْ يسارَكْ تُبيّنْ لِلْخِيامْ

شَبْلَهْ وَلامو وموسيها وَصَلْ   وذاك لى فى مُرادَكْ قدْ حصَلْ
وإنْ كان شى يا فتى معكم يصلْ   هُنا وفى زِنجبار السّحرْ قامْ

الصُّبحْ نِشْمُرْ على جودَ الكريمْ   فى السِندبار المُغيّبْ با نقيمْ
نَصيفةَ الوضْح والدّايمْ مُديمْ   رَجّعْ سُهيلَ الخضَرْ يقدمْ أمامْ

اقْبُظْهْ باقِى نهارَكْ وَٱتّبعْ   نَصيفةَ ٱللّيلْ وَٱلآخرْ دَفعْ
إلى الحِمارَيْن وَٱلعقربْ رجعْ   تُصابحِ ٱلبرّ زامًا بعد زامْ

تِصْبَحْ كَلِيفِى وإلّى حَوْلَهَا   عَلَامتهْ كِتْف عالى طُولُهَا
أَشْجار فَوْقُهْ طَوِيلهْ زُولها   مِن بعْدها قُورمُمْبَاسَهْ تَمَامْ

وَالْقَبْط هٰذى إلى وَاسِين فِيهْ   نِجْم الحمارَيْن كُنْ حافِظًا عَلَيْه
وخَلّ كُلِّبِنْ يَجِدْ ما يِشْتَهِيهْ   وبعْد أَقْبُطْ بِسَهَيْلَكْ لَا تَنَامْ

اقْبَطه لُمَّان مَازِيوَهْ تِبَانْ   إنْ شُفْتها يا فَتَى حِزْت الأَمَانْ
القَبْط أَلسِّنْدبارى لا يِبانْ   راس الجزيرهْ أُشَايِرْها قِيَامْ

أُمّ الدِّجاج الجزيرهْ والشجارْ   وفوقها هى جزيرتْ زِنْجبارْ
وكُنْ مُسَايِرْ معا ضَوْءِ النَّهارْ   إلى مُتُونى ودام الْمُلْك دامْ

خَلِّ الجزايرْ يَمِينَكْ والجميعْ   وكُلّ ما قُلْت يا صاحِبْ تَسِيعْ
عَبْرُور فيبِهِنَّ ذى طُولُهْ رفيعْ   وَادْخُلْ بِلاد الْمَقَالِى وَالْخُدَامْ

وَخَتْمها بالنبى يا سامِعِينْ   والآل وَالصّحْب ثُمّ التابِعِينْ
ومَنْ تبعْهم على مَرّ السِّنينْ   عِداد ما قدْ سَجَدْ ساجِدْ وقامْ

تَمَّتْ وبالله التوفيق

THE QAṢĪDAH OF SA'ĪD SĀLIM BĀ ṬĀYI' FROM SAYḤŪT TO ZANZIBAR

Introduction

1 My Lord, Thee I beseech, ease what comes hard
Upon Thy servants amid land and sea
Through the Prophet Aḥmad/Ṭāhā's esteem, best of Mankind.
Holy, Clement, Thee I beseech, Sublime.

2 Now listen, shipmate, what I have to say.
Whenever the sun sets bless the Apostle.
O God, I say, repentance bring and grace.
Forgive, forbear, make the words come easily.

Compass bearings

3 Sayḥūt, from it at even we set sail
When we'd finished what there was to be done,
Course set true on 'Aqrab of the Ascension,
For a night entire wherein we keep each watch aright.

4 Set a day's course upon the star Ḥimārayn
And, on the night that follows, it will make clear signs to you.
Into the stars, after half-days three, comes
Suhayl of the Acension. Set course on it and keep to it.

5 Ahead of you appears 'Abd al-Kūrī – those
Are rugged hills. I'll show you what they're like.
On the starboard side 'Pharaoh's Balls' [Ki'āl Fir'awn]recede,
So too Gardafui [Jurduf], as my words intend.

6 Alter course now to Sindibār.
Bear westward, mate, in the broad daylight
[Toward] Binnah, Bunhā [its son] and Khutha [its sister] – look out for them!
Watch out! With clouds about, be above all vigilant!

7 Up rises [Cape] Ḥāfūn before you, tall, prominent.
Once it has fallen behind, set course
Upon Suhayl the Green, of the Setting. Give heed
To the helmsman on duty: watch he doesn't fall asleep!

8 For half-days three [follow] the coast of Barr al-Khazāyin:
Its first [landfall] is [Cape] Ma'bar. By it goes.
Rashād too, Ayl of the sweet water.
As that one disappears the next comes in full view.

9 The third half-day Rās al-Khayr lies before you,
Sticking out [to sea] and by morning the coast is alongside you.
The star Ḥimārayn, may [Allah] hear your prayer,
If you set course on it you'll come to no harm/ you need fear no harm.

10 [Al]-Sīf al-Ṭawīl [the Long Beach] is a prominent landmark;
Qar'ad you have, and Hubyah [Obbia], dropping to your wake.
Taking, my friend, two half-days clear,
Then Hīrāb [last] of these places, after which [come] seven watches.

11 If you should say, man of eloquence, it needs longer,
This may [happen so]; [but] if there be a constant wind with you
You will arrive at the reefs there in good time.
You and I must keep a close lookout [for dangers].

12 Set this course of yours on the star Ḥimār,
I mean Ḥimārayn, you master yarn-spinner of the nights.
Go along with the coast, follow it on
To Murūtī and landmarks will come into your sight.

13 Here a tall tree, then the sand dunes.
Two sand dunes one sees, red both of them,
A low projecting peak, too, you'll see overlooking [them].
After these we'll divide them into shares.

14 'Aqrab now is your star. Follow it.
Where it goes we accompany it.
It covers all the Banādir coast
As far as Barāwah – after this lies safety.

15 [Now onwards] we keep company with Ḥimārayn.
When we pray let our prayer be met with an answer.
Set a true course on it, mate of mine, and all will be well.
By morning you'll be at al-Jubb [the Juba] or thereabouts.

16 These, man of insight, are the Kiyāmah islands,
Ṭūlah, Ṭiwālah. Follow them, keeping straight ahead.
Umm al-Khawādir comes up on the course.
You see there trees standing out on the coast.

17 The star is now that [Ḥimārayn] of which we spoke.
If you alter course take the rhumb of it in its direction.
Voyage along with the coast, keep close to it
And you will arrive at Kuwāyūh where this coast disappears.

18 [Now comes] Khawr Yāyāh to starboard, with its reef.
Beware of putting in [to it] like those who go [there].
[To starboard] observe the seven[islands], all separate, appear,
And Māndā headland rise up there.

19 Shaylah's sand-dune is by you – strike out for it.
Follow it [Shaylah], turning [coastwards], approaching it.
Then alter sail and run parallel [with it] but take avoiding action.
If you follow the port side you will mount onto the shoals.

20 Keep to the middle of the pleasant gulf
Where your mind will be happy and relaxed.
If, man of eloquence, you be a reader and writer,
Look to your port side and you will see [Lamu's] huts.

21 Shaylah and Lamu, now their shipping season has arrived,
What you were hoping has now happened.
If, lad, you've any onions with you,
The price of them has gone up here [in Lamu] and in Zanzibar.

22 At dawn, by God's bounty, we shall set sail
Fixing [course] on Sindibār of the Setting
For half of a half-day – for the Eternal alone is immutable
Then alter[course] to Suhayl the Green and go ahead.

23 For the rest of the day set your course on it: follow
It for the first half of the night, but in the latter [half] alter your course
And bear over to Ḥimārayn and 'Aqrab.
In the morning you will be opposite the coast, watch after watch.

24 By morning you'll be at Kalīfī and the country round it.
Its landmark is a high sand dune running the whole length of it
With tall trees on top. Pass on from there.
After it Mombasa's peaks come fully [into view].

25 To Wāsīn the course to set is upon
The star of Ḥimārayn – keep to it.
Let everyone find what he wishes.
Then set your course on Suhayl – don't fall asleep!

26 Set your course on it till Māzīwah appears
And after that/When you see it, lad, you've won to safety.
The course is on Sindibār until there comes into sight
The headland of the Island, [and] its signs rise up.

27 [Then comes] Umm al-Dijāj the island with the [coconut] palms.
Beyond it lies the Island of Zanzibar.
Keep running along [between it and the main island] in daylight
To Mutūnī [Island]. [God's] sovereignty lasts for eternity.

28 These islands [well] to starboard leave.
And listen, mate, to all I have to say.
About them is a long projecting shallow reef,
And enter the land of the masters and the slaves/
And with safety enter the land of cloves.

29 Listeners [to me], conclude [these verses] with the Prophet,
[His] Family, Companions, then the Followers
And those who have followed them throughout the passage of the years,
As many times as one at prayer prostrates himself and rises.
Concluded, and in God is success.

Commentary on the verses of the translation

1 The Prophet Aḥmad is of course Muḥammad, and the term rendered into English as 'esteem' is *jāh*, which is difficult to translate exactly. The poet is beseeching Allah by the standing that the Prophet has with him. *Jah* also means the Pole Star; the play on words is doubtless intentional.

3 Sailing vessels from the Ḥaḍramī and Mahrah coast proceed first to the Mahrī port of Sayḥūt which is also a good area for fishing. Thence they sail to Africa, thus avoiding the winds of Cape or Rās Gardafui, also known as Rās 'Asīr (Bā Maṭraf). 3/2 means that all the gear and equipment required for the voyage has been taken aboard.

4 *Rusūm* was explained to me as *ishārah*: signs or indications. Bā Maṭraf says that when this star Ḥimārayn "makes clear its signs to you" it means that when we survey our progress with the astrolabe (he means the sextant) we shall be able to determine the distance we have come from Sayḥūt to our present position.

5 *Ki'āl/ka'āl*, testicles, is a Ḥaḍramī word (cf. Serjeant 1957: 17). This feature was described to me as two small islands (*ṣayyilayn*). Al-Quṭāmī (1964: 65) calls it *Qarāqir Fir'awn*. I was told there is much dust (*ghubār*), cloud (*ghamām*) and dew (*ṭaylul*, dimin of *tall*) at Gardafui.
I quote from my "Maritime and customary law off the Arabian coasts" (1970: 199): "When a vessel arrives at Rās Ḥāfūn, from Africa, they i.e. Ḥaḍramīs told me, the crew prepare a coconut for the Jinn to pacify them – after this there is *fawlah*, i.e. *salāmah*, safety. Sometimes in return for a safe arrival they sacrifice an animal. At Rās 'Asīr or Gardafui the sailors place some food in a little box and throw it into the sea so as to ransom themselves from the wrath of the Jinn who inhabit the mountains of Gardafui."

6 In the Gulf it is called *al-Silibār* (vocalisation uncertain) as in al-Quṭāmī (1964: 179). Mariners have to be careful at Rās Binnah because of the islands and hidden shoals (*shi'āb*) under the surface.

7 Sailors call Suhayl 'the Green' because of the greenish light it gives, as they call al-Ḥimārayn 'the Red' (*al-Aḥmar*) because of the reddish colour of its light. If the order "*Imsik al-najm al-khaḍar*" be given, you know Suhayl is meant. The helmsman on duty, literally the 'master of responsibility', must be consulted, according to Bā Maṭraf, as to whether or not it will be possible to hold the course to Suhayl since the winds and currents at Rās Ḥāfūn may at times force a temporary diversion. Two other hands must be stationed with the helmsman in time of need, or to take over when he is drowsy or in need of rest; those in training usually take this role. He quotes Badr b. Aḥmad al-Kasādī, *al-Qāmūs al-baḥrī*, p. 76 (presumably printed but unknown to me),.as saying that the helmsman (*al-sakūnī*) has authority over the crew in the matter of altering the sail from one direction to another and in lashing or unlashing certain ropes during the ship's progress (cf. Tibbetts 1971: 524, also quoting al-Quṭāmī). The helmsman's assistant is called *qabḍī*, plural *qabḍiyyah.*

8 Barr al-Khazāyin extends from Rās Ma'bar to Rās al-Khayr, the latter often called Rās al-Khayl. Ayl lies at the estuary of the Najāl River where the water is known to be sweet should they wish to take on water (*yimizzirū*), a place for watering being called *mamzar* (al-Quṭāmī 1964: 22).

9 Southwest of this Cape (*Rās*) is a sandy coast. After changing course to Ḥimārayn you are safe because there are no further difficulties at sea. With the main danger past, it may be at this point that ships travelling in convoy (*sinjār*) will release each other from the obligations that convoy entails.

10 (Al)-Sīf al-Ṭawīl extends nearly to Mogadishu. Rās Hīrāb is a black coloured cape upon its two northern thirds, and white upon its southern third – a small mountain on the sea coast washed by the waves (al-Quṭāmī 1964: 106). *Hīrāb,* a keel, illustrates the long shape of this landmark.

Bā Maṭraf says that when the building of a dhow begins, the blood of a goat is poured onto the keel to divert the envious eye (*'ayn al-ḥasūd*), to drive away evil spirits and to gratify the jinni (*mārid*) of the sea.

11 My notes state that a *misbār* is a *rīḥ mustamirr qawiyy*, a strong continuous wind or *rīḥ ṭāris*; when the wind is *ṭāris* it is full.

Fishṭ / fushṭ / fisht (all used) is defined by al-Quṭāmī (1964: 42) as a sort of soft rock, *al-ḥajar al-hashsh*, though by Bā Maṭraf (1972: 38) as *qishār*, which is a rocky area under the sea like the foreshore of al-Mukallā.

12 Al-Ḥamar/Ḥimār is probably simply al-Aḥmar 'the Red', epithet of Ḥimārayn (cf. note to verse 7).

The crew is divided into the first and second night watch (*zām awwal; zām ākhar*), each with a ship's cook (*ṭabbākh*) to prepare and distribute coffee. Each watch lasts six hours. With each watch also is a *ṣadr al-samar*, a man known to be adept at spinning yarns and who receives an extra supplement to his wage by keeping the sailors from boredom and sleep by telling his tales. Bā Maṭraf lists the type of tales he heard: stories of the Jinn; of World War I; of Adam; of Sa'īd Bāshā (the Turkish general from the Yemen) and his days in Lahej; of the punishment of the grave; exchange of popular verse (*muṭāraḥāt shi'riyyah sha'biyyah*): Yūnānī (Greek medicine); the Pharaohs; magic on Socotra Island; the spermwhale (*ḥūt al-'anbar*); winnings from lotteries; the virtue (*faḍīlah*) of clipping the fingernails; the raids of Badr and Uḥud; 'Antarah b. Shaddād; the terrors of the Day of Resurrection; men's wiles; adventures of some sailors in large ports such as Bombay, Basrah or Mombasa; traditions of the Prophet on eating meat, wheat (*burr*) and dates; Anti-Christ (*al-Masīḥ al-Dajjāl*); popular folk dances (*al'āb*); gambling; the general conduct of the nākhudhā, the serang and the clerk (*kātib*) of the ship – exciting much laughter and mockery.

Murūtī is a dangerous coast with many submerged rocks. The signs or indications of Murūtī are the waves breaking over its shoals (*shi'āb*), shining with phosphorescence at night.

13 My notes record that a tree on the coast is a sort of landmark, and the fish found off the coast of al-Sawāḥil is the *tirnāk (Scomberomorus commerson*). Bā Maṭraf speaks of groves of *nakhl Iblīs*, literally 'The Devil's Date Palm', but in fact, according to Landberg, a Yemeni name for the dwarf fan palm which is called, in Ḥaḍramawt, *naws* (*Arabica* V, Leiden 1898). Al-Quṭāmī speaks of a reef the indication of which is "*nakhlat Iblīs ṣulbuyn.*"

In the vicinity of the groves are two mounds (*aqwāz*) of red sand, then you come to a small projecting peak with a black volcanic crest overlooking the two mounds of sand (Bā Maṭraf).

Miraykūz, explained as *mustaṭīl*, long is the diminutive of *markūz*, passive participle from the verb *rakaz, yarkiz/yarkuz*, ficher dans le sol . . . qu'il y reste debout (Landberg 1920–42).

Nāf means to be high, tall. 13/4 according to Bā Maṭraf means "We shall substract them (the two sand hills) from the account" after we have passed by them.

14 Bā Maṭraf defines the Banādir Coast as stretching from Mogadishu to Lamu where it ends. My informants said that the *banādir* or ports, are Mogadishu, Wār Shaykh, Markah, Barāwah and finally Kismāyū.

15 As you are now crossing the Equator you cannot use the *kamāl* (sextant), and you must come to the Juba using reckoning and the star Ḥimārayn. When you pass the mouth of the Juba River you will be opposite Kismāyū (Bā Maṭraf).

16 After Kismāyū you see the Kiyāmā Islands one after another in a long string; small islands, situated near the coast in a southwest direction. Three of these islands are taken by sailors as landmarks (Bā Maṭraf). Umm al-Khawādir, I was informed, is one of the islands and Bā Maṭraf says it is marked by groves of tall *nakhl Iblīs.*

17 After Umm al-Khawādir, if you want to put out to sea to avoid the reefs (*shi'āb*) along the coast, especially at night, you take a new course to Suhayl Mughīb, the new rhumb lying between Ḥimārayn and Suhayl. You have to watch the coast all the time till you arrive opposite Kuwayūh Island (Fig. 3). At this point the coast falls away from you. *Hām* is probably from *hamā*, Fr. *cacher*, "to hide" (Landberg 1920–42) (Bā Maṭraf).

18 You will now see the land on the horizon gradually coming up and Khawr Yāyā and its reef will appear to starboard. Near this *khawr* (lagoon) you will see small fishing boats over its reef because fish abound here. Your ship is too large to go to the shallows where the fishing boats are, so set a course far away from them. As you pass Khawr Yāyā you will see to your starboard seven islands separate from each other, the largest of which is Battah (Pate). Then there rises in front of you the headland of another island, Māndā (Bā Maṭraf).

Verse 18/3 was said to me to refer to seven sand-dunes (*qawz*).[5]

5 Editors' note: Captain Owen, in his *Narrative of Voyages. . .*, published in London in 1833 and thus

V

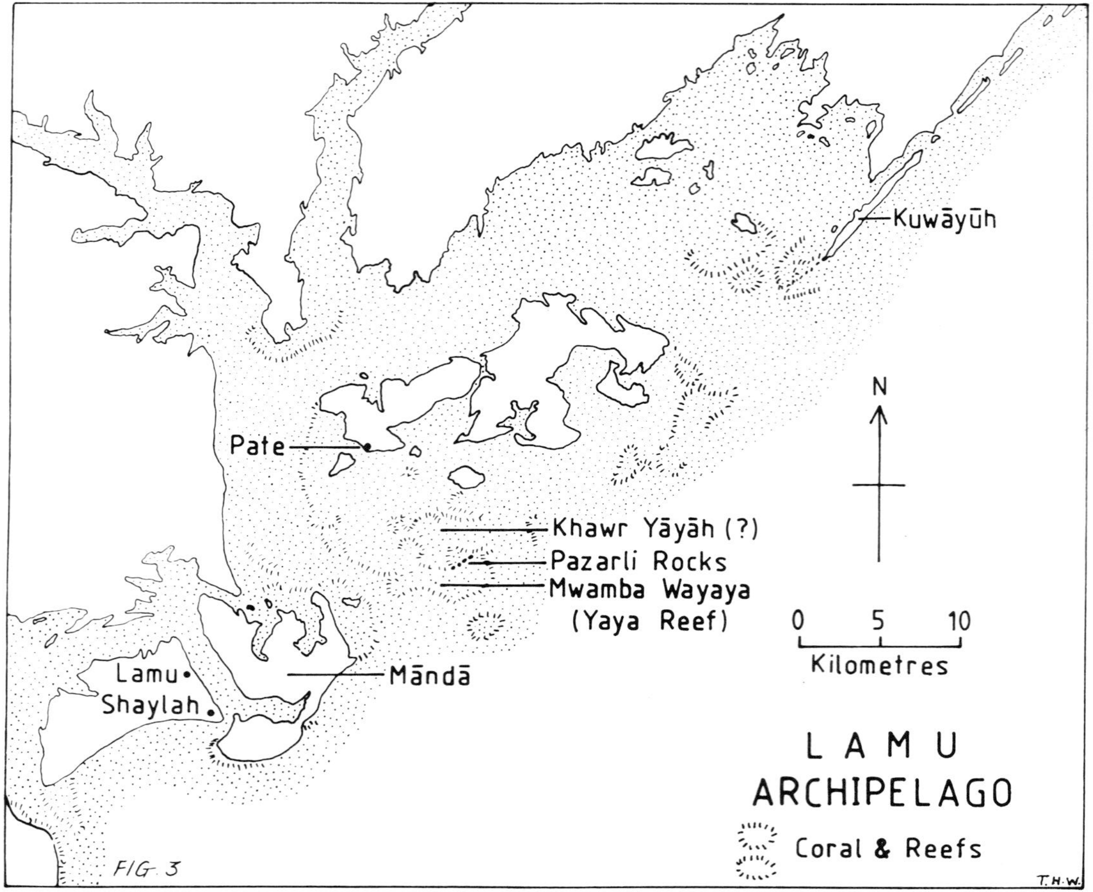
Kuwāyūh
Pate
Khawr Yāyāh (?)
Pazarli Rocks
Mwamba Wayaya
(Yaya Reef)
Māndā
Lamu
Shaylah
N
0 5 10
Kilometres
L A M U
ARCHIPELAGO
Coral & Reefs
T.H.W.
FIG. 3

19 You should approach Māndā Island until you can see the sand-dunes of Shaylah/Shīlah (in modern Swahili, Shela) village on the eastern tip of Lamu Island, which loom over the low cliffs at the southern end of Māndā Island. When you get near the headland alter direction (*dawwir*: turn round) because your course will now be to the northwest so that Māndā will be on your starboard side and Lamu to port.
19/3 means that you should go slowly using the lead (*buld*) to ascertain the depth of the sea. This should enable you to avoid the submerged reefs extending from Māndā and Lamu Islands, and to take the direction of Lamu (Bā Maṭraf).

20 In the middle of the *khawr* of Lamu you will avoid the rocky submerged reefs and progress slowly towards Lamu harbour. On the port side, on the south confines (*aṭrāf*) of Lamu you will see the 'tents (*khiyām*)'. Bā Maṭraf interprets Bā Ṭāyi' to be alluding to the fact that *khiyām* in classical Arabic means not only a tent, but any house built without stone (i.e. *'ushash*). To know this sense of the word you have to be *faṣīḥ* 'eloquent', and literate!

21 The sailors from al-Ḥāmī of Ḥaḍramawt, trading with East Africa, used to take with them articles of merchandise such as salt, dried salted fish, sardine-oil (*ṣīfah*) and onions (*baṣal*). These last, being the principal crop of al-Ḥāmī village, seem to have fetched a good price in Lamu and Zanzibar. However, my informants said that by 'onions', merchandise in general was intended. Lamu also provided relaxations after the hardships of the voyage (Bā Maṭraf). For some of the delights of Lamu, see Serjeant 1974: 167.

22 We cannot continue long with our course set on Sindibār in a southerly direction because of the waves dashing against the side of the ship, causing it to roll. It will be followed for six hours only and then we shall return to the course on Suhayl. The poet is trying to alleviate for the voyagers the distressing effect of the conditions while on this course, and he points out that neither hardships nor ease (such as they had in Lamu) endure for ever – Allah alone is Enduring. When they take the new course, the waves will dash against the stern of the ship (Bā Maṭraf). 22/2 uses the form *mughayyib* for *mughīb* to fit the metre.

23 For six more hours until dusk, the ship's course is set on Suhayl, then for yet another six hours till midnight. As they are some distance from land in this Formosa (Ungwana) Bay, the course is altered from Ḥimārayn to 'Aqrab to draw nearer the coast. Malindi seen in the distance is passed the next morning.

24 Kalīfī or Kalīfah, in modern Swahili Kilifi.

25 The voyagers have a pleasant view of the coast with its many trees. Rās Wāsīn is opposite Mombasa harbour. The ship passes on to Jāzī (Gazi) village (Bā Maṭraf).

26 After Jāzī comes Māzīwah (Maziwi Island). There are many shoals in the sea here. It is now time to prepare for entering Zanzibar and after a while there appears Unjūjā

written only slightly later than the composition of the pilot-poem, wrote of approaching the Lamu area from southern Somalia: "We now steered along shore for Lamoo, passing several islands, but none of sufficient importance to require a description, until we saw what we called the Pesanby Rocks, known to the Arabs as the Seven Islands of Eryaya, which is also the name of the great channel through the reefs into Patta" (Owen 1833, I: 363). This channel is today known as *Mlango Pazarli* (Swahili: Pazarli gate), and the reef *Mwamba Wayaya* (Sw. Yaya Reef). The Pazarli Rocks rise prominently above the reef in this area, are about seven in number, and appear on the starboard side of ships sailing southwards, just before reaching Manda Island. Although Khawr Yāyāh was unknown to several skilled fishermen questioned in the Lamu-Pate area, they stated that just inside the reef is calm water with excellent fishing, and it is probably in this area south of Pate Island where Khawr Yāyāh is to be found.

(Unguja) Island, the largest of the Zanzibar islands containing the harbour and town. The ship now makes for the northern cape. Val Hinds has informed me that the Arabs are believed to have planted many casuarina trees on for instance Maziwi Island, and other places along the Kenya and Tanzania coasts to act as landmarks for sailing vessels.

27 Umm al-Dijāj, says Bā Maṭraf, is at the foot of the cape of Zanzibar Island (*asfal rās jazīrat Zanzibār*); it appears this must mean Tumbatu Island. You proceed slowly till you reach Mutūnī Island which you keep on the starboard side.

28 You keep both Umm al-Dijāj and Mutūnī (Mtoni, a little north of Zanzibar) well to starboard for fear of the *'ayrūr*, the reef, directly under the surface of the sea and which encircles these two islands, then bear to port and safely enter Zanzibar harbour.

28/4 Clearly in Bā Ṭayi''s time, slaves were an important article of trade and in Ḥaḍramawt, Sawāḥil slaves are distinguished from Nūbah slaves. *Mawālī* (masters) can also mean 'clients' but this is unlikely here.

29 The poem concludes with the *Ilāhiyyah* mentioning Allah, the Prophet and his House, and the Companions and Followers in the traditional way (Serjeant 1951: 6).

Note: I have not found, on the maps and charts consulted, Rashād and Umm al-Khawādīr, while Umm al-Dijāj is only provisionally identified. The two latter Arabic names may disguise local African names.

Glossary to text

3/1 *Shamar, yishmur;* to sail
3/3 *Qabaḍ, yaqbaḍ, qabḍ;* set a course, bearing on
4/2 *Rusūm;* signs
5/3 *Ki'āl/ka'āl;* testicles
6/1 *Raja';* to turn, change from one bearing to another (T)
7/1 *Irtafa';* to go directly out to sea (T)
7/2 *Tawakhkhar;* to fall behind, *dafa';* to set course to
7/4 *Mawlā/ṣāḥib al-darak;* helmsman
10/1 *'Alāmah* plur. *'alām;* signs
10/4 *Hirāb;* keel, *zuwām*, plur. of *zām;* watch
11/2 *Misbār;* continuous strong wind
11/3 *Fishṭ*, plur. *fushūṭ*; reef (T)
13/1 *Kitf*, plur. *kutūf*; sand-dune. Syn. *qawz*, plur. *qīzān*; sand-dunes or hillocks covered by sand
13/3 *Miraykūz*; see commentary
17/2 *Khann*, plur. *khunūn/akhnān;* rhumb (T)
18/1 *Shi'b/sha'b*, plur. *shi'āb*; reef (T)
19/3 *Dawwar (al-shirā')*; to swing round (the sail)
19/4 *Al-jawsh;* port, opposite of *dīmān*; starboard
'Aẓm, plur. *'iẓām*; submerged rocky area or shoal
24/3 *Zall*; to pass over a shoal (T), using Bā Maṭraf's reading
26/1 *Lummān*; up to, until
26/4 *Ishārah*, plur. *ashāyir;* signs, landmarks (T), sea signs.
28/3 *'Ayrūr*; reef, shoal

References to the glossary in Tibbetts' *Arab Navigation* are indicated by (T).

References Cited and Select Bibliography

Chittick, H. N., 1969: "An archaeological reconnaissance of the southern Somali coast." *Azania* 4: 115–130.

–, 1976: "An archaeological reconnaissance in the Horn: The British-Somali Expedition, 1975." *Azania* 11: 117–133.

Consociazione Turistica Italiana, 1938: *Guida dell'Africa orientale Italiana.* Milano.
Ferrand, G., 1924: *L Elément Persan dans les Textes Nautiques Arabes.* Paris.
–, 1921: 28: *Instructions Nautiques et Routiers Arabes et Portugais*, 3 vols. Paris: Geuthner.
Grosset-Grange, H., 1978: "La côte africaine dans les routiers nautiques arabes au moment des grandes découvertes." *Azania* 13: 1–35.
Koji Kamioka and Hikoichi Yajima, 1979: "The inter-regional trade in the western part of the Indian Ocean: the second report on the dhow trade" (English summary). *Studia Culturae Islamicae* 9. Tokyo
Landberg, C. de, 1920–42: *Glossaire datînois.* Leiden
Bā Maṭraf, Muḥammad 'Abd al-Qādir, 1972: *Al-Rafīq al-nāfi' 'alā durūb manẓūmay al-mallāḥ Bā Ṭāyi.'* Aden.
Ommaney, F. D., 1957: *Isle of Cloves.* London
al-Quṭāmī, 'Īsā, 1964: *Dalīl al-muḥtār fī 'ilm al-biḥār*, 3rd ed. Kuwait.
Serjeant, R. B., 1951: *Prose and Poetry from Ḥaḍramawt.* London.
–, 1957: *Saiyids of Ḥaḍramawt.* London.
–, 1970: "Maritime and customary law off the Arabian coasts." In *Sociétés et Compagnies de Commerce en Orient et dans l'Océan Indien: Actes du Huitième Colloque International d'Histoire Maritime* (1966), ed. by M. Mollat, pp. 195–207.
–, 1974: "The ports of Aden and Shihr (Mediaeval period)." In *Recueils de la Société Jean Bodin*, 32, *Les Grandes Escales*, 1, Antiquité et Moyen-Age, pp. 207–224. Bruxelles.
–, 1980: "Customary law among the fishermen of al-Shiḥr." In *Middle East Studies and Libraries: A Felicitation Volume for Professor J D. Pearson*, ed. by B. C. Bloomfield, pp. 193–203. London.
Shumovsky, T. A., 1957: *Tri niezvestnuie lotsii Akhmada ibn Mādzhida, arabskogo Vasco da Gama.* Akad. Nauk. SSSR Institut Vostokovsdeniya. Moscow. (Portuguese translation by Myron Malkiel-Jirmounsky, *Três Roteiros Desconhecidos de Aḥmad Ibn-Mādjid*, Lisboa 1960).
Tibbetts, G. R., 1971: *Arab Navigation in the Indian Ocean before the Coming of the Portuguese.* London.
–, 1974: "Arabia in the fifteenth-century navigational texts." *Arabian Studies* 1: 86–101. London-Cambridge.
Villiers, A., 1940: *Sons of Sindbad.* London.

First published in *Paideuma*, Mitteilungen zur Kulturkunde 28 (1982), Franz Steiner Verlag, Stuttgart (formerley Wiesbaden).

VI

A *MAQĀMAH* ON PALM-PROTECTION (*SHIRĀḤAH*)

THE importance of the date-palm to the traditional economy of Ḥaḍramawt is abundantly evident in the literature of the country and when living there one is impressed by the reliance of the people not only on the date harvest but on all parts of the palm for building, fencing, fuel, the manufacture of domestic articles, and the like. Some sixty years ago Paul Popenoe[1] quoted an estimated figure of 200,000 date palms for Ḥaḍramawt, contrasted with only 100,000 for the Yemen. No doubt, reports[2] to the erstwhile Aden Protectorate and South Arabian Federal Governments might contain more precise statistics. In those days also V. H. W. Dowson,[3] the expert on the date culture of Iraq, visited the country and has published an excellent account of the use of dates there. It was a British objective to promote and extend palm cultivation, and I recall hearing the late Sayyid Sir Bū Bakr b. Shaykh Āl Kāf tell the British Resident that owing to this policy the danger of famine had been largely removed due to the relative abundance of dates. It is a popular saying that in times long ago one could travel the whole length from the top of the Wādī down to the Prophet Hūd's Tomb under the shade of palms, and judging by the clear signs of cultivation in many areas now desert, this may well have been true.

Raiding the date crop of a neighboring district has doubtless gone on from time immemorial—it is frequently recorded in Ḥaḍramī chronicles. Shanbal, for instance, speaks of a Banū Ẓannah/Ḍannah raid in 635/1237-38 in which they cut the *kharīf* crop.[4] Ḥaḍramīs discovered that palms could be destroyed by pouring kerosene (*ghāz*)[5] into the center top of the tree—a wanton act in a country where years of labor and investment are required to create this valuable asset. In Ruḍūm of the Wāḥidī Sultanate in 1947 I saw dead palms belonging to the Mashāyikh, who said this was done by the tribesmen, but these latter said they were hungry and had asked for dates, which the Mashāyikh refused to give them. According to an article in *Al-nahḍah*, "the first to make use of this method to destroy palms in Dawᶜan were men of the Dayyin tribe (east of Wāḥidī territory), more than twenty-five years ago (from 1951 on) when the struggle between them and the Dawlah (the Quᶜayṭīs?) was at its fiercest."[6]

Palm protection (*shirāḥah*) by armed tribesmen doubtless remounts to antiquity and not less so the malpractice and exactions of the tribes who undertake the

1 Paul Popenoe, "Scale-Insects of the Date-Palm," *JAOS* 42 (1922): 205-6; idem, "The Pollination of the Date-Palm," *JAOS* 42 (1922): 343-54. Cf. *Lughat al-ᶜArab* (Baghdad, 1926), vol. 4, p. 298.

2 *The Department of Agriculture Report for 1958-63:* Protectorate of South Arabia, Aden, gives production for the whole area in 1963 as 7,500 tons, but this is clearly a guess.

3 V. H. W. Dowson, "The Date and the Arab," *Journal of the Royal Central Asiatic Society* 36 (1949): 34-41; idem, "To Arabia in Search of Date Palm Off-Shoots," *JRCAS* 39 (1952): 45-66.

4 This appears in my as yet unpublished edition of this Ḥaḍramī chronicler.

5 *Qazzaz* is the verb derived from *ghāz*.

6 *Al-Nahḍah* (Aden), III, 105, 21 December 1950, p. 2.

protection. *Al-jawhar al-shaffāf*, composed in the first half of the ninth/fifteenth century, alludes to palm guardians taking the fruit from the trees that do not belong to them to give to others "as is the custom of the guardians of the palm trees of Ḥaḍramawt (*ka-mā huwa ʿādat ḥurrās nakhīl Ḥaḍramawt*)."[7] It is against the abuses of this protection that Sayyid Ṭāhir inveighs in the *maqāmah* I published thirty years ago.[8]

The word *shāriḥ* (plur., *sharaḥ* and *shurrāḥ*) is basically used of those protecting, for example, millet against the depredations of birds, usually children armed with slings (*naḍaf*). Al-Yāfiʿī under the annals for 694/1294-95 quotes the following passage: *Mā lī wa-li-ʾm-ḥirāsah; anā anzil min am-mishbāb wa-atruk am-zarʿ*, "What have I to do with guarding (crops)? I shall get down from the *mishbāb* and leave the crop."[9] He describes the *mishbāb* as "pieces of wood set up in the middle of the crops" upon which is constructed an *ʿarīsh*, "hut," in which the guardian (*ḥāris*) sits. The protection, in this case in Shāfiʿī Yemen, is against thieves and wild animals. The ground or area which comes under the protection of an armed tribesman is called in Ḥaḍramawt *mishrāḥ*, and *māl* means there "property in palms." Any man who possesses *māl* is called a *ṭabīn*.[10] The *ṭabīn* hires a laborer skilled in date-palm culture to look after his property, and we even have the text of a typical contract between them in *Al-nubdhat al-muḥarrirah* of Baḥraq.[11] In Ḥaḍramawt the *ṭabīn* is very often a Sayyid, but he may belong to one of the other classes; and my informant Raḥayyam who owned a few palms was himself a shaykh. It might be that in ancient south Arabia the *MŚWD* (Ḥaḍramī Musawwad, i.e., Sayyids) could at the same time have contained in their ranks individuals who were *ṬBNN* or *ṭabīn*s.[12] At any rate, there were, until fairly recently, three main classes interested in the palm-groves of Ḥaḍramawt: the owner, the share-cropping cultivator, and the tribesman-protector.

Of course palm proprietors, even Sayyids, would not necessarily invariably hire laborers to look after their palms, especially before the nineteenth century, for the pious Sayyid Muḥammad b. ʿUmar al-Saqqāf[13] used to recite the *sūrat* Yā Sīn every time he planted a palm, notwithstanding the number he planted and his continuous activity in this direction. Evidently he then attended to his palms, at least to some extent, in person.

7 Cf. my article "Materials for South Arabian History II," *BSOAS* 13 (1950): sec. iii, p. 582 (SOAS photocopy II, 279).

8 Ibid.

9 Al-Yāfiʿī, *Mirʾāt al-janān* (Haydarabad, A.H. 1337-39), vol. 4, p. 226; al-Sharjī, *Ṭabaqāt al-khawāṣṣ* (Cairo, 1321/1903), p. 44.

10 The *ṭabīn* is usually a person who advances money and/or items in kind to a sharecropper. Popular verses such as the two obscene allusions to the *ṭabīn*'s wife (see my *Prose and Poetry from Ḥaḍramawt* [London, 1951] p. 164, nos. 51, 52) show the dislike the sharecropper feels for his *ṭabīn*, while another (no. 55/3) says, "The well-worker runs away and the *ṭabīn* is demanding." On the other hand another (166, no. 76) says,

> As for me, nobody has a *ṭabīn* like mine,
> He's brought my "feast," Barbarī sheep
> tethered by twine.

By "feast" the feast gift is intended.

11 Baḥraq, "Forms of Plea, Šāfiʿī Manual from al-Šiḥr," *RSO* 30 (1955): 10 ff.

12 Cf. Jacques Ryckmans, *L'Institution monarchique en Arabie méridionale avant l'Islam* (Louvain, 1951) 22: "Selon N. Rhodokanakis (in D. Nielsen, *Handbuch der altarabischen Altertumskunde* [Copenhagen, 1927], p. 122), le terme *mśwd*, que l'on retrouve à Saba et Qataban, où il s'oppose respectivement aux termes *msḫnn* et *ṭbnn*, qui désignent l'ensemble des propriétaires terriens, se rapportait aux membres d'une classe privilégiée parmi ces derniers."

13 Sayyid ʿAbdullāh b. Muḥammad b. Ḥāmid al-Saqqāf, *Tārīkh al-shuʿarāʾ al-Ḥaḍramiyyīn* (Cairo, 1353/1934), vol. 2, p. 173.

Perhaps under certain conditions palm protection was a reasonable affair, but this had apparently not been the case for a very long time. Raḥayyam himself could plainly remember the days before its abolition, which in 1948 were not yet remote. He said that the tribes from Tarīm down to Hūd were poor and hungry and, so, especially practiced palm protection, but those between Tāribah and Saywūn, better off, engaged little in palm protection (this part of the Wādī as far as the Wādī Dawᶜan received remittances from Java and the Far East). According to Van den Berg,[14] however, the Sultan of Saywūn had persuaded his tribes to accept the re-purchase of protection-fees, i.e., those prior to 1886, and it is the Kathīrī perhaps to which Raḥayyam alluded. The palms in the wādīs about Saywūn belong to *ᶜabīd*, "slaves."

Van den Berg states that the fee, originally 10 percent, sometimes amounted to over half the crop. Raḥayyam was more explicit. The tribes exacted a fifth on crops (*ḥawāṣil*), especially dates, and tithes (*ᶜushūr*) on millet-cane (*qaṣab*) and grain (*ṭaᶜām*). There are in fact, he said, two types of protection, *shirāḥah buṭl wa-ẓulm*, "palm-protection of invalidity and oppressive (taxation)," namely a fifth of the crop; and *shirāḥah ṣamīl*, "palm-protection by club," meaning by agreement, namely a tenth or twentieth of the crop. The former is a forced contract, and the latter is at the discretion of the owner. The term "protection by club" arises from the fact that the man who looks after the crop bears a club as a weapon—the characteristic arm of the *miskīn*,[15] here the peasant—in contradistinction to the *qabīlī* who bears a rifle. In the case of grain or fruit the *qabīlī* takes his fee at the time of harvesting.

It was probably the representations of the proprietors who suffered under this regime, supported by reference to Islamic *sharīᶜah*, that persuaded the British Resident of the time, W. H. Ingrams, to press for the termination of *shirāḥah buṭl*.

In the case of *shirāḥah ṣamīl*, a *qabīlī* or a *miskīn* may be hired; he then makes himself a hut (*ᶜarīsh*, plur., *ᶜurūsh*) of palm-frond (*saᶜf*), and, if a *qabīlī*, he leaves his gun there. If anyone steals the dates this man deals with him. Should the thief be known, the case is one which is dealt with by the customary law of the protecting tribesman. If the thief is not known the palm warden informs the palm proprietor that he did not commit the theft himself, but he gives the owner no compensation for his own failure in his duty.

The *qabāʾil* divide up the palm growing lands into sectors. Each sector comes under the protection of a certain tribal group or individual, or at least so the tribes claim, though the palm proprietors may think differently. The tribes may even have recourse to the owners of the land in question to define the exact extent of ground the latter own so as to support their own intertribal disputes about the land which comes under their protection.

14 L. W. C. Van den Berg, *Le Ḥadhramout et les colonies arabes dans l'archipel indien* (Batavia, 1886), p. 79.

15 For the distinction between *masākīn ḥaḍar*, merchants, craftsmen and petty traders, and *masākīn ḍuᶜafāʾ / ḍaᶜfā*, workers in clay, see my "Social Stratification in Arabia," in *The Islamic City: Selected Papers from the Colloquium Held at the Middle East Centre, Faculty of Oriental Studies Cambridge from 19 to 23 July 1976* (Paris, 1980), p. 130. Sayf b. Ḥusayn al-Quᶜayṭī, *Al-amthāl wa-ʾl-aqwāl al-Ḥaḍramiyyah* (unpublished MS in Haydarabad, Deccan), cites a proverb, "*lā tusāyir man silāḥ-uh ṣamīl-uh*," "do not travel with a companion whose weapon is his club." This is because being unarmed, he will not be able to join in defense against an attack on the party.

The rights to protection of palm-groves became known as *shāʾim wa-lāʾim*[16] or simply as *shāʾim*. The *shāʾim* developed into a sort of investment or stock of which one person could dispose by sale to another, and, in recent times at least, the palm-proprietor could buy back the *shāʾim* on his own land and install his own warden to guard his crop without paying the tribes for protection. The British Resident, or perhaps more exactly the local governments under his advice, took action to reduce the *shāʾim* on all produce to 5 percent, and the new measure was enforced. The tribesmen had in the past, as we have seen, taken very much more than 5 percent, and it seems that the Kathīrī Sultans were powerless to stop them. When the *shāʾim* was restricted to 5 percent, the measure was naturally not altogether pleasing to those who had money in it.

There seem to have been three stages at which the palm proprietors had to suffer the depredations of the palm warden.

1. About the time the covers are placed on the date clusters when they are just beginning to redden and are known as *faḍah*, and later at the *munāṣif*[17] stage, dates are taken by the Bedouin.[18]
2. When the palm frond (*saᶜf*)[19] dries and changes color the Bedouin come and cut it.
3. The Bedouin send to the owners to tell them it is time to cut the date crop (*kharīf*) and take their portion of what is cut.

These three despoilations are described in Sayyid Ṭāhir's *maqāmah*[20] translated below, but he also alludes to other devices by which they rob the palms. The Bedouin take anything that falls to the ground; they shake the dates at the green (*bisr*) stage and collect anything that falls, for, unpalatable as these may seem to a Westerner at the earlier stages, the diet of the Ḥaḍramī is so monotonous that even *bisr* is welcome. Often the Bedouin take more than the share of the crop due to them by custom or contract. For instance, they will perhaps take a cover (*khubrah*) with its content of dates with the words, "*Baghayt ghadā l-ummī/li-khāltī*," "I want 'elevenses' for my mother/mother-in-law."[21] This bare-faced theft is covered by the decent term of *ghadā* (lunch), which is often supplied by employers to employees as a part of their wage. In this way sometimes as much as nine-tenths or nearly all of the crop is lost to the palm proprietor. Indeed some lost so much of their crop that they did not even bother to harvest it.

In the Wādī Dawᶜan, from which hail many wealthy Ḥaḍramī merchants settled in foreign parts, palms were purchased for what seem purely prestige or sentimental reasons. In January-February 1948, for example, at al-Maṣnaᶜah, palms are recorded as selling sometimes at 3,000 rupees and bringing in a harvest of dates worth 13, 15, or

16 Cf. my "Two Tribal Law Cases (documents)," *JRAS*, April 1951, pp. 38, 44: *shāyim wa-lāyim* is that which touches a man's honor. Muḥammad b. Hāshim, *Tārīkh al-dawlat al-Kathīriyyah* (Cairo, 1367/1948), p. 134, says, discussing *shirāḥah*: "The custom (*ᶜurf*) was current that each tribe had the right of protection of the properties (*amwāl*) which their area covered and the protection was called in their *ᶜurf*, *shirāḥah*, and the area of (their) influence *shāyim*, and the owners had never any right to transfer the *shirāḥah* to whomsoever they wish." Cf. ibid., p. 106, for a raid by a Shanfarī group on the *masāriḥ* of Al-Tamīm at Dammūn and on cattle and donkeys going out to work (*sawāriḥ*) in the fields.

17 *Munāṣif* dates have one side ripe, the other half-ripe, deep brown and bright yellow.

18 Bedouin simply means tribesfolk, not nomads.

19 *Saᶜf* is used for fuel, basket making, and other purposes.

20 See again my *Prose and Poetry*, pp. 164 ff., Arabic text, secs. 16-24.

21 Lit., maternal aunt, used euphemistically for mother-in-law.

20 rupees; and from this the cost of maintenance had to be deducted. Dung (probably from the privies) cost a quarter rupee per *jibl* (plur. *jubūl*), "pannier," two of which make a donkey load. A letter dated 16/xii/1365 (11/xi/1946) to the Quᶜayṭī Government from Khuraybah in Dawᶜan concerning the water-channels and protected date palms (*sawāqī wa-mashāriḥ*) of the village complains that people are still paying the taxes they used to pay in "the period of oppression (*ᶜaṣr al-ẓulm*)" (i.e., prior to the reforms sponsored with the Ḥaḍramī states by the British Residency) for which they see no justification except that those who collect the taxes have succeeded to them from their fathers. The writer cites the case of the villagers of al-Ribāṭ nearby who used to make payments to the Marāshidah (sing. Murshidī), Qatham/Qitham, and Sumūḥ (sing Sawmaḥī) Bedouin, all of the Saybān tribe, but who have now recovered their rights. He says that the Āl al-ᶜAmūdī Mashāyikh, the people of Dayjūr, have withdrawn from such (things) as this (*takhallaw ᶜan mithl hādhā*) at the instance of the Nāᵓib so they cannot see why the Khāmiᶜah and Marāshidah tribesmen should be exempted from giving up taxes they exact. "It is not justice that we should pay three taxes (*ḍarībah*), one to the Khāmiᶜah and Marāshidah, one to the Government (*Dawlah*), and one to the palm wardens of the peasants (*shurrāḥ al-ḥaḍar*) who watch, to look after our property (*amwāl*, "palm groves"). The writers, Aḥmad b. Muḥammad Bā Rās (who signs it) and all the palm proprietors (*mawwālah*) of al-Khuraybah, ask for protection against this payment to the Marāshidah and Khāmiᶜah."

The author of the *maqāmah* which is the subject of this article, al-Sayyid Ṭāhir b. al-Ḥusayn b. Ṭāhir al-ᶜAlawī, was born in Tarīm in 1184/1770 and died in 1241/1825-26 at Masīlat Āl Shaykh, to which his father had moved along with his family when Sayyid Ṭāhir was twenty-four years old. Sayyid Muḥammad b. Hāshim,[22] who dwells on his biography at some length, sees in his career part of the effort of the ᶜAlawī Sayyids to set up a state (*dawlah ᶜāmmah*) and suppress disorders (*fitan*). The chiefs of Tamīm and the Āl Kathīr agreed to making him ruler as Imām, and he took the title Nāṣir al-Dīn and gave the address at the Friday Prayer girt with his sword and carrying a *bunduq*. As the ᶜAlawī Sayyids, or most of them, had given up the bearing of arms[23] for several centuries to devote themselves to religion, this was a new departure. Sayyid Ṭāhir was, however, at the same time a serious scholar whose works include responsae (*ajwibah*) to questions on *sharīᶜah*, and in a practical sense he acted to promote conciliation, but in effect arbitration (*iṣlāḥ*), between parties at variance with one another. A curious treatise of his is cited by Muḥammad b. Hāshim, *Taḥrīm al-mashṭ bi-ᵓl-tamr al-mamzūj bi-ᵓl-ward* [The prohibition of combing with dates mixed with roses], which he avers was a custom at that time.

Two neighboring sections of Tamīm, Āl Mirsāf and Āl ᶜAbd al-Shaykh, quarreled over protection of crops (*shāyim shirāḥah*) of a group called Āl Ḥamtūsh. The latter had decided to transfer the *shirāḥah* of the land they owned from Ibn ᶜAbd al-Shaykh to Āl Mirsāf, but the former objected on the grounds that Āl Ḥamtūsh were his protected persons (*rabīᶜ*), and he did not assent to his *shāyim* being withdrawn from them. Imām Ṭāhir wrote to Āl ᶜAbd al-Shaykh at Bā ᶜItīr supporting the rights of proprietors to act as they wished over their property. Āl ᶜAbd al-Shaykh tried to

22 Muḥammad b. Hāshim, *Tārīkh al-dawlat*, pp. 127ff.

23 See my *Saiyids of Ḥaḍramawt* (London, 1956), p. 19.

convince Ṭāhir that Āl Ḥamtūsh were under their protection (*rabāᶜah*), and to remove the *shāyim* was a violation of their protection and a treading (*daws*) upon their honor, which would bring disgrace (*ᶜār*)[24] and humiliation among the tribes. They ask him to take this into consideration and look at both sides "with a single eye." He replied that it was all right to have Āl Ḥamtūsh as a *rabīᶜ* to defend, "but if you protect him (*rabaᶜta-hu*) to gain power over the Sayyids and Miskīns by beating, force, abuse, reviling, and destruction of their properties, then this is a (kind) of protection to which none of the tribes of Ḥaḍramawt have resorted before you, even Yāfiᶜī or Shanfarī." (Ṭāhir was opposed to the Yāfiᶜī group.)

Ibn Yamānī, the chief (*muqaddam*) of all Āl Tamīm, entered into the case and eventually he summoned all the Tamīmī tribes to meet him at his village Qasam at Ṭāhir's instance. Ibn ᶜAbd al-Shaykh accepted his decision (*ḥukm*) on condition that he should make a critical examination of it (*yunāqida-hu*) with the chief of the Bā Jaray tribe of the Shanāfir.

In 1225/1810-11 and 1226/1811-12 the chief Sayyid families of Ḥaḍramawt, including the Āl ᶜAydarūs, the family with whom lay the hereditary chieftanship known as *naqābah*, the B. Shihāb, B. Sahl, ᶜAydīd, Ḥabshī, Jifrī, Āl Kāf families, and others agreed on giving him their allegiance (*mubāyaᶜah*); the Āl Tamīm also adhered to this. They were to promote the *sharīᶜah* and were of course opposed to customary law. Nevertheless, there were further troubles between some of the Āl Tamīm and certain land-owners, and the Tamīmīs in question would not accept the decision on the issue because it was outside the customary law (*al-ᶜurf*) between the two parties. Furthermore, the notables of the tribe hung back from enforcing the decision as, on account of the treaty between them and Ṭāhir, they should have done. Eventually a meeting was arranged between the Tamīm chiefs and Ṭāhir at Masīlat Āl Shaykh and an agreement drawn up between them. It stipulated that *shirāḥah* and *biqār*[25] were not to be restricted to certain recognized people among them and that proprietors (*ahl al-amwāl*) were in charge of their properties in the matter of *shirāḥah wa-biqār* and the like, engaging whomsoever they wished for this purpose. The palm wardens (*sharaḥ*) assented to this *fatwā*.

The efforts of Sayyid Ṭāhir and the ᶜAlawī Sayyids to establish a sort of Shāfiᶜī imāmate in Ḥaḍramawt did not meet with adequate support in the country, and he eventually disengaged himself from any active participation in affairs. The Sayyids later corresponded with Muḥammad ᶜAlī Pasha of Egypt and with the Imām of the Yemen to persuade them to intervene in Ḥaḍramawt, but neither was sufficiently interested or able to do so.

It is more than likely that Sayyid Ṭāhir composed his *maqāmah* about the time of the dispute between Āl Mirsāf and Āl ᶜAbd al-Shaykh and that he intended it as

24 How greatly the protectors esteemed these rights over the palm plantations is perhaps to be seen in the poem of Bā Gharīb where he expresses his longing for Ḥaḍramawt while living abroad (quoted in my *Prose and Poetry*, p. 124):

Ḍayyaᶜtu ʾl-amwāl wa-ʾl-shāyim
Lī fi ʾl-Masīlah wa-fī Bashᶜar.

I have lost my estates and palm protection rents
Which I had in the Masīlah and Bashᶜar.

Bashᶜar is a place south (*baḥrī*) of Tarīm.

25 *Biqār* is not known to me, but *baqara* means "labourer la terre" (C. Landberg, *Glossaire daṯînois* [Leiden, 1920-42]), and I can only suggest that it may have something to do with ploughing by the *baqqārah* or "cattle-men."

propaganda against the abuses by the tribes of their position as protectors of the palm proprietors, among whom of course many of the Sayyid class would be numbered. He displays great literary skill in describing and delineating the habits and haughty, overbearing manner of the tribesman to the non-arms-bearing groups whom he considers *farth*, "dung," as opposed to *dam*, "blood," i.e., the tribal element.

16 This Maqāmah Concerns Those Engaged in "Palm-Protection" in Ḥaḍramawt

Composed by the Blessed Sayyid[26] Ṭāhir b. al-Ḥusayn b. Ṭāhir al-ᶜAlawī (God aid us through his virtue, Amen)

Praise to God, Lord of the Universe. One of his[27] dicta, God receive and rest (his soul), is a pronouncement of his. "Palm-Protection" comes into the category of basically reprehensible actions, devoid of goodness or grace, (characterized by) unscrupulousness, associated with oppression and insolence, and accompanied with harshness and haughtiness.

Viz., the Tribes of Ḥaḍramawt, for whom Death would be seemlier than Life, have obtained mastery over "the property of the people by corruption" (Qurʾān 4:159, 9:35),[28] gaining possession thereof without any right or analogy. They have not the least semblance (of a claim) to it, be it much or little, nor have they any proprietary rights, not even a grain or date-stone pellicle.[29] The proprietors never hire them for the palms at all, nor for one instant did they permit them to enter their property, neither did they ask them to watch (over the palms) or for help. But the palm warden considers that he has more right to the palms than their true owner, saying, "This is mine," without respect for the Noble Creator and his Creation;[30] for "Satan has made their works seem good unto them" (Qurʾān 8:49), making their natures fair in their own eyes, and their actions seem excellent to them.

You find that they will not listen to advice nor turn from what they have resolved, "for when it is said unto them 'do ye not evil in the land!' they say, 'we do naught but what is good.' 'Verily they are the evil-doers but they perceive it not' (Qurʾān 2:11). 'God destroy them, how they are made to lie!' " (Qurʾān 9:30).
17 They persist in wrongdoing, and inflict great loss on (God's) servants. "When it is said unto him,[31] 'Fear God,' pride[32] makes him sin. Gehenna shall suffice him, and a bad bed that is" (Qurʾān 2:202). Yet, withal, would only that they did not steal, and not take an unreasonable wage,[33] or that they would be satisfied and content with what the palm-proprietor leaves (aside) for them, treating the

26 *Al-ḥabīb al-barakah*, a quite unusual form of epithet even for a Sayyid still living.

27 The author, Ṭāhir b. al-Ḥusayn.

28 These Qurʾān passages both contain references to the Jews, and one refers to the monks—this seems intended to insult the tribes. *Māl*, "property," means palm groves.

29 I.e., not a jot or tittle. Cf. al-Wāqidī, *Al-maghāzī*, ed. Marsden Jones (London, 1966), vol. 2, p. 732: give alms, "*wa-law bi-shiqq tamrat-in.*"

30 I.e., mankind.

31 The palm warden.

32 The Qurʾān refers to tribal pride, an apt quotation here.

33 Palm wardens usually took much more than the fifth customarily paid them under agreements they had forced on the proprietors.

proprietors with politeness and (due) humility. "Nay, on the contrary! What they were doing (sinfully) has won sway over their hearts!" (Qur᾽ān 83:14).

When one of them wishes to go out to walk around the palm-plantations under his protection, you will find that he tricks himself out in all his finery and girds up his kilt,[34] putting his right foot up against his left leg and leaning on a stick,[35] winding a sash around his waist, binding a black wool cord[36] (around his turban), drawing it tight, and tying a (black) wool thread around his wrists and calves, puffing himself out boastfully, ruffling it proudly, swaggering, and strutting.[37] Whenever he has walked a little way, he clears his throat roundly[38] and blows out his nose with his fingers,[39] imagining that he has succeeded well or that he possesses the kingdom of Chosroes and Caesar. He never passes by a peasant[40] but he bullies him, nor a piece of palm-branch without breaking it off, nor a palm owner, but he scolds him abusively, saying, "what are you up to—putting the covers on the ripening date-clusters[41] or cutting before you come to ask me. At the moment God does not will[42] that you should cut (them)!"[43]

(So) he pours forth a torrent of bad language and unpleasant abuse, while the palm owner follows in his steps behind him, trying to ingratiate himself with the tribesmen and beseeching favors.[44] The palm warden remains (the while) frowning, silent, and full of wrath, (merely) putting his hand out (to say, "Be off with you!"), and walking away with loud threats. The life of the palm owner is spent

34 The tribesman wears his kilt or waist-wrapper quite short, whereas Sayyids carefully observe the injunction to cover the knee. The *fūṭah* is girt up so that the tribesman can move quickly to stop anyone cutting the dates. When wearing the *fūṭah* short, one is said to be *mutakashshil*; when wearing it long with the foot trailing on the ground, one is *mutabarrij* (the latter is also applied to a woman who decks herself out). A knot tied at the top of it about the waist to hold some small object is called *ᶜaṣb*.

35 This is common in Ḥaḍramawt, mostly among the Bedu and is regarded here as a sign of insolent power.

36 A cord (*ḥabl*) is wound round the turban to be ready for action, it is said, and to serve as protection against blows. The *ḍafīr* cord is tied around the wrists, under the knees, or around the ankles. Raḥayyam maintained this is done for reasons of pride, but I have seen such a thread on a Tarīm Sayyid's wrist, and have been told that it is worn if one has a pain (*ṣāniq*) in the leg or arm—in the latter case it should be above the elbow or wrist. An Arab of al-Bayḍā told me it is worn around the calf because it holds the flesh compactly together and keeps it from wobbling (though these people are spare enough), preventing soreness and stiffness. In some parts tribesmen wear a row of these threads around the top of the calf linked into silver buckles called *ḥubūs* (sing. *ḥibs*).

37 A true if satirical description of the tribesman's behavior.

38 To indicate disapproval one clears one's throat and says, "*ham ham*," among many tribes, thus giving warning of a matter touching one's honor. At a Sayyid house in Qaydūn I recall the teasing of an ᶜAmūdī shaykh who had taken a young girl newly to wife, but for whom, it seemed, he had not proved altogether adequate—*tishtakī minnuh, yādhī-hā*. His Qaydūn friends kept his temper simmering while he tried to induce the Sayyid to take him seriously enough to provide a *duwā* to meet his condition. When he felt the fun was touching too nearly on his honor he would clear his throat in this way, moving his hand to his dagger to indicate things had gone too far. Cf. Aḥmad b. Ḥanbal, *Musnad* (Cairo, A.H. 1313), vol. 1, p. 381, *tanaḥnaḥa wa-bazaqa*. If in the *ṭahārah* and another attempts to enter, the occupant clears his throat in this pointed way!

39 In the more tribal districts Arabs blow the nose with the fingers and shake the product out of the window or under the palm-mat on the floor. This was not considered unmannerly in Wāḥidī country, but I never saw it done in Sayyid houses in Tarīm or Saywūn. In the context here it is insulting.

40 *Miskīn ḍaᶜfā*. See n. 15 above and also my "South Arabia," in C. A. O. van Nieuwenhuijze, *Commoners, Climbers and Notables* (Leiden, 1977), p. 232.

41 Open basket-work covers are placed over the date clusters to keep the birds off them.

42 Lit., "has not written."

43 With the curved toothed *sharīm* sickle.

44 Such as to be allowed to cut his own dates!

in attendance on the palm warden, in coaxing and flattering him, even through his children and women, servants and slaves—whose favor he tries to win by little gifts,[45] stories, and all sorts of amusing tales, so that they may perhaps laugh to him or smile in his face, though they scoff and deride him (all the while). This is if they are pleased and mollified by him.[46]

Some of the palm wardens make a show of scrupling to avoid forbidden things and of pretensions to righteousness, but they are not aware that they are eating of forbidden foods[47] and are deluded thereby. Say to them, "so ask those who have
18 Books of Monition[48] if you do not know" (Qurʾān 16:43). You see them carrying rosaries,[49] praising God and "reckoning that they are guided aright" (Qurʾān 7:29). "They try to deceive God and those who have believed, but they deceive only themselves and are not aware" (Qurʾān 2:9). Man, hast thou ever heard of so reprehensible an action as this, or have thine eyes ever beheld such an evident (sin)?

How many vicious practices[50] there are in "Palm-Protection," and what a volume of deadly sins lies therein, engulfing bodies and souls and angering Him who knoweth all mysteries, such as the abominable brokerage fee[51] involving shame[52] and infamy. Viz., whoever it be who sells palms belonging to or owned by him, the palm wardens take part of the vendor's money and a part from the purchaser which they (must) pay immediately. It might seem as if he (the tribesman) had sold the most expensive commodities[53]; yet he concluded no contract between the two parties nor exerted himself in any way,[54] nor went to any trouble. Then he takes over the palms at once and gives the purchaser a taste of the harshest oppression, mulcting (him) of the larger share in everything,[55] while condescending to give to the purchaser as if he were conferring the greatest of favors upon him, though he has no responsibility for the cultivation[56] or any of the expenses.[57] Notwithstanding this he says, "why don't you look to the cultivation of the palms and see to them? Look to them[58] before they wither! Attend to them before they die!" because he knows if they die the loss will be his,

45 Such as fish, sugar, a *thawb*, etc.

46 Through the gifts given by the *ṭabīn*, or "palm-owner," to the tribesman and his retainers.

47 Cf. Qurʾān 5:45, 65, 66.

48 So Rodwell, after al-Bayḍāwī.

49 This may refer specifically to the Tamīmīs who carry rosaries wound around their dagger-hilts, but one sees this also in other parts of South Arabia. For rosaries of *yusr*, cf. Ibn al-ʿAydarūs, *Al-nūr al-sāfir* (Baghdad, 1934), p. 272, a sort of blackish coral; it is found among other places off the coast of the Wāḥidī sultanate.

50 *ʿAyb* can mean "something that brings shame upon (one)." One says, "*ʿAyb ʿalayk taqūl hākadhā*," "it is shameful of you to say so." To stop a child or adult from an unseemly action one says, "*ʿAyb!*"

51 If you purchase, for example, 400 palms, you give the *qabīlī*, tribesman, two and a half or even as many as five palms. The vendor does the same, and until such payment is made, the *qabīlī* will not permit the new owner to enter into possession. The broker's commission is of course well established all over southern Arabia, but the author complains that this is an exaction, not payment in return for a service. The actual transfer fee is given by Van den Berg, *Ḥadhramout*, p. 79, as 5 percent.

52 *ʿĀr* has much the same sense as *ʿayb*.

53 Because the fee is excessive in proportion to the value of the goods sold.

54 Unlike the broker (*dallāl*) who works for his money.

55 According to my informant, in the harvest of fruit and grain. In my unpublished collection of documents from the Thibī *ḥawṭah* (sacred enclave), lists of these imposts are to be found.

56 Such as digging round the roots of the palm, irrigation, etc.

57 Such as the wage of the *khaddām al-nakhl*.

58 To irrigate them.

for their fruit brings him profit. The palm owner is compelled[59] always to work for others, exposed to loss, humiliation, misery, regret, chagrin, and sufferance, poverty, and distress.

What an amazing thing it is that anyone should sell free men in return for serfs,[60] and in lieu of a tranquil heart and ease one should take quarreling and altercation, abandoning the pleasures of solitude to mix with Hypocrites. The respectable man is he who would remain in the house[61] without coming into contact with those blackguards.[62] Fie upon property which humiliates Man to this degree; perish worldly chattels which bring him to this pass!

19 These words I say extempore, hoping only that one of them may hear them with attention, humility, and submission, so that he may repent of his sin and return to his Lord, God have mercy upon him who shows my faults and points out my sins to me. So give warning, for warning profits the Believers, and the mercy of God is near unto the righteous. God loveth those who repent and cherisheth those who avoid sin. But penitence has certain provisions and a fixed rule; it is no mere clacking of the tongue[63] or wagging of the head and forefinger. God watcheth over the hearts (of men), and each must soon meet Him face to face. So let him hear what the Lord of Mankind[64] has said concerning forbidden things.

He (God bless him) said, "God has an angel over Jerusalem who proclaims each night, 'whosoever consumes forbidden things, from him God will not accept intercession or a propitiary gift, i.e., neither a super-rogatory nor an obligatory action.'"[65] He (God bless him) said, "whosoever buys a piece of cloth for ten dirhams, and a single dirham of the price is a forbidden thing,[66] his prayer God will not accept so long as he continues thus." He (God bless him) said, "any flesh which has grown from forbidden things, Hell best befits it."[67] He (God bless him) said, "he who cares not whence he acquires his money, God will not care whence he makes him enter Hell."[68] How much there is in the verses (of the Qur᾿ān) and

59 The Arabic word has the sense of *corvée* (*sukhrī*), common throughout the Middle East at one time and which still existed in the Yemen probably up to the 1962 coup d'état.

60 *Ariqqā᾿*, properly "slaves."

61 It would be better to sell your palms and not endure so many insults and so much oppression, along with the labor of tending the palms for so little profit, as do many of the religious Sayyids who have retired from the world.

62 In Sayyid writings the word *ashrār*, "blackguards," is regularly used of the Arab tribal groups which held Wādī Ḥaḍramawt at their mercy, exacting the maximum from the land-owners and peasants. The Sayyids and other Ḥaḍramīs of religion are known by them as *al-Akhyār*. Cf. ᶜAydarūs b. ᾿Umar al-Ḥabshī, *ᶜIqd al-yawāqīt al-jawhariyyah* (Cairo, A.H. 1317), pp. 17-18. The Abyssinians in pre-Islamic times are called al-Ashrār, *li-mā aḥdathū fī ᾿l-Yaman min al-ᶜayth wa-᾿l-fisād wa-ikhrāb al-bilād.*

63 This seems to mean the clicking of the tongue on the palate, a sound of assent in Ḥaḍramawt when accompanied by tossing back the head, but of dissent if the head be shaken from side to side. Another gesture is lifting the right hand palm outward shaking it from side to side to indicate the negative, while to put the hand to the brow denotes assent.

64 The Prophet.

65 For *ṣarf wa-ᶜadl*, see my article "The *Sunnah Jāmiᶜah* . . . Documents Comprised in the So-Called 'Constitution of Medina'," *BSOAS* 41 (1978): sec. i, p. 24. The tradition does not seem to be in Wensinck's *Concordance*.

66 Aḥmad b. Ḥanbal, *Musnad*, vol. 1, p. 346: *man ishtarā thawb-an bi-dirham-in wa-fī-hi dirham-un ḥarām-un*, etc. By "forbidden thing," goods unlawfully acquired are meant.

67 Ibid., pp. 321, 399: *lā yadkhulu ᾿l-Jannah man nabata laḥmu-hu min suḥt-in.*

68 Al-Bukhārī, *Ṣaḥīḥ*, *Buyūᶜ*, vol. 7, though not verbatim as quoted here.

Traditions which confirms this; how many an admonition do they contain—which only he who listens and testifies can comprehend. Whosoever maintains that "palm-protection" is lawful, without (his) having been hired by the palm owner—a genuine transaction without question, has joined the band of the Erring. His link with (the faith of) Islam has vanished into shame and pain, retribution and punishment, fetter and chain, because he has transgressed the Law of Almighty God and disobeyed the command of the Apostle when he said "Allah, Allah![69] Let God's servants avoid this corruptness."[70] For God knows "the deceitful eye and what (men's) breasts conceal" (Qurʾān 40:19). "So let not the present life delude you, nor let the Deceiver delude you concerning God" (Qurʾān 31:33). For soon you will perish, and you shall harvest what you sow, and "you shall be rewarded only (according to) what you do" (Qurʾān 52:16). "And they who have acted unjustly shall learn how they shall be overturned" (Qurʾān 26:227). If one of them should say that they do not listen to good counsel, nor follow the truth, nor take warning from admonitions, I shall say, "for (our own) excuse with your Lord, and it may be that they will act rightly" (Qurʾān 7:163).[71] "To my Lord alone shall their account be rendered, if you but know" (Qurʾān 26:113). "Far is your Lord, the Lord of Glory, from what they attribute (to him)" (Qurʾān 37:180). Bless the Apostles, and praise to God, Lord of the Universe.

Chapter on the Stage When the Dates First Begin to Redden[72]

And (the season when) the palm wardens take their ease, (the season of) slaughtering and supping,[73] of setting the covers on the date-clusters, and of cutting, of annoyance and intimidation, and concerning what they take by trickery and theft from the palm owner.

Some of this I have already explained, and I beseech God for strength to tell of their evil ways and what they do, saying, "God aid me to arrive at the truth" by

69 Allāh Allāh, fear God in respect of such a thing (Lane).

70 This tradition does not seem to be in Wensinck.

71 The Qurʾānic context has an apologia for the Prophet's seeming to waste time on admonishing hardened sinners—so also Sayyid Ṭāhir justifies his admonishing the tribes engaged in "palm protection."

72 Or turn yellow, depending on the variety in question. Specific technical terms distinguish stages in the ripening of the date. In ʿUman, S. B. Miles, *Countries and Tribes of the Persian Gulf* (London, 1919), vol. 2, p. 396, records nine stages, but they do not correspond exactly to the Ḥadramī terms which are: a) *bisr*, green; b) *faḍah*, turning red or yellow; c) *qarʿ*, half red and ripe (vb., *qarraʿ*, to become *qarʿ*), syn. *munāṣif*; d) *maḥū*, syn. *ruṭab*, described as *manfūkh*, swollen; ripe; and black, yellow, or red, according to the variety; e) *bāligh*, syn. *maftūl*, dry and wrinkled; f) *khaṭiʿ*, rotten dates, not eaten, sour, gone bad. A term given me elsewhere was *maḥil*, *bisr*, which becomes partly black but is not *maḥū*. Some of these terms seem unknown to the lexicons.

73 Lit., "ladling out" of soup, always made from a slaughtered animal and drunk from the common bowl after eating the meat. It is at the time of the early reddening that the tribes come from their houses to the Masīlah and make their *ʿarīsh* of palm-branches in which they stay to protect the ripening date-crop. There are even small *dārs* in the fields outside al-Hajarayn in which the palm wardens used to stay at the time of the *kharīf*, but by 1947 all this had ceased. The tribesmen cut the *faḍah* dates, leave them exposed to the sun, then shake them into the cone-shaped basket known as *maqsham* or an ordinary palm-leaf basket to make them ripen quickly, then sell them and with the proceeds buy a sheep to kill and eat.

aiding you and me with the means to repent, for He is generous and freely giving. I seek refuge in God from Satan the accursed—in the name of God the Compassionate, the Merciful, seeking to reach the Mighty King through the Holy Prophet, the Sacred House, Zamzam, and al-Ḥaṭīm.[74] (I beseech God) that He remove this vile wicked thing, a reprehensible affair under no restraint or limitation.

The first thing by which they begin[75] and upon which they contract and concur, (is that) when the reddening dates begin to show on the palm-trees, impudence
21 and laughter begin to appear in them, and a great deal of babble and abuse. They let loose their women and children, saying to them, "deal roughly with the peasants,[76] block up the paths with thorns.[77] Don't cover up the stranger if he falls by the wayside;[78] but if it is a man of arms[79] speak to him tactfully. Eat all you can which is still without covers before those greedy[80] men who give no charity[81] of it, nor entertain the poor from God's portion, and who do not (properly) understand the worth of what God has preserved and protected for them." The (palm wardens) often miscall[82] the palm owner with far worse words than these, insulting his honor in front of rascals and beggars, never thinking that for this there will be a reckoning (with God some day), nor do they realize the inevitability of catastrophe, stopping, and quakings of the earth.

When the palm owner comes to put on the covers,[83] he sends to them, great and small, politely consulting them, with the words, "we are going to put on the covers, if you are not too busy, but otherwise we shall go away."[84] After an hour[85] they answer the palm warden, saying, "you think that we're keen to leave it.[86] We

74 Al-Ḥaṭīm is merely employed for the rhyme, a well a few feet west of the Meccan Kaᶜbah.

75 This word means also "to undertake." This opening sentence is a parody of the style of tribal documents and their rather pompous phrasing. For example, one of the Thibī agreements reads, "the noble Sayyid ᶜAlawī b. Muḥammad b. ᶜAlawī al-ᶜAydarūs and the Headmen of the Āl Shaybān have agreed, undertaken, resolved, and decided by their honor. . . ."

76 "Beggars" might be a better rendering of *masākīn*, since at the time of the *faḍah* stage beggars come to collect fallen dates, but the palm wardens drive them away. In this connection, Aḥmad b. Yaḥyā b. al-Murtaḍā, *Al-baḥr al-zakhkhār* (Cairo, 1366/1947-1368/1949), vol. 4, p. 284, states that if it is the custom of owners of fallen fruits (*sawāqiṭ al-thimār*) to make them free, it is permissible to take them since custom (*ᶜurf*) has an influence on the like of this. So in Zaydī Yemen that was evidently the practice.

77 *Salam*, acacia, thorn would be laid down to keep people out of the palm-groves.

78 Scold a beggar or bedouin if he takes the dates. Some less severe palm wardens would only warn pilferers.

79 I.e., a tribesman, who is not to be treated in this insulting fashion. Care must be taken to create no incident (*bādī*) that might lead to a feud, but the tribesman counts all others, Sayyids included, as *farth*, "dung," the slaying of whom will not give rise to a feud between tribesmen.

80 By "greedy," perhaps to be rendered as "envious," the palm proprietors are intended.

81 *Zakāt*.

82 As in other Arab countries expressions of abuse might be, "*umm-ak*," "your mother"; "*aysh abūk bā*," "what does your father want?"; "*aysh umm-ak batt-uh*" (for *baghat-uh*), "what does your mother want"? See also my article "Sex, Birth and Circumcision," in Adolf Leidlmar, *Hermann von Wissmann Festschrift* (Tübingen, 1962), p. 194.

83 The covers are made mostly in Madūdah and in 1954 cost 40 East African cents each. They are sold in batches of 40 (*qarn*), one packed on top of another to form a bundle. Brokerage is 40 cents per *qarn*.

84 Without putting the covers onto the date-clusters.

85 The palm warden pauses a while without giving an answer, out of tribal haughtiness and contempt for the palm owner.

86 The palm warden means that this is not the case as he cannot offer the excuse to any person who comes to ask him for a few dates that the covers have now been put on them and they cannot be touched.

had, as a matter of fact, thought of sending for you before today because this year the date-crop is a very little only,[87] and if it is left (without covers) much blame will fall on us."[88] (In such ways) do they make out that the date-crop is poor, belittling and despising it.[89] They add that it requires watering and that the fruit is dropping off,[90] making little (of it) so that the reckoning of what they are going to take to bring away and eat will tally. The palm owner says to his laborers, "let one man go to cover each palm while I stay here to see which of you is getting on with the job."

When the laborer goes (to do so), they (the palm wardens) say to him, "Keep off our 'bounty',"[91] stopping him, so the peasant[92] says, "point out exactly what is yours." "Each small child-palm,"[93] says the palm warden, "leave it alone, and
22 take care you don't go near it!" "Whatever you say," answer (the laborers), "in whatever you mark out."

But, after all that, the (tribes-people) go up to the father-palm[94] and pull off its fruit. Their women and children bear down (on the dates) stripping them off with their fingers,[95] and their friends too, close and distant relatives, come racing up to it, while all the owner can do is to look on and stay sitting still, all (the palm owners) preoccupied and silent together, as the (tribes-people) quarrel and snatch (anything they can get) from each other. When all this is done, the palm owner can scarcely believe that (the trouble) is over, for he gets but little out of the date-crop, and the cares and vexations he suffers are without number. O God, make haste to bring retribution to the palm wardens, by the right/truth of Muḥammad, Lord of Mankind. There is no gainsaying that most of it falls to the share of the palm wardens, in the way of (what is taken) at the ripening of the fruit in "bounties," or fraudulent practices. Yet despite this they say, "the palm owner has

87 The rhymed terms *shawm* and *lawm* at once recall to the Ḥaḍram so Ḥaḍramī the phrase *shāyim wa-lāyim*, rights over palm-plantations of the palm warden. See "Two Tribal Law Cases," I, 44-45. But *shawm* here means "very little"; cf. classical *shuʾm*.

88 As long as the dates are left without covers many people will come to ask for gifts of them, and if the palm warden refuses them he will be stigmatized as *bakhīl*, "mean," and his honor will suffer.

89 The palm warden represents the crop as meager because he has already taken some and wants to cover up the deficiency.

90 They imply that the owners have not seen to the oiling of the date clusters with castor oil earlier in the year to prevent the dates dropping prematurely.

91 The *ṭarḥah* is a handful or two of dates given to a beggar who asks for them as one is climbing the palm to pluck one's crop. This is never refused, but the *khaddām al-nakhl* also cheats his master and gives the date-bounty to his relatives so that, indeed, one way or another, the beggar may receive more than the owner. Here the palm warden is telling the cultivator not to give the "bounty" to his own women and children.

92 Arabic *ḍaᶜīf*, who is called *jaᶜīl*, laborer.

93 Each small palm standing by itself is called *walad*, child. The following are the various types of growth: *khalᶜah*, plur. *khilāᶜah*, small young palm which has reached the stage of bearing fruit; *manqar*, a tall, single-stemmed palm; *qarīn*, a palm with two trunks stemming from a single root; *dawwārah*, a palm with three or more trunks; *mallah*, a large palm like the *dawwārah* but planted near a well. For the term *khāliᶜ* and the saint known as Khāliᶜ Qasam, see my article "Hūd and Other South Arabian Prophets," *Le Muséon* 47 (1954): 157. A *baqlah/bajlah* is a self-sown palm, the opposite of *khalᶜah*. See also my article "Some Irrigation Systems in Ḥaḍramawt," *BSOAS* 27 (1964): sec. 1, p. 66. The Ḥaḍramī says: "*Lī qid-ak bā takhlaᶜ tibāq lābāh*," "If you are going to plant, look at its (the palm's) fathers" (*tabāqā li*, to look, examine, etc.). This is also used metaphorically of human beings.

94 The *ab* or "father" is a tall palm. The varieties of palm are divided, arbitrarily it seems, into male and female—that is apart from the true male palm used to fecundate the date-bearing palms. A "male" variety such as *Zār* or *Madīnī* is known as *ab*; the "female" variety such as *Ḥamrā* and *Hijrī* is known as *umm*, and one speaks of *al-umm* and *al-abū*.

95 *Kharīṭ*, pulling the date-stalk through two fingers to strip off the fruit, not pulling off each date separately.

left us nothing[96] (though, in point of fact, he set neither locks or keys upon it),[97] but we shall take away what we want without any bother (from him)."

When the date-crop ripens and becomes "medium-ripe"[98] they bring baskets and bags[99] to it and begin carrying it off in goat-skin bags[100] or plaids.[101] After they have got all they wish and think they have done well, and profitably they return to their places to meet, agree and undertake,[102] to conceal what has taken place. For about ten days or a week after this they stay resting, killing fat sheep in great luxury and ease, as if they had upheld the Faith and the sanctions of Islam! Nay! On the contrary, they are demolishing all its very Pillars, treading the paths of Satan, little thinking they are on their way to Hell-Fire—God defend us from the same, deliver us and you (also) from abiding therein and make us safe from such fears and tribulation.

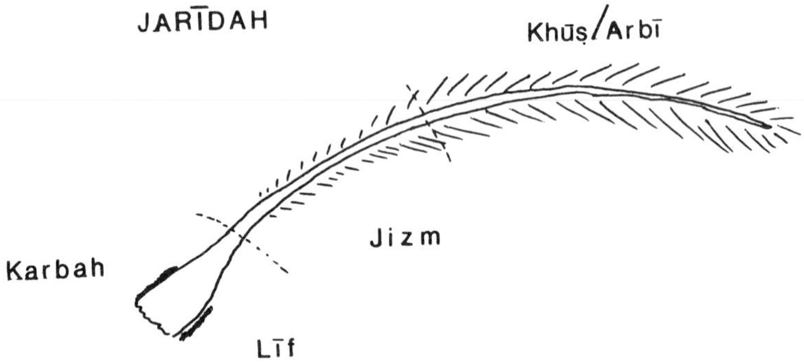

FIG. 1.—*Jarīdah*, palm spathe; *karbah*, pl. *karab*, butt or stub of palm branch; *līf*, palm fiber used for ropes or for kindling (*rushun*; *jizm*, pl. *juzūm*, or *quṣm*, thorns used to dry meat upon or for making awnings (*shubbak*) for windows (now curtains are used); *khūṣ*, palm leaf, also known as *arbī* (no plural) and used when dry for kindling (*rushun*)

96 This statement is made to counter any objections the palm owner may raise.

97 The peasant's viewpoint is expressed in the *rajaz* (See my *Prose and Poetry*, text 166, no. 80):

> I wish all land-lords (*ṭabānah*) fever and disease,
> For four years long not carrying their keys.

98 *Muntaṣif* = *munāṣif*.

99 The *maghdafah* is anything carried over the shoulder like a sack, so it might be a cloth used to wrap up and carry the dates.

100 A whole goat-skin used as a bag, carried over the shoulder, as well described in C. Landberg, *Etudes sur les dialectes de l'Arabe méridionale*, vol. 1, *Ḥaḍramoût* (Leiden, 1901-1913), p. 257. I have the impression it is more used by tribes than peasants.

101 The *malḥafah* is a black indigo-dyed fringed cotton cloth which the Bedu carry on either shoulder as they feel inclined; it may be draped over both shoulders or used as a waistband. My informant said it was 7 by 3 *dhirā*ᶜ and Baḥraq, *Nubdhat*, p. 6, gives the size as 7 by 4 *dhirā*ᶜ which is also exactly what it is today. Baḥraq was writing in al-Shiḥr, a famous weaving center. Cf. Landberg, *Ḥaḍramoût*, p. 709.

102 The author again parodies the style of tribal documents. By "undertake," he means the Bedouin gesture of drawing the index finger down the brow and nose, with the words, "*shallayt bi-wijhī an af*ᶜ*al*," "I undertake by my honor (face) to do." Cf. n. 74 above.

Then when these days have passed by and they have engaged submissive
workmen, they cut down all the yellowing palm-leaf, stalk and stem,[103] may God
cause them to perish, fearing not the punishment of the Angel[104] who knows all,
without respect for shame or blame or flame, as if indeed, they had never entered
23 into the Faith of Islam and knew not the taste of Death.[105]

After they have committed these abominable deeds and unseemly actions,[106] taking in their grasp a formidable handful of sins and misdemeanors, they send heralds to the palm owners saying, "come (here), for the date-crop is ripe and this Ramaḍān month, wherein lieth profitable merchandise,[107] has come upon us unexpectedly. In it we wish to be free[108] to devote ourselves to acts of worship and the telling of beads."

So up come the palm owners at once because if they were to tarry punishment would be meted out to them on that account. When they come to the palms and set about cutting, up come the tribal headmen to them with swaggering gait and say, "why do you not send to the palm warden to let him know? Hold your hands until you send (for him)."

In this way is the palm owner kept back from his palms, waiting to see how his affairs will turn out, apprehensive of the vexations and insult that he will encounter. After a short while the palm warden arrives like a fiend, keeping an eye (on the people at the palms) and grumbling till he comes up to them, scolding (all the time). The palm owner gets up, makes excuses to him, humbly begging of him and kneeling on the ground in front of him, but the palm warden is in a temper, shouting abuse, and waving his hands. After a while the people[109] there go to him and ask him to let them cut, (declaring) that if they have erred, they acknowledge (their fault). After a great to-do and quarreling, accompanied by volleys of abuse and rowdiness, he does give them permission, but the palm owner is humiliated and treated with contempt.

However, they get up and cut what they manage to get, saying nothing as though they were dumb (animals) unable to speak, or blind and unable to see. Each laborer amongst them is followed by a woman and child with basket,

103 The different parts of the palm branch are distinguished by special names. The whole is known as *jarīdah*. The *karbah* (plur., *karab*) is the broad base or butt of the branch. Around it is fiber (*līf*) used for kindling (*rushun*). This narrows and the leaves begin to appear in the form of small spikes, this portion being known as *jizm* (plur., *juzūm*) but also as *quṣm* (plur., *aqṣām*), probably basically both the same word. The *jizm* was used for sun-drying pieces of meat and was also at one time used for bars (*shubbāk*) for windows where now curtains are employed. The upper part of the branch with long leaves (*khūṣ*, sing., *khūṣah*) is called *arbī* (no plur.) and is used dry as *rushun* with the effect of producing almost instantaneous heat when desired, for instance, to boil water in a kitchen.

104 Perhaps *malik*, "king," i.e., God, should be read for the *malak*, "angel," of my informant.

105 An allusion to such Qurʾān verses as *sūrah* 3:185, etc.

106 The text here is perhaps a little dubious.

107 Reminiscent of the Qurʾān. The merchandise is God's reward for acts of devotion such as the Fast.

108 From the duties of palm protection.

109 My informant said these would be beggars and other palm wardens acting as intercessors for them. It seems, however, that the palm owners and their laborers are meant.

322

pannier, and cone-shaped straw hat.[110] Whenever any dates fall they pick them up quickly, but if a cover falls they throw it aside with derision.[111]

24 When they have finished cutting and picking the dates off the stalk into the cover, (accompanied by) the hard treatment (to which they are exposed) the chief palm warden gets up and comes to them. Leaning on his stick he blows out his nose and clears his throat. Then he says, "divide and set aside the *zakāt*-alms and bring it to me to distribute amongst the people without contention (from them)." Nor can they possibly disobey any advice of his. He then takes it like a mad bull,[112] distributing it in inappropriate quarters without referring to anyone for advice.[113] He singles out his relatives, allies, friends, servants, and favorites to be its recipients, without fear or regard to the condign punishment and retribution of God that day on which he will be given his book in his right hand[114] and none of them shall avail him by what they have acquired,[115] nor shall any of his reproof and reprehension be averted that day upon which he comes face to face with the shameful and sinful actions to his account.

May God make me and you my hearers, among those who listen to admonition and do not what angers their Lord and those who amass good deeds to please their Creator, those who abandon acts of disobedience, for the best of admonitions lies in verse and in speech, who distinguish the Signs[116] (given) to those of wit and understanding, with the words of those who distinguish between what is licit and what is forbidden. God aid you and me to lay hold on what is lawful and to avoid what is forbidden and sin. May He deliver and cleanse us of this "Palm-Protection" for it is one of the greatest of sins. May God pardon your sins and mine, by virtue of Muḥammad, child of ʿAdnān; may He guard you and me against the evil of dissension and tribulation. May He set us, of His mercy, in the highest of the Gardens of Paradise, for He is Generous and All-Bestowing. God suffices us, and how excellent a Guardian He is. God bless our Lord Muḥammad, his Family, and Companions. Praise to God, Lord of the Universe.[117]

110 Of the type worn by peasant women. There are several basket types which might be used, a *quffah hadiyyah*, gift basket, in which small presents, usually of food, are sent, *quffah ṭaḥīn*, for flour, and the *maḥmalah*, which contains six to seven *muṣrās*.

111 *Huqlah* was said to mean derision and clapping of the hands. They take out the dates from the covers which act as a sort of bag and throw them away though presumably they would normally be kept for another season.

112 *Tazaywam*, to be filled with a sort of wild frenzy such as affects a bull before copulation or attempting to gore.

113 *Zakāt* cannot of course be dispensed to merely anyone, but the palm warden pays no attention to what *sharīʿah* law lays down in this respect.

114 Reminiscent of Qurʾān 17:71.

115 Presumably as intercessors for him on Judgment Day.

116 I.e., Qurʾānic verses.

117 For further material on dates see my *Prose and Poetry*, Arabic text, secs. 126-33, for a poem on the varieties of date in Ḥaḍramawt and their properties; al-Malik al-Afḍal the Rasūlid, *Bughyat al-fallāḥīn*, eighth/fourteenth century, for date cultivation in the Yemen (MS, my edition in preparation); and for *ʿuhdah*, pledging of property in palms, etc., see my "Materials for South Arabian History," sec. iii, p. 591; a MS on this subject is now available to me.

VII

THE CULTIVATION OF CEREALS IN MEDIAEVAL YEMEN

(A Translation of the *Bughyat al-Fallāḥīn* of the Rasūlid Sultan, al-Malik al-Afḍal al-'Abbās b. 'Alī, composed circa 1370 A.D.)

Agriculture is the occupation of the great majority of Yemenis, and even the foreign experts who have been coming to the country since Imām Aḥmad's days admit that the country is, by and large, well farmed. One cannot fail to be impressed, on the one hand, by the magnificent sweep of broad fields in the fairly open country near Ibb, layered one above the other, and beautifully farmed and kept, while, on the other, every tiny parcel of land in the high and stony rocks of the steep mountain sides, terraced from bottom to top, seems to be utilised. Nevertheless I think I detected, in some of the higher northern districts, that certain terraced lands of marginal nature had gone out of cultivation. This is likely to happen more frequently where the returns are poor, the rains scanty, and the population are attracted to work elsewhere.

The pre-Islamic inscriptions naturally reflect the agricultural activity upon which the economy of the Yemen is based. It was however the Rasūlid monarchs of the thirteenth and fourteenth centuries A.D. who were not merely interested in the practical sides of husbandry, stock-rearing, agricultural almanacs and calendars, as the Rasūlid MS.[1] of mixed contents shown me by Qāḍī Ismā'īl al-Akwa' shows, but took the trouble to record in writing their personal observations, side by side with their gleanings from earlier Arab writers. The *Bughyat al-fallāḥīn*,[2] completed not before 773 H. (1371 A.D.) may be described as the quintessence of their researches. The author, al-Malik al-Afḍal al-'Abbās b. 'Alī (flor. 764 H./1363 A.D.–778 H./1376 A.D.) quotes extensively, among Yemeni sources, from *al-Ishārah fi'l-'imārah* of his father, al-Malik al-Mujāhid 'Alī b. Dāwūd (721 H./1321 A.D.–764 H./1363 A.D.), but he also uses materials he found in the *Khaṭṭ* or handwriting of his father – recorded by his father on the authority of his own father, on the authority of his (Dāwūd's) grandfather. The extent of the agricultural tradition in the Rasūlid house is apparent – our author's grandfather, al-Mu'aiyad Dāwūd b. Yūsuf (696 H./1297 A.D.–721 H./1321 A.D.), is said to have abbreviated an agricultural encyclopaedia,[3] and Dāwūd's grandfather, al-Malik al-Muẓaffar Yūsuf b. 'Umar, ascended the throne in 647 H./1250 A.D. Our author also frequently quotes *Milḥ al-malāḥah* which, he states, was written by 'my grandfather (*jadd*) al-Malik al-Ashraf'.[4] The *Milḥ al-malāḥah* is extant, though so far in a single defective copy only,[5] but no copy of *al-Ishārah* is yet reported. It is possible that a few pages in the Biblioteca Ambrosiana[6] might come from *al-Ishārah*, for the section on millet (*dhurah*) has sentences nearly identical with the *Bughyah*, as might also the section entitled *Ma'rifat al-matālim*, though the latter might be from a missing section of *Milḥ al-malāḥah*, or both might be the notes of an earlier Rasūlid monarch.

Of the *Bughyah* itself two copies have been at my disposal. The first, a copy made available to me in Tarīm by the kindness of the historian Saiyid Muḥammad b. Aḥmad al-Shāṭirī, I transcribed in 1953–4; it is dated Muḥarram 9, 1197 H./15 Dec., 1782 A.D. The second[7] was shown me by the late Fu'ād Saiyid on my return via Cairo from Ḥaḍramawt: he provided me with a microfilm – it is undated. It is said that another copy exists in the Yemen but it has not been available to Yemeni scholars or to me. Since I commenced my edition of the text, Ibn Baṣṣāl's *K. al-Filāḥah*[8] has been published in Tetuan, and Professor Iḥsān 'Abbās has drawn my attention to a published edition of another Spanish treatise, *al-Falāḥ fi 'l-filāḥah* of Abu 'l-Khair[9] of Seville, both of which are utilised by our author though he only once cites Abu 'l-Khair by name.

Establishing and Interpretation of the Text

It might be thought a simple matter to establish a text of the *Bughyah* from the MS. and printed sources cited above, to which must be added the abbreviated version shown me by Qāḍī Isma'īl, but the contrary is the case. Our author does not quote *verbatim* from either his Yemenite or Spanish sources, but abbreviates or paraphrases. I incline to think he had the fuller version of Ibn Baṣṣāl at his disposal, and while the latter author is a useful aid, the *Bughyah* is quite often more correct than the printed text of Ibn Baṣṣāl, which is poorly edited. My objective has been to attempt to preserve the text as nearly as al-Malik al-Afḍal presented it, including his mis-readings of, in the main, Ibn Baṣṣāl, provided these make sense. From time to time however it is necessary to interpret the text of the *Bughyah* from one of the sources upon which it draws. With this translation I have not included the marginal notes containing further technical vocabulary for which I must go to Zabīd to establish. My text is now pretty satisfactory, but a few passages remain about which the Qāḍī himself is unsure.

As long ago as 1954 I had some help from Qaḥṭān al-Sha'bī, later first President in an independent Aden, but at that time an agricultural assistant. My debt to Qāḍī Isma'īl al-Akwa' is overwhelmingly great. Not only have I questioned him to the point of weariness – if he could be wearied – concerning Yemenite technical terms and usages, but on our journeys together he has neglected no opportunity of showing me in the fields practical illustrations of what the *Bughyah* describes. His brother Qāḍī Muḥammad also helped from time to time. The *Bughyah* describes technical practices strange to me, but the greatest difficulty of interpretation lies in the many technical terms peculiar to the Yemen, not merely unknown to the lexica, but particular to the Lower Yemen where the Rasūlids held their court, farmed and taxed the countryside. In the northern districts of the Yemen quite different technical words are currently employed, and Qāḍī Isma'īl himself is not acquainted with some of the Tihāmah vocabulary.

In view of the liberal policy of the Government, after the end of the Egyptian occupation, in opening the Yemen to research I have decided to anticipate my edition and translation of the *Bughyah* by rendering into English the most important and purely Yemeni chapter – that on cereals, the staple diet of its people. I may later follow this with the chapter on pulses. To the dates of seasons of the agricultural year, given according to the Rūmī months and/or agricultural stars, I have added, in brackets, the corresponding date in our solar year, following the almanac of Shaikh Muḥammad Aḥmad Ḥaidarah, *Ṭawāli' al-Yaman al-zirā'ī*[10] for the year 1391 H./ 1971 A.D., published in Ta'izz. If this rough and ready method should introduce a margin of error upon which historians of the calendar would be informed, it can be but slight. Indeed it is surprising how many entries in the Ḥaidarah almanac reproduce data figuring in the *Bughyah* – this implies a continuity in almanac literature which I could in fact demonstrate.

Present-day Investigation of South Arabian Agriculture

For the former Aden Protectorates I have seen agricultural reports going as far back as 1947, but the best source for the type of work carried out up to 1967 is Dr A. M. A. Maktari's[11] study of irrigation. I am informed that insofar as further development has taken place since 1967 it follows the general schemes laid down by the British. In the Yemen external organisations had already begun to interest themselves in agricultural development in the latter days of Imām Aḥmad.[12] The United Nations Development Programme has been active of more recent years and a number of F.A.O. reports were put at my disposal through the kindness of Annika Bornstein. Though these reports are informative where experts deal with their own specialisation, since, obviously, they have no knowledge of Arabic, they can be misleading. One such report, for example, makes the bold assertion that 70 per cent of the land is owned by merchants and nobles. I am highly sceptical of the truth of so contentious a statement, for the plain fact is that we know extremely little about land tenure in the Yemen. In certain districts in South Arabia such as Laḥj/Lahej[13] this may be nearly true, but so diversified is the terrain that one cannot generalise. How much land is mortgaged under the '*uhdah*[14] form of contract in the Shāfi'ī districts – and is this the practice in Zaidī territory? Are the share-cropping contracts similar to those in Ḥaḍramawt or Lahej or Jīzān,[15] and do they differ in mountain land from the Tihāmah? These and many other questions require investigation. An indication of the complexity of land tenure and water rights may be found in the U.N.D.P./F.A.O. *Survey of the agricultural potential of the Wadi Zabid*,[16] the area to which the *Bughyah* so frequently refers. Ismā'īl al-Jabartī (ob. 806 H./1404 A.D.)[17] is credited with establishing there the water-law followed at the present day.

For the Wādī Jīzān, nowadays in Saudi Arabia (but in the mediaeval period part of the Rasūlid domains), there are two important F.A.O. reports – the C. O. van der Plas[18] study contains invaluable sociological, linguistic and other data on what is probably a typical Tihāmah valley. The forthcoming report of the Sir William Halcrow team of which Dr Maktari and I were members includes Maktari's study, 'Land tenure and water rights', with a useful technical glossary – to say nothing of carefully worked-out agricultural and physical surveys relevant to this present study, executed by various experts in the team.

Agricultural Calendars and Almanacs

Muḥammad Ḥaidarah's edition of his almanac for 1365 H./ 1945–6 A.D. I have already translated,[19] adding much additional information about South Arabian calendars. Now also I have at my disposal the Jīzān agricultural calendar which we worked out from local informants in December 1971. In the Wādī Jīzān there is much variation as to the precise dates, by several days in fact, upon which the period covered by the twenty-eight stars falls. I relied on Shaikh Aḥmad b. Manṣūr al-Sa'dī of Abū 'Arīsh who was stated to be the best authority in the Wādī. The Jīzān stars correspond exactly to Glaser's 'Sun Stations' and 'Morning Ascension', though there is a wide discrepancy in the dating of the two.[19] Both these differ again from the Ta'izz calendar, which last, however, does exactly tally with that of Wādī Ḥaḍramawt. In point of fact there is considerable variation from district to district in Southern Arabia, apart from other Arab countries, and it has to be established what the local usage is in each place. Not all countrymen who use the star calendar to guide them can even list the star periods, and anyway this lore is probably beginning to be abandoned nowadays in some districts. In all the calendars mentioned the star period is thirteen days, with one star of fourteen days' duration.[20]

For the author of the *Bughyah* the year commences with Rabī', and the Ta'izz almanac gives the dates of the seasons as follows:

Rabī' (Spring) 22 March, Ḥamal (Aries)
Ṣaif (Summer) 22 June, Saraṭān (Cancer)
Kharīf (Autumn) 23 September, Mīzān (Libra)
Shitā' (Winter) 23 December, Jady (Capricorn)

Aries (al-Ḥamal) was found convenient to use as a term to a letting in Lahej, for Bā Makhramah[21] was asked for his opinion concerning 'a person who rents (*ista'jar*) from another, land in Lahej up to the sun of Aries (*shams al-Ḥamal*), according to their custom (*'ādah*)'. His reply commences with the assertion that most of the (contracts of) lettings (*ijārāt*) are improper (*fāsidah*).

The agricultural year in Jīzān, on the other hand, commences a little before mid-June with the sowing of *shabb* millet, though in the mountain districts there this last takes place a little earlier. This is the opening of the season of Kharīf (12 June).[22] So in Jīzān the fiscal year during which the *zakāt*-tax is paid upon grain produce, runs, approximately, from Kharīf to Kharīf, or as it may be expressed, 'from *shabb* to *shabb*', since *shabb*-millet is planted at the stars al-Dhirā', the last star of Ṣaif, and al-Nathrah the first star of Kharīf. Al-Dhirā' is actually often called Shabb as a proper name because of the association of millet sowing with it. This Jīzān practice would seem to me to be ancient.

A Russian Arabist, Anatoli Agaryshev,[23] has given a not very accurate version of the agricultural star calendar republished by me from Glaser's earlier work, this being used among the Zaidī tribes, but perhaps not in the districts in which the Rasūlid sultans interested themselves in agricultural economy. It is probably in use in the district east of Kuḥlān for I heard the saw in July, *'Man ṣulum mā ẓ/ḍulum*, He who plants Ṣulmī (i.e. grain sown at the star Ṣulm) will not lack a good crop.'[24] We also saw 'Allānī[25] being sown there on 17 July presumably ripening in 'Allān (September) – this is wheat and barley, but it is also called 'Alibīyah since it is sown at the Star 'Alib (about 12–24 July). On the same day we passed men sowing Khāmisī, mostly barley, and *'alas*-wheat, but there were also broad beans (*fūl*). The ploughman was followed by another man a few paces behind throwing the seed in the ground and covering it over with earth with his foot, just, in fact, as the *Bughyah* describes (pp. 46–7). Khāmisī will ripen at the star Khāmis 'Allān (2–15 September) and I noted when in al-Qārah a proverb, *'Idhā dakhal al-Khāmis tajid al-ḥāmī*[26] *kāmis.* When the Khāmis (star) enters you find the field guard (also called *shāriḥ*) in ambush (to catch persons trying to rob the crop).' All over the fields in the north are dotted little stone cabins (*miḥrās*) where the armed tribesman watches for the robbers at harvest-time.

It seems apposite at this point to quote a couple of ditties sung about harvest-time. As the peasants tread out their grain (*yadūmūn*) with their cattle they sing,

Tread you out a treading, *Dūm lak yā dawm*
O duff, o milk gruel. *Yā 'aṣīd yā zawm.*[27]

Qāḍī Ismā'īl quoted a saying, *'Ya lait Ṣan'ā' 'aṣīd, wa-'l-baḥr zawm, wa-Qā' Jahrān malūjah wāḥidah*, Would that Ṣan'ā' were duff, the sea milk gruel, and the Plain of Jahrān all one round of barley/wheaten bread.'[28] On which Ḥusain b. Sa'īd commented that the man who said this must have been very hungry! Jahrān is a large fertile plain. At the winnowing on the mountain of the 'Ayāl Yazīd when one winnows the grain (*yimdhaḥ al-ṭa'ām*) I heard in late November 1966, the song.

Blow O Wind, O strong blowing wind,
Hubbī yā nawd yā nawwādah
O daughter of goodness,[29] O noble wind.
Yā binta 'l-khair yā jawwādah

At the star Sihail (Suhail) in northern Yemen (25 July), in fact at Mashāf on the low Tihāmah foothills, I heard the saw, *'Sihail, fī yawm-uh miyat sail, min ghair sawāri 'l-lail*, Sihail during the day-time of it are a hundred floods, not reckoning the rain-storms of the night.'[30]

There is a hot period in the Yemen known as al-Jaḥr which, it was said to me in Ṣan'ā', lasts *manzilatain*, i.e. two months, and the 9th day of it in 1972 was 27 May. In 1964 at al-Qārah I was however told that, '*Al-ghubār yijī fi 'l-Jaḥr, shahr 'Alib*. The dust-storms come in al-Jaḥr, in the month of 'Alib (12–24 July).' The *ghubār* must be identical with the *ghubrah* of Jīzān, a north wind bearing dust and sand commencing about 22 June and abating about mid-August. I endured the same wind in Ṣubaiḥī territory when stationed there in the hot months of July and August 1940. Glaser considers al-Jaḥr would fall in late May, June and July. There is a crop-season named after it (p. 50).

The Rūmī months have, I think, been used in the Yemen from pre-Islamic times. They are mentioned by the ninth-century Hamdānī;[31] distinguished authors like Nashwān b. Sa'īd and 'Abdullāh b. As'ad al-Yāfi'ī have composed poems of a semi-didactic type upon them. On two occasions Khazrajī[32] refers to them, noting rains falling in Aiyār and earlier, from the 1st of Nīsān, in 795 H./1392 A.D., and heavy floods reaching the sea at Zabīd in Tammūẓ 802 H./1400 A.D. For convenience I repeat these Rūmī months:

Tishrīn al-Awwal 14 October	Naisān/Nīsān 14 April
Tishrīn al-Thānī	Aiyār or Mabkar
Kānūn al-Awwal	Ḥazīrān
Kānūn al-Thānī	Tammūz
Shubāṭ	Āb
Abhār	Ailūl

Ḥaidarah's Ta'izz almanac expressly states that the 1st of Shubāṭ (14 March) is the opening of the '*ām al-Rūm al-zirā 'īyah* (sic), which I take to mean the agricultural year of Rūmī months.

Apart from the *hijrī* calendar, two more systems may be mentioned. Glaser describes a system which starts with a period known as Lailat wa-lā-sh (The Night of nothing) when the Sun and the Pleiades are in conjunction, falling about 18 May, though the period itself would appear to start some ten days earlier, and it lasts three months – after this follow nine months of thirty days, Tis'at-'ash, Sab'at-'ash, Khamst-'ash, Thalāt-'ash, Ḥad-'ash, Tis', Sab', Khams, Thalāth. The other is the Ḥimyarite months mentioned in the *Bughyah* which Professor Beeston discusses on pp. 1–4. A less sophisticated type of reckoning I chanced upon is that, on the right of the Ṣan'ā'-Ṣa'dah road not far from Raidah, is a hillock called Kawlat Nā'iṭ, a sort of landmark. When at dawn the sun shines (*tishriq*) halfway up or about the head of the *kawlah* (hill) this marks the faṣl al-Ṣaif and time for sowing millet (*madhrāt al-dhirah*).[33]

32

Rain-making (*tasqiyah*)

Invoking Allah for rain is not only countenanced and approved in Islam, but, as I show in my *South Arabian Hunt*, was, for example, linked to the chase of the ibex in the pre-Islamic ages – it takes many forms in Arabia. In the regions around Ṣan'ā' there is a rain-ceremony known as *tasqiyah*. On 21 July 1972, driving down the Ṣan'ā'-Hodeidah road, we came upon, somewhere before Manākhah, a group of men and boys walking down the road with some black goats and singing. Our driver Ḥusain b. Sa'īd of Jabal 'Ayāl Yazīd told us they were invoking rain (*yitsaqqī*). The goats are slaughtered but not eaten, being left to the vultures (? *nusūr*), wild animals and dogs. In northern Yemen usually they make a tour of the town or village and then slaughter the animals, in no specific place, but usually near the Jāmi' mosque or else above the *birkah* (large cistern) which will probably be empty of water. It is, as a rule, he added, only when the camel or bull is worn out and old (*ta'bān*) that they leave it to the *nusūr*, for they usually eat the younger animals. In the Jabal they usually take a bull (*thawr*) or calf (*'ijl*) of one to one and a half years.[34] Of what is said or sung at the *tasqiyah* he could only remember

Give us water, o Allah	*Isqī-na yallah*
Pity us/Give us rain, o Allah	*Irḥam-na yallah*
Have pity on the beasts	*Irḥam al-'ajmah*[35]
Thirsting for water	*'Āṭishah li'l-mā*
Hungry for fodder	*Jāwi'ah li-'l-'alaf*

etc.

Again, on 5 August, driving down into Wādī Ḍahr from Ṣan'ā' we came upon women and boys walking up the road from the wādī, performing a *tasqiyah* but without any goats or other animals for sacrifice. My diary records rain on 9 and 10 August.

Gifts of First Fruits at Harvesting

On our way from Ṣan'ā' to Ḥadā country in mid-June we were offered gifts of many a bunch of green lentils (*'adas*) which the farming folk, men and women were clearing off the fields. This custom is said to be dying though we saw no sign of that! A gift of first fruits of peas (*'atar*) or beans (*qillah/qillā*) is known as *ghasūs*. As you pass the harvesters they say, '*Ta'āl – lak shuwaiyat ghasūs*, Come and you'll get some peas/beans.' *Lasīs* is a gift of wheat (*birr*), and *sa'īf* one of any millet crop eaten green (*sa'af/ista'af*, to eat green *zar'*). In Wādī Sirr at mid-July a woman with a tiny vineyard offered us *ṣabūḥ* of grapes, and a little later Shaikh Aḥmad Faraj of Bait

Hanamī (Banī Ḥushaish) in Wādī Rijām heaped a generous gift of grapes upon us, saying it would be shameful (*'aib*) to take money for them. Saiyid Muḥammad 'Abd al-Quddūs tells me the word *ṣabūḥ* is the term especially applied to the first fruit gift of grapes.

Abū Makhramah[36] alludes to the ripe dates and heads (of *dhurah* probably) given to the poor at harvest-time in Ḥaḍramawt, and the Shāfi'īs even laid down what it is fitting to say at the time of the first fruits.[37]

Bread

Bread, its ingredients, types, methods of preparation in south-western Arabia could in itself be the subject of a separate monograph. Hamdānī's[38] brief remarks on the bread used in northern Yemen about 900 A.D. have an historical interest for us. 'The people of Ṣan'ā'', he says, 'have *ruqāq* [thin rounds of bread] which is in no [other] town, so thin, wide and white, because of the way in which the firm consistency of the wheat comes into play. The wheats [*abrār*] of the Yemen are 'Arabī talīd [perhaps by *talīd* he means the 'Arabī variety introduced to and grown in Yemen], Nusūl [and] *'alas*-wheat [*burr*], this latter being the nicest of them as bread and lightest [in consistency]. In Ṣan'ā' the round (*raghīf*) of bread is not broken, but folded over and rolled up like a scroll.' Ṣan'ā', he adds, has many sorts of bread, and this one can remark at the present day in the old markets of the town. Of the types of grain Hamdānī[39] says, though his text is not always easy to understand, there are *burr 'Arabī* which is not *ḥinṭah* (wheat), Maisānī, Nusūl and Halbā'. Either Halbā', or the three last-named, the text being dubious at this point, is/are only to be found in Najrān. Two other varieties he mentions appear to be a smooth black and white kind, and a coarse red type. There is Ṭahaf, and white, yellow, red, and dust-coloured (*ghabrā'*) millet (*dhurah*). Sesame, he says, is of a quality unapproached (elsewhere), notably that of Ma'rib and the Jawf which is pure and shining. With these may be sown chick peas, broad beans, cummin, etc.

A pattern of grain consumption is perhaps suggested for the first half of the sixteenth century by Bā Makhramah,[40] though it is not certain how far the *fatwā*-decision he gives represents actual contemporary conditions. It seems to imply that wheat and barley might each be eaten for two months of the year, millet for three or four, and *kinib* for four – I imagine this refers to the ratio of the available supply of these cereals. Despite the large imports of foreign flour into present-day Yemen one might find districts where the ratio of consumption of local cereals could be studied.

Regarding the planting together of two crops, I observed on June 15, south of Ṣan'ā', but later in many other places, fields in which at the time of sowing (*badhr*) of Ṣaifī millet, shallow trenches or furrows (*mandab*, pl., *manādib*) a little raised above the field level had been made across the furrows in which the main crop of millet had been planted (Fig. 1). These *manādib* were intended to catch the water from a channel leading from the hill slope behind the field. They had french beans (*fāṣūliyā*) sown in them for use as animal fodder (*'alaf*), but the whole field benefits from the run-off conveyed along these cross-furrows.

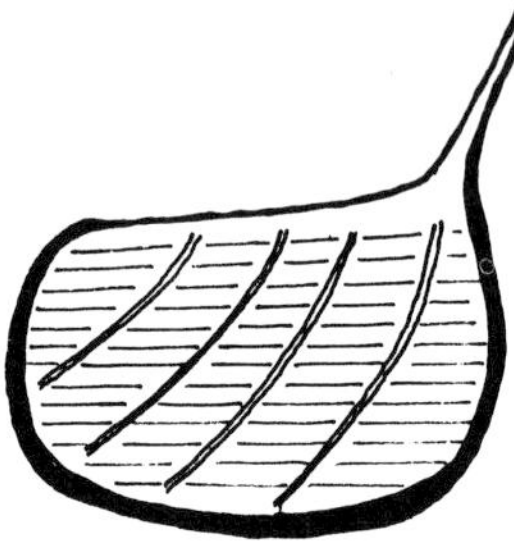

Fig. 1. Field with irrigation channels cutting across ploughed furrows.

Hamdānī on Cereal Cultivation

'One of the marvels of the Yemen', Hamdānī[39] continues, 'is that most of its cereals [*zurū'*] are grown on rain-land [*a'qār*] – therefore dough made from them has a firm consistency and bread[40] made from them is soft. This is because a field [*jirbah*] drinks at the end of Tammūz [14 July–13 August] and beginning of Āb [14 August]. It is then ploughed in Ailūl [14 September–13 October] when it '*jammat*',[41] i.e. has drunk/absorbed its water and its surface has dried. Then it is ploughed once, then again in Tishrīn I [14 October], then a third time in Tishrīn II [14 November]. Then it is sown in Kānūn I [14 December] and the crop stands till Aiyār [14 May] when it is harvested no water having reached it. At al-Qarārah and al-Hujairah it is cut promptly in Nīsān [14 April] and the close of Ādhār [13 April]. The field will still retain much of its wetness [*jamm*], so it is ploughed and resown, producing grain [*ṭa'ām*], before full time because of the heat of the season, which is cut in Ḥazīrān.'

'As for Ma'rib, the Jawf and Baiḥān, the *widn* [i.e. the field (*jirbah*) and, in Tihāmah dialect, the *dhahab*] fills from the *sail*-flood, and, when full, *ṭahaf* and bulrush-millet [*dukhn*] are sown in it. The water is absorbed [by the soil], its shoot [*nabt*] springs up, and for the duration of a month and days the field does not drink – until it [the crop] is cut and [the field] is [re-] ploughed for the sowing [*zar*ʻ] I have mentioned. Often there is sown [*ṭuriḥ*] in the field along with millet seed, sesame[42], cow-pea [*Lūbiyā*ʼ][43], green peas [ʻ*atar*], cucumber [*qiththā*ʼ], melon [*biṭṭīkh*], and pumpkin [*qar*ʻ].' He states that this is the practice in such districts as Najd (the area east of the Sarāt of the Yemen), Najrān, the Jawf, Baiḥān, and the whole of Tihāmah. In certain places in Najrān, he says, there is a variety of millet cane (*qaṣabah*) which has two, three, or more *miṭw*, heads (*sunbul al-dhurah*).

Rasūlid Gardens

In a number of places the author of *Bughyah* alludes to horticultural experiments made in the royal holdings (*al-amlāk al-saʻīdah*) such as his father's successful trials of rice at al-Jahmalīyah and another place (*infra*). One wonders if he followed Ibn Baṣṣāl's instructions which he quotes.

The Rasūlid sultans, says Qalqashandī,[44] resorted to Taʻizz as a summer place (*maṣīf*) and to Zabīd for wintering (*mashtā*). In the little village at the foot slopes of Jabal Ṣabir, overlooking Taʻizz, called Thaʻabāt from the spring (*ghail*), there was a garden with a royal dome and sultanic residence looking on to trees brought from everywhere so that in it the fruits of both Syria (al-Shām)[45] and India grew together. He also speaks of a park (*muntazah*) called al-Ṣahlah[46] above Taʻizz into which the Lord of the Yemen had led water, having built large buildings there in the midst of a garden. Khazrajī's chronicle of the Rasūlid dynasty certainly gives the impression that the Rasūlid court regularly commuted between Taʻizz, Zabīd and al-Mahjam – these with the Rasūlid village of Thaʻabāt were, incidentally mint cities. In 708 H./1308–9 A.D. the Rasūlid sultan al-Mu'aiyad built the palace known as al-Maʻqilī[47] which had a pool (*birkah*) 100 by 50 cubits in size with brass birds spouting water on the sides, and a fountain as described by G. R. Smith (p. 121). In the same year the Sultan also built a second palace in the garden of Ṣālah of Taʻizz where the late Zaidī Imām Aḥmad also had a castle in which he was besieged in 1955. This, I think, must be the same as al-Jahmalīyah[48] garden of *al-Bughyah*. A certain Bustān al-Shajarah at Taʻizz is noted in 726 H./1326 A.D.[49]

I first visited both Thaʻabāt and Ṣālah in December 1969 and found a small piece of celadon in the former, as also a rectangular structure, forming a field, that might have been the pool now full of earth. A large circular wall was being bull-dozed away at our visit in 1972.

In the Wādī Zabīd the Rasūlid sultans also had properties – Bustān al-Rāḥah[50] seems to have been at Zabīd itself, and al-Bustān al-Sharqī[51] as its name indicates probably lay to the east. The histories[52] often simply mention al-Bustān at Zabīd, perhaps the former. Khazrajī[53] also refers to a district at al-Tuḥaitā of the Wādī Zabīd called Saryāqūs al-Asfal which the Rasūlid al-Malik al-Ashraf had purchased in 798 H./1395–6 A.D. and started to work, but there may have been Rasūlid holdings here before that. There is also a passing reference[54] to Bustān Manṣūrīyah between al-Qurtub and Zabīd.

The Ṭāhirid monarch who died in 882 H./1477–8 A.D., like his Rasūlid predecessors, 'planted palms in numerous parts of Wādī Zabīd, al-S ḥarī and Mawshaḥ [?].[55] He planted sugarcane in numerous districts and rice [*aruzz*], etc.'

Some Notes on Millet

It need scarcely be said that experience in growing a cereal so basic to the economy of the Yemen as millet has brought the farmers an accumulated store of practical wisdom and knowledge on every aspect of its cultivation, varieties, qualities and uses, accompanied with a large technical vocabulary with numerous local variations, just as methods of treatment also differ from district to district. For this reason it has seemed worth reproducing some few scattered entries from my diaries.

On 23 May 1972 we saw Ṣaifī millet being sown at Ḥizyiz, a little to the south of Ṣan'ā'. There are three processes in preparing the ground for the Ṣaif crop, and only for the Ṣaif crop. One harrows (*shabbar*, verbal noun, *tashbūr*) with a harrow consisting of a wooden bar in which iron teeth are set, a week or so after rain, so that there is

Fig. 2. Harrow, wood with iron teeth.

no mud (*khulb*). After that a man runs (*yaḍbuṭ*) the plough (*ḥalī*) through it – then he runs a wooden beam over it (*yakamm-ah*, vb. noun, *kumūm*) to flatten it. Now comes the time for sowing (*waqt al-madhrā*). They split the ground with a plough, and two men go sowing handfuls down the plough-line. They only do this in earth holding moisture (*ḥimmah*).[56] We noticed the soil was soft and easily turned, but the upper part of the field was not being treated, as it had not retained enough moisture.

A type of scraper often seen is called *maṣabb* (*masabb?*) *al-ḥadīd.*

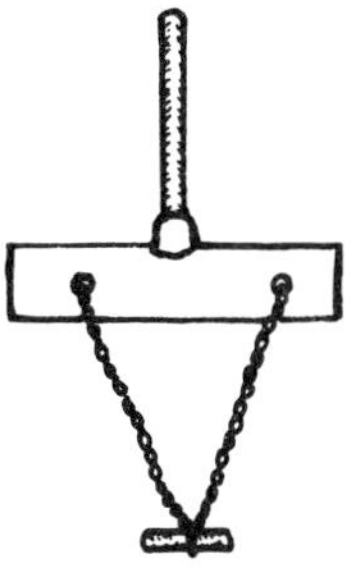

Fig. 3. *Maṣabb al-hadīd.*

One of these at Jahānah consisted of a rectangular iron plate, perhaps 3–4 feet wide by over a foot high, with a metal socket in the top side of which had been fixed a wooden pole. Two holes to the right and left of the plate allowed a chain to be threaded through from the front and back again, the ends meeting together in a wooden handle. One man holds and manipulates the pole of the scraper, and one or two men draw the handle of the chain along. As in the previous case the field is ploughed, then flattened with this instrument, then a second ploughing follows. I recall seeing it used also to make small field banks.

On 2 June near Nakhlat al-Ḥamrā' in Ḥadā country, the period described as aiyām al-Thawr, i.e., *aiyām al-midhrā* or sowing days, we saw people sowing everywhere. After working from early morning they break off at mid-day, to restart in the later afternoon and work on until nightfall. If they work in the middle of the day the seed does not do well. En route we saw the *saqlah* variety of barley growing – the other type of barley is a black variety. For four years there had been little rain in this district, and now and then the Ḥadā shaikhs remarked on fields with patches where the crop had withered – this they called *'aṭash* (lit. thirst), but most fields were bright green and *rāwī*, watered. The corresponding terms in Jīzān are *ḍāmī* (thirsty) and *sāqī*, watered. On 15 June we saw sheep pasturing among the young millet – at this stage about a foot high. They do not eat it because it tastes bitter, but they eat the weeds in between the stalks. In this region south of Ṣan'ā' we saw many fat-tailed sheep. In Jīzān in the Ṣaif season which there runs from about mid-March to mid-June it was maintained that if millet which had had no water were eaten by cattle during one star of that period it would kill them. Ḥaḍramis told me that the millet cane is at the short stage called *ṭamal*, then *khawlah*, then *qaṣab* – as the *ṭamal* is bitter (*qārr*) it kills sheep (*ghanam*). The terms used in northern Yemen for the parts of a millet plant – or at least those known to me, are *'idhq* for the head at all stages, *sabūl*, sing., *sabūlah* for the ear, *qam'* for the envelope round the grain, *'ajūrah*, plur. *'ajūr* for the stalk, called, it seems, in some mountainous districts *jazab*. The curved piece of stalk between the *'idhq* and the top knot of the stalk is called *miḥjān*.

There are three sowings of millet in the Wādī Jīzān, all of them having a second and usually a third growth. These are, with their sowing times:

Shabb – end of May
Khalf shabb
Khalf khalf shabb

Kharīf – last week of August
Khalf kharīf
Khalf khalf kharīf
Jinnīyah or Khalf khalf khalf kharīf

Makhraṭ – end of October and early November
Khalf makhraṭ
Sa'ūdāt/Su'ūdāt

In Jīzān Zi'ir can be sown at any suitable time, but it is especially planted in Ṣaif and called Ṣaifī. It has a second growth (*khalf*) but Gharb-millet has not. Makhraṭ millet has the best stalk for foddering but it is little in quantity. Khalf al-kharīf is the best second growth and next to Makhraṭ produces the best stalk. *Dukhn*, bulrush-millet, is sown at any suitable time, but particularly at the end of the Kharīf season (up to about mid-September) if there is rain.

In Jīzān I was able to collect a number of weeds from the growing *dhurah* in December and these were identified as far as possible by Dr Said H. Mawly of Tendaho Plantations, Ethiopia, but I shall only mention of these *wabal*,[57] and *'udār, striga hermonthica* which has purple flowers and grows about the root of the millet, a bad weed. Many of the other weeds are used for animal fodder and allowed to grow on the field banks. In Jīzān insect and other pests are known as *ṭair*, (properly 'birds'). The worst is *'usāl*, aphids. They appear about the time of the heat and floods there, about the end of March, and sometimes at the end of Kharīf, i.e. the first two weeks of September. Even on 12 December 1971 there were some in part of the Jīzān district. *'Usāl* was described as a white worm black and sticky on the millet, from which it makes water run down. Animals will eat the stalk in this condition but it is not good, and there is no second growth (*khalf*) to millet once this pest has attacked it. In Jīzān rain was said to put an end to this pest, and Qāḍī Ismā'īl also volunteered the same statement. *Al-dūdah*, the stalk-borer, a black worm, eats the heart (*qalbah*) of the stalk and damages it. *Irḍah*, the white ant, is a well known pest, and *khurmuj* is smut, the black disease of the ear of the millet. *Jārish*, probably a sort of grass-hopper, was mentioned, and an unidentified insect called *al-ḥuṭām*.

It is a little odd that the *Bughyah* makes no allusion to the planting of other crops with millet such as *dijrah*, cowpeas, which I noticed growing with Ṣaifī millet between Kibs and Ṣan'ā' on 10 August.

* * *

The English translation of al-Malik al-Afḍal's account of the cultivation of cereals now follows. Originally it was my intention to append to it a glossary of technical terms culled from the text, a few of which are discussed in my annotations, along with many others which I have collected in South and Western Arabia, especially those from the Saudi Arabian province of Jīzān. Since however I am not yet able to include the additional vocabulary, mainly from Tihāmah, that is provided by the marginal notes to the Cairo MS., this seems best left to another occasion, for it will necessitate an inquiry in the Wādī Zabīd and elsewhere.

The relevance of this section of the *Bughyat al-fallaḥīn* to any survey of agricultural practice in the Yemen and even parts of Saudi Arabia at the present time will be obvious from the very fact that one can explain al-Malik al-Afḍal's treatise by reference to present day usages and language, and actually see, from day to day, these usages enacted before one's eyes in the fields of South Arabia. This emerges from my annotations to the English rendering of this section of the *Bughyah*. *Everything* described by our author there can be seen today. Nevertheless the *Bughyah* is little more than a sketch applying to certain districts within the experience of the Rasūlids – an introduction, as it were, to a very wide subject indeed.

The sixth chapter on cereals, divided into nine species

1. The first is wheat,[58] *it having a number of varieties.*

In *al-Ishārah*, my father, God rest him, said: 'If wheat be sown with us in Aiyār [14 May–13 June] and Ḥazīrān [14 June] up to the time when the sun falls in Cancer [23 June] which, by general testimony is summer [*ṣaif*] then it stands four months and is harvested. Some may be sown following the summer of *wasmī*-rain [*qaiẓ al-wasmī*],[59] and Yemenite folk call it Dithā' and Qiyāẓ, it being the poorest variety of wheat. Varieties of wheat in the Yemen are many – 'Arabī, Halbā, Wasnī/Wisnī, Nasūl – which is 'Alas, and Ḥabashī [? Abyssinian].'

The learned Democritus[60] said: 'The cultivator must not sow on a day when the north wind is blowing because it spoils the ground, nor does the seed flourish and establish firm roots in it.' He (also) said: 'Sowing of wheat must not be delayed beyond its [proper] days, and early [sowing] will bring it better fortune [*barakah*].' He (further) said: 'Wheat and barley seed must not be washed, for, if washed, its grains turn out thin and do not fill out much. When the cultivator sows at the waxing of the moon and the ascension of the preponderant (?)[61] Zodiacal signs which cause generation, that seed will thrive and produce abundantly.' Democritus said: 'I have [on occasion] sown at the waning of the moon and [had no cause] for regret.'

Wheat is (a cereal) of great utility, so care over its preservation is a matter of importance. One of the ways whereby wheat lasts, even over a long period, is that it be taken (to the granary) in its ears.

I say: 'In the Holy Book,[62] in the story of Joseph al-Ṣiddīq – peace be upon him – and the interpretation of the King's dream, occurs [the passage]: "And whatsoever ye harvest, leave it in its ear."[63] That is to say, he ordered them to store it against the years of drought so that it might be safe from pests.' They say that *jāwars*[64]-millet, if taken (to the granary) in its ear, will keep a hundred years. I say: 'That cannot be applied to every country without restriction, for hot countries cause the perishing of grain together with its ear.' They say: 'A way whereby wheat is kept safe from rot and alteration of its taste is that papyrus or Persian reed be put down for it and it be placed on it.' If wheat or other seed be placed in earthenware pots and covered with a hyena-skin in such a way that the smell of that skin can pervade it then it will stay free of all pests.

Now concerning the varieties of wheat already mentioned, one of these is Wasnī/Wisnī,[65] the best variety of wheat, its grains being thick, red, with a tail[66] to them, heavy in weight. The places for sowing it are the temperate districts, and the time for sowing it in Mikhlāf Ja'far and the adjacent districts is *ṣurāb*.[67] This is the variety used in the mountain country of the land of the Yemen watered by rain at the beginning of Tammūz (14 July); it stands three and a half months and is then harvested. It is sometimes sown in al-Quṣaibah and parts of the Ta'izz[68] district like it that are watered by running streams half-way through Tishrīn I (29 October), sowing of it continuing until the middle of Tishrīn II (28 November), this being the best of Wasnī/Wisnī wheat, and it too is harvested after three and a half months. A sowing of it called Qiyāḍ[69] is sown in the places watered by running streams, this coming after the millet-harvest, from the beginning of the winter (*shitā'*) season, in Kānūn I (14 December) up to Kānūn II (14 January–), standing three months and then being harvested.

Another (variety) of wheat is 'Arabī[70] which is white and fine of grain, the places for sowing it being the cold districts. It is sown at the beginning of the month of Ḥazīrān (14 June–), and continues being sown until the first of summer (*ṣaif*) time, i.e. when the sun falls in the Zodiacal sign of Cancer (23 June). This is the time in the mountains and mountainous districts for sowing it as *ṣurāb*[67] (autumn cropping) – it stands four months and is then harvested. In *Milḥ/Mulaḥ al-malāḥah*[71] he said: 'Some may be sown there[72] (though) rarely, following the *wasmī*-rain, it being called Dithā' in the Ṣabir country[73] and some regions of the Mikhlāf.[74] Some called *'aqar* (rain-land)[75] may be sown in the mountains of al-Sha'ir, al-Ḥaqūl, al-Shawāfī, and districts adjoining them. All that consists of winter (*shitā'*) sowings according to the amount of *wasmī*-rain occurring that year. It is the weakest variety of wheat.

Ḥalbā is another (variety) of wheat. In *al-Ishārah* he said: 'It has short thick white grains without a husk as there is upon the other varieties of wheat. The places where it is sown are (the same as) those where 'Arabī wheat is sown, but it is sown at the close of the sowing of 'Arabī wheat, half-way through Ḥazīrān (28 June), standing three and a half months and then being harvested.'

Another (variety) is Ḥabashī (? Abyssinian). In *al-Ishārah* he said: 'Its grain is middling between long and short, white and red, while in quality it stands somewhere between 'Arabī and Wasnī/Wisnī, there being a type of it called Samrā' [brown].[76] Sowing of it commences half-way through Ḥazīrān [28 June] as the sun falls in the Zodiacal sign of Cancer, and the last [sowings] are up to the [first] of Tammūz [14 July].'

The way wheat is sown, as he says in *Milḥ al-malāḥah,*[77] is 'that the ground is ploughed three or four times, or (even) more than that, in accordance with the goodness or poorness of the land and its hardness or softness. It is manured with good manure and ploughed over, that which is watered by running streams being levelled so that it may be irrigated evenly and the water settle in it evenly covering all of the ground. What is watered by rain-water requires no levelling. When the time for sowing comes it is sown by hand-scatter while the ground is moist from rain – yet without being runny[78] from heavy rain, and ploughed over so that it becomes covered by earth. The sowing by hand-scatter[79] should be middling in such a way that seven grains drop on the space of a single foot-step or what more or less approximates to that. Over good ground sowing by hand-scatter is less dense because the crop [standing] on it would be crowded together, be overmuch, and spoil itself; while poor or middling good land should have a middling hand-scatter [of seed], without it being [too] thin. Scatter-sowing means that it [the seed] is tossed by hand – one takes a handful of wheat and tosses it in front of, and about oneself, after marking out a marker for it with which to distinguish the place [already] scatter-sown from the place which one has not [yet] scatter-sown, lest the scatter-sowing of the seed over it be repeated, or some remain unsown. It is [then] ploughed over, and any scrub, grass, roots or the like on the ground are cleared, nor should one leave the ground until it is cleared of any weeds on it. If however, it is land which is watered by running streams, the ground, after scatter-sowing, is divided up into sections like troughs. This division is made through its (the land) being ploughed with the ploughing implement, one even furrow, alongside which one ploughs an additional furrow in such a way that a side of the (second) furrow stands up [so as to form a bank along with the side of the first furrow] – beyond this is left a piece [of ground] [of a size] conforming with the known force, or lack of force, of the running stream employed for irrigation,[80] and one ploughs [yet] another furrow with a second furrow beside it, so that a side of it stands up. One continues in this way until the ground is divided up into troughs/plots so that each plot may be watered separately with the water of the running stream and it may cover all the ground, plot by plot.'

Green grain for parching (*farīk*)[81] begins first to be taken from it after three months. He said, in *al-Ishārah*: 'It [*farīk*] stands in the same place as *jahīsh* with the folk of the Tihāmas with regard to millet' – and (it begins to be taken) from Wasnī/Wisnī after eighty days. When one wants to make parched grain of it, let him cut from half-way up the stalk when the crop has just begun to turn yellow and the ear has filled. It is bound into sheaves; each sheaf what the hand can grasp, and one lights a fire with a flame but no smoke, and burns off the hairs [awn] of the ear in it until they are burnt up and the grain in its ear is cooked. It is rubbed with the hand on a palm-leaf basket-work tray, or beaten with a stick in a goat-wool bag,[82] then winnowed. If only partially cooked it is toasted lightly in a piece of pot-sherd, an open-necked jar or a frying pan, without too strong a fire on it so that it dries up.

When (wheat) has ripened, turned white, and its heads have begun to bend over, it is reaped with sickles about a handspan above the ground or less, according to the amount of straw wanted.[83] It is then loaded up and stacked (in a circle) one (sheaf) on top of another, the ears placed towards the centre of the pile and the cut edges (of the stalks) to the outside; some of the ears along with (their) stalk are used to cover it over.[84] It remains stacked up in this way for eight days, more or less. It is not stacked up while there is any damp in it from dew or rain so that it would rot and perish. On the contrary it is stacked while dry, then it is separated (by hand or wooden fork) from the stack, and trodden out by oxen, or with a stone attached to the implement (the yoke ?) of the oxen, the stone having either a hole in it through which the rope is tied to it, or a groove round the middle by which it is tied. It goes on being trodden out until it is freed of all the straw, and then its straw is beaten.[85] Thereupon one tosses it (the grain) in the winds with a stick that has fingers like those of the hand until the grain is free of straw. When it is free of it (straw) but some ears (still) remain in it (? the straw), one repeats the treading out with the stone and oxen until all is free (of it),[86] and it is winnowed with a pole to the top of which is nailed a board of about half a cubit, until it is cleansed of straw and anything else. As often as one winnows it in the wind one brushes the surface of the grain with (a piece of) brushwood resembling a broom to remove any ends of straw[87] still remaining on it, and ears not trodden out, and one makes a single heap of it. It (the grain) is then measured[88] and taken (to the granaries), but it is not lifted until after it is cool of the sun's heat lest the worm come to it, and stored in a store with apertures facing the north wind.[89] It should not be below a dwelling, kitchen, fire, lamp, beasts or straw; nor should there be any aperture to the east or south. It will then be free of the worm – (this has been proved by experience).'

Among the varieties of wheat also, as known to the folk of Ṣanʿā' of the Yemen – among these is excellent Bawnī[90] wheat, *qiyāḍ* (cropping in winter [*shitā'*]), *ʿaqar*[91] (grown on rain-land), and *ṣurāb* (cropping in autumn) – of the *ʿaqar* open Egyptian bread is made – white Ḥaḍūrī[92] wheat, called after Ḥaḍūr of the Banū Shihāb country, of which cake[93] is made. Other varieties of wheat also are the Dhamārī and Maisānī[94] kinds.

2. *The second species is* ʿalas-*wheat.*[95]

In *al-Ishārah* he said: 'It is of two varieties, white and red, containing six grains in envelopes, each envelope holding a couple of grains –

when good there may even be as many as three grains in an envelope. It is of the wheat category. Its grains (can) only be detached from their envelopes when it is crushed between the stones of the quern [*raḥā*], the latter being called *mijashshah,*[96] or pounded with a wooden pestle and mortar as is done with rice. When freed of its envelopes its grain is called Nasūl.[97] When flour of it is made into bread it is firmer than wheaten bread, bread made of it resembling that made of Wasnī/Wisnī wheat except that in quality, utility and goodness it is inferior to it; in the form of flour it exceeds its measure when in the form of grain. With us sowing of white *'alas* takes place at the beginning of Tammūz [14 July–] and it is sown like Wasnī/Wisnī wheat in the places where the latter is sown also, it being scatter-sown just as [Wasnī] is. It stands three months and is then harvested, being called *Abyaḍ* (White) because its husk is white, though its grain is red.

'Now concerning the second variety which is red, the places where it is sown are the same as those of 'Arabī wheat, it being sown exactly as the latter is, and harvested after three and a half to four months. In the wādīs and places where Wasnī/Wisnī is sown it does not do well. It is called *Aḥmar* (Red) because its husk is red, but its grain is nearer white. It is sown in accordance with the rain at the time mentioned above, as *ṣurāb* [cropping in autumn], and in places with running streams as *qiyāḍ* [cropping in winter (*shitā'*)] at the beginning of the month of Kānūn II [14 January–].[98] From the time it is harvested it is left exposed to the sun and not stacked – then it is trodden out with oxen and the stone till it is crushed, and winnowed, as wheat is, in the winds until the grain is free of straw, but each pair of grains stays in a cover until it is crushed with the *mijrāsh*[99] – I mean the stones already mentioned – and pounded with the wooden pestle and mortar until it is freed of its husk.'

3. The third species is barley.

It is of two varieties, one with a husk – the best known among folk, and a huskless variety called *sult,*[100] called Ḥabīb[101] by most people. In *al-Ishārah* he said: 'It is sown in the cool places where 'Arabī wheat is sown, the time for sowing it being from half-way through Ḥazīrān [28 June] to the first of Tammūz [14 July], and it is closer to the nature of wheat than the rest of the varieties of barley. It is sown in the same way as the husked barley, and each stands three and a half months and is then harvested as wheat is, and stored.' I say: 'Scholars differ with regard to the description and nature of *sult*. The 'Irāqīs, al-Baghawī[102] and al-Qāḍī Ḥusain, say: "It is a grain resembling wheat in colour and softness, but resembling barley in its cold nature [*ṭab*'],[103] and the one tastes the same as the other." To this al-Shāfi'ī testifies in *al-Umm*,[104] as does al-Buwaiṭī,[105] but al-Ṣaidalānī, al-Fūrānī and others take a contrary view to his, saying: "It has the appearance of barley though its nature is hot like wheat, but it has no husk." Al-Nawawī[106] said: "The sound, indeed the correct view is what the 'Iraqīs say, and what the generality of people [i.e. scholars] have resolved, it being what the lexicographers remark. So then there are three views on the matter, the soundest, which is also the deposition of al-Shāfi'ī according to al-Buwaiṭī and *al-Umm*, being that it is an independent species (*aṣl*) in itself, but it has been lumped together with others. The second is that it is lumped together with wheat because it supplies its place, and the third (view) is (that it is lumped together) with barley because it is a sort of it." Al-Qāḍī Ḥusain said: "Nay indeed, it is ordinary barley." This positive view was also taken by the shaikh Abū Muḥammad, and Abū 'Alī in *al-Īḍāḥ*, and by al-Māwardī in *al-Aḥkām al-sulṭānīyah*.'[107]

4. *The fourth species is millet* (dhurah).[108]

He said, God rest him, in *al-Ishārah*: 'It is specially [common] in the land of the Yemen, and the bulk of what is sown in Tihāmah consists of it, though there is less in the mountains. Its origin is from the land of the Blacks [al-Sūdān], and it has [several] varieties. That variety of it which is sown in the mountains is Raisī,[109] an intensely white grain with widely spread out heads. In the mountains the proper time for sowing it is in the ten chosen days[110] of Nīsān/Naisān,[111] and the places for sowing it are the temperate districts and the valleys near the heat.' In the Writing [*Khaṭṭ*] of my father,[112] God rest him, he says: 'In the year 737 H./1336–7 A.D. I saw in al-Ḥuṣain district a yellow millet stalk upon which were forty-three heads, apart from those beginning to burst out of their covers, about to come out.' He said (also): 'And, in the year 740 H./1339–40 A.D. I saw a stalk at al-Ḥ w lī/J w lī[113] of the lands of al-Janad thirteen cubits high.'

There is a variety of it (millet) called Shuraiḥī with a harder grain than the two former varieties, between white and yellow in colour. The places where it is sown are in the mountainous districts – in those that are temperate and nearer the cold. The time for sowing it is the same as that of the two former varieties, and it is harvested after full seven months, its heads being smaller than those of the yellow (variety) and its grains overlaying one another in the head because of their abundance. From the land of Abyssinia sometimes comes millet resembling this variety, with, perhaps, three grains in one envelope.

A variety of it called Ja‘aidī,[114] somewhere between the Shuraiḥī and Ṣafrā’ (Yellow) (varieties) is sown in the Saḥūl and ‘Annah districts. The time for sowing it is the first to the middle of Aiyār (14–28 May);[115] it is harvested after four months, the last of it five months from the day it was sown. Its grain is extremely hard, with a fatness to it not found in any other varieties of the Baiḍā’ (White) sort.

Another variety of it called Gharibah[116] closely resembles Baiḍā’ (White) in colour and grain, but it is not of the same excellent quality. The places where it is sown are the mountains and valleys of the hot districts, and the time for sowing it the first of Ḥazīrān (14 June–), it being harvested after four months or less – it is the first crop of millet to be harvested in the mountains.

There is a Ḥamrā’ (Red) variety of it sown in the cool mountains at mid Ādhār (28 March) and harvested after nine months.[117] Belonging to the Ḥamrā’ (Red) type is a variety called Baḍ‘ah[118] with a sharpness to its taste, sown in valleys (irrigated) by running streams and in cool places with much dew. It (too) is sown on the first of Nīsān/Naisān (14 April) and harvested after nine months.[119]

Another variety of millet is called Ṣawmī.[120] The time for sowing it is at the rising of the Sābi‘ (seventh star) of Banāt Na‘sh (Ursa [Major])[121] up to half a month after its rising, it being harvested after four months. Then it has a second growth which is harvested, then a third which is harvested, no crop of the mountains but this variety being harvested thrice. Its grain is reddish in colour, and it is coarse and dry, scarcely attractive to eat. If (grown) repeatedly on the (same) ground it exhausts it.

All the varieties of millet in the mountains – during the days of winter (*shitā’*) the ground is ploughed thrice for them and manured. The way it is sown is after it (the ground) has dried and is not runny, but middling between wetness and dryness – the best of conditions for sowing. The ground is ploughed in straight even furrows, each furrow alongside the next, not running over[122] on to it (?). The grain is sprinkled between the ends of the fingers into the bottom of the furrow, from two to three grains being dropped (at a time) according to the goodness or poorness of the ground. Following the track of the oxen and plough one takes a pace, dropping (the seed) as it was dropped first. Dropping the seed in the plough track is so that the grain will be covered with soil – one takes a pace also and drops the grain. So one's way goes on until one finishes (sowing) the ground. Some folk drop the seed-grain and tread it with the foot, some cast it without treading it, according to the lightness or heaviness of the soil. When the crop rises and it is forty days old the tops/ridges of the furrows are ploughed with oxen so that the crop rises from the bottom of the furrow, and the bottom of the furrow becomes empty with none of the crop in it – this they call *kaḥīf.*[123] When a month or less has elapsed following the *kaḥīf*-ploughing, the middle (lines of ground between) the crop, bare of any crop before the *kaḥīf*-ploughing, are ploughed. Some folk clear the crop of bush and grass before *kaḥīf*-ploughing it so that it may grow strong and flourish – in accordance with the quality of the ground and the greater or lesser area to be irrigated (?).[124] Then after the *kaḥīf*-ploughing the soil is turned back onto the roots of the crop so that it will do well, and it is cleaned of any bush or grass among it; this ploughing is called the *khilfah* because it is vagarious (*mukhtalif*)[125] not in straight (lines) like the ploughing of furrows or the *kaḥīf*-ploughing. Then twenty or thirty days later the furrows in which the growing crop was first are ploughed with a straight not a crooked ploughing so that the crop is turned back to the tops/ridges of the furrows, and no (ploughing) work is done on it after that. It is cleared of its leaves – whenever one of its leaves turns yellow one clears it away from its (stalk), until it forms a head in which the grain shows.

When one wants grain to parch (*jahīsh*)[126] one cuts the ears before it (the grain) hardens and kindles a fire without smoke for them, beating out the fire with a stick or stone till it no longer burns fiercely yet is intensely hot, it being called *mallah.*[127] The ears are placed in this fire and covered over with some of the burning embers – when they are cooked one takes them out and they are rubbed together in the hand – which is best – or beaten out with a stick in a coarse cloth, and one clears them of twigs and stems and cleans them.

If *jahīsh* is not wanted, or the crop exceeds the *jahīsh* (taken to parch), the crop stands until the ears harden and no moisture remains in the grain and it ripens – whereat it is harvested with sickles at the points[128] for cutting off the heads and collected together at the threshing-floor (*baidhar*),[129] i.e. the *jurn/jirn*, and then trodden out with cattle and the stone, as before in the case of wheat, and winnowed in the wind. Each time one winnows it, one ends by collecting together the husk from among it with brushwood bound together like a broom. When the heap is cleaned and in good (state) one measures it out and takes it to the storage-silos[130] hollowed out of smooth rock, customarily (used) to bury grain, for, in the mountains, this is the best way to keep it. If one wants to store seed-grain one should select the cleanest ears, those with the largest grain, the best quality of them, and lay them in the sun until the moisture in them dries out, beat them out with flails, clean them of the débris of the millet-heads, and take them, after drying them well, to a good well-ventilated place where neither the sun's heat nor damp can reach them, so they will not get wormed[131] for the passage of a year – when one sows with them.

As for Tihāmah one of the varieties of its (millet)-crop is Baiḍā' (White), called (also) Budaijā (?).[132] Its grains are small, with a hardness to them. It is the best of the Tihāmah crops, and they depend on it for their food. Its sowings vary, one being Thālithī (Third), called after the rising of the third (star) of Ursa Major, i.e. the first of Āb (14 August).[133] This special name is given it in Zabīd and Rima'. That which is sown at this time in the districts of Mawr and Surdad they call *shabb*.[134] Then it is sown at the rising of the Khāmis (fifth star) of Ursa (Major) and they call it Khāmisī (Fifth) – this being the first of Ailūl (14th September); this name is used in (Wādī) Zabīd, Wādī Rima', Wādī Surdad and Wādī Mawr. They (also) call it Bainī (Between) because it (is sown) between the Thālithī (Third) and Sābi'ī (Seventh [stars of Ursa Major]).

A (sowing) of it (*dhurah*) is Sābi'ī (Seventh). In *al-Ishārah* and *Milḥ al-malāḥah*[135] they said, 'It is so called after the Sābi' (Seventh) or Ursa (Major) and is sown on the 19th of Ailūl (2 October). It forms the bulk of the crop of Baiḍā' (White) millet in Tihāmah, and is the most abundant, cleanest and most fortunate (*abrak*) of that. In Mawr and Surdad it is called *shabb*, and in Lahej and Abyan *bukr*[136] (first crop); Sābi'ī (Seventh), in the Ḥais area, Rasbān, Mawza', and the Ḥāzzah[137] (area at the juncture of the mountains with the Tihāmah plain) of the Wādī Zabīd is called al-Ḥaḍḍār.[138] Its sowing sometimes precedes the sowing of Sābi'ī (Seventh) of Zabīd by a short interval, about ten days, or they may be sown together. In Mawr and Surdad the time for sowing it is the 19th of Ādhār (1 March).

A variety of it (*dhurah*) is a crop called Zi'ir. In *Milḥ al-malāḥah*[139] he said, 'It is sown in Wādī Surdad and Mawr, the time for sowing being the 29th of Ādhār (11 April).

From all the Baiḍā' (White) millet crop in Tihāmah one takes grain for parching (*farīk*), i.e. *jahīsh*, after two and a half months. It ripens in eighty days and is harvested three months from day it was sown.'

After its being harvested it has a second growth again, producing a good yield called *'aqb/'uqb*[140] and a third growth follows that known also as *khilf*. On outstandingly good ground yet a fourth growth, once more after the *'aqb* (third growth), may follow the White crop of it (millet) and give a yield which they call Jinnīyah,[141] but when eaten there is no pleasure in its taste.

The top of an underground silo (*madfan*) just when the filling has been completed, the top of the pile of millet grain showing at the centre of the foot of the picture. The wooden measure (*qadaḥ*) has an iron band round the top and bottom. The measure when filled with grain is smoothed over level with the top with the hand, whereas in the Ḍāli' district it is heaped up on top conically, the surplus spilling over. Mabyan, where this photograph was taken on 10 November 1966, is a small town in the Zaidī country, a little north of Ḥajjah, but its mediaeval buildings were badly destroyed by Egyptian bombing, though not the *madfans* which are at least mediaeval if not much older; they are constructed on arches with roofs of fitted masonry. The hat worn by the men is the woven bamboo cap (*khaizurān*) of the Yemen.

(*Photo R. B. Serjeant*)

In *Milḥ al-malāḥah*[142] he said: 'For all of them the ground is ploughed until it is rendered good from bush and grass, and the valley-land is watered by floods or running streams, the rain-land (*ḍāḥī*) by rain-water. When the ground has drunk and the watering of it is completed, ploughing or sowing is put off until it dries then it is ploughed again. When one wants to sow, the seed-grain is distributed over it (the ground) by spreading [it] in the middle of the trench [*shaṭṭ*],[143] one lot opposite another. One puts down two to three grains [at once][144] next to each other with no space between them, quite different from the sowing of millet in the mountain districts. Between the places[145] where the grain [is sown] will lie the extent of one pace, and each one of the furrow-bottoms[146] for containing the said sown grain shall have spreadings[147] [of grain] at their said times. When the crop comes up and is forty days old it is ploughed with oxen in the same way as the millet crop is in the mountain districts – which [operation] is called *kaḥīf*, and in the Tihāmah *shitāh* [?].[148] In Tihāmah it is not re-ploughed after this ploughing, neither does the crop receive another working after that; none of its leaf is used as fodder, nor is it cleared [of weeds] until the harvesting of it is over. Once it is harvested it is taken to the threshing-floor where it is threshed with flails but not trodden out with oxen or the stone – so it is with the Tihāmah crop, contrary to [the way in which] the mountain crop [is treated].'

Another (variety) is Ḥamrā' (Red). In *al-Ishārah* he said: 'It is sown at the same times as Baiḍā' [White], and they call it Ḥujainā' because its ears come up crooked [*ḥajnā'*],[149] bent over. It is dry and hardly attractive to eat. It is sown also, at the time for sowing Baiḍā' [White millet], in the valleys of Zabīd, Rima' and districts near-by. The way in which it is sown is that in which Baiḍā' [millet] is sown, and it is harvested just as Baiḍā' is, and threshed as it is threshed. In Rima' valley of the Tihāmahs some is sown at the evening [*'ishā'*] rising of al-Thuraiyā [the Pleiades] [6 November],[150] called [on this account] 'Ashawī/'Ishwī, it being sown on the six[teenth] of Tishrīn I [29 October]. After it comes Nasrī, so called from the rising of al-Nasr [al-Wāqi'(?)],[151] Baiḍā' [White] and Ḥamrā' (Red) (varieties of millet) being sown; it is sown in Lahej especially on the sixteenth of Tishrīn II [29 November], and Ḥamrā' follows it five days later. After it comes Jaḥrī,[152] sown during the first ten days[153] of Kānūn I [14–23 December]. There is some that is sown in the districts already mentioned during the first month of winter [*shitā'*], called al-Ḥaḍḍār al-Tamrī [that which is ready along with the dates].[154] Baiḍā' [White] and Ḥamrā' [Red] [varieties] are sown that come in the days of the palms [i.e. date-harvest]. This appellation after the dates is in Wādī Zabīd. All of these [varieties of millet] are harvested three months from the time of sowing.'

One of the varieties of millet is Ḥ r jī/J r ḥī.[155] with a larger grain than Budaijā (? Bujaidah[132]). Its husk is black and its ear comes up crooked, bent over, *like Ḥamrā' (Red), but it is sweeter than Budaijā* (Tarīm Ms.) / *like the husk of the eggplant (badhinjān) and it has a larger grain than Budaijā* (Cairo Ms.[157]), but its grains have not the hardness of Budaijā. It is sought after, not unpalatable, coming next to Budaijā in quality, nor is it any less valuable than it. It forms the bulk of the Wādī Nakhlah crop at Ḥais as has been stated. It is sown ten days after the Sāb'ī (sowing of 2 October) of Budaijā, and in good fat land only. It is sown also as 'Ashawī, i.e. at the rising of al-Thuraiyā (the Pleiades) in the evening as I have mentioned earlier, and is so called after that; this (rising) is on the six(teenth) of Tishrīn I (29 October). The places where it is grown are (all in) al-Sharj[158] al-'Ulyā of Wādī Zabīd.

Then (comes) Jaḥrī (sowing ?) (when) the Baiḍā' (White) and Ḥamrā' (Red) (varieties of Budaijā are sown, the period for sowing it being the first ten (nights) of Kānūn I, (14–23 December), after the Sābi'ī (crop) has gone. The most suitable (crop to sow) at this time is Ḥamrā' (Red millet). As for Budaijā/Bujaidah it gives a poor crop and there is no good in its stalk. Jaḥrī[159] is sown in whatever land it be of the upper or lower parts of Wādī Zabīd, but as for Sābi'ī it is only sown in the best ground. The harvesting of this (all of it) takes place after three months. That Budaijā sown in the lower part of Wādī Zabīd as Sābi'ī is harvested after seventy nights and (even) less, down to as little as sixty nights.

Summer (*ṣaif*) (begins) when the sun first falls in Ḥamal (Aries) (22 March–) – what is sown at that time is called Ṣaifī, most of the sowing then being Ḥamrā' (Red) millet and bulrush (*Pennisetum*) millet (*dukhn*).[160] The time for sowing it is from the first of Ḥazīrān (14 June) and it is harvested after four whole months. It is sown in whatsoever ground in the hot regions it be, and in the mountains also in land near the heat, though as for the cold districts it is not sown there. In Surdad and Mawr they call what is sown at this said time Wasmī (Wasmī-rain crop) because it is sown at the first rain of the year – which latter for the cultivator starts when the sun falls in Ḥamal (Aries) (22 March).

My father, God rest him, said: 'Rūmī[161] bulrush millet [*dukhn*] has a fruit[162] like that of the bulrush millet of the Yemen except that it is finer and looser of ear. Its grain is white in colour, small, with a delicious taste to it [when] prepared with milk and bread. There is some better than any bread I have (ever) eaten, and this is made with milk, duff [*'aṣīd*][163] and such-like things. [There is] a plant which is brought from Abyssinia and its seed is abundant there.'[164] My father said: 'I ordered it to be planted, and as to its grain it is like broad beans with their skin, and the taste of hazelnuts and chestnuts that they eat for dessert – the smell of it used as dessert is like all [other] dessert, but with a delicious taste, sweetness and excellence to it. Also if it be toasted it is nice to eat. In a negro language it is called *Ḥ ṣ b.*'[165]

'As to the bulrush millet of this district in the Yemen and Najd (the area E. of the Sarāt of the Yemen) the ground, any suitable ground, is ploughed for it, but it only does well in hot and temperate districts, and as for cold (districts) it is not sown there.'

'The way of sowing it,' as he said in *Milḥ al-malāḥah,*[166] 'is that the ground is ploughed for it twice or thrice before sowing it and the seed cast, three or four grains to every two paces, the seed-grain being cast from below[167] the ploughing gear, behind the plough, and the earth turned back over it. Each time one casts three or four grains one treads them down well with one's foot so that they stick in the ground and grow strong straight shoots. Ground to receive it is not ploughed up unless it be dry of moisture, but at the time for sowing it is sown while the ground holds moisture that will make the grain sprout. It is sown in ground watered by running streams and in that watered by rain-water, those lands watered by rain being best when the autumn (*kharīf*)[168] rain comes to it (the bulrush millet). When it is forty days old it is ploughed (*kuḥifa*) with oxen in such a way as to bring the bottom of the furrow to the top and the top/ridge of the furrow to the bottom, as one does with millet. Half a month after *kaḥīf*-ploughing it the field is ploughed lengthwise between the tops/ridges of the furrows of the (standing) crop so that it can absorb the rain[169] and its crop be singled, though it is not ploughed with a thorough-going application at the time when there is dew for fear of (the leaf) turning yellow, but (only) when the sun has played on it and the dew on it has dried away. When it has had rain on it that will be enough for it, but if rain be scant and it is (watered) by running streams, then it is watered by the stream when it has received the final ploughing that follows the *kaḥīf*-ploughing. The time for sowing it is on the first of *Ḥazīrān* (14 June) and it stands four months then is harvested. In Tihāmah it is sown at the summer sowing season (*matnam al-ṣaif*)[170] on the twentieth of the month of Aiyār (2 June) and is harvested after three months.

When it is wished to make parched grain (*farīk*) of it a fire is set to blaze with a fierce flame and the (bulrush millet) ear parched over the smoke-free flame of it. Once it is ready two of its stalks with their ears are stripped by being rubbed together, the grains of it (the millet) fall down and are tossed up and down on a basket-tray (to remove them) from their husks, and one partakes of it.

When one wants to harvest it it should be harvested only when its ear is dry, i.e. when dry of rain or dew – should rain or dew reach it – then it is put in a place where neither rain or moisture can get at it. It is beaten out with flails, i.e. curved sticks, removed from its husks and lifted. Should there be a large quantity of it, it is threshed/trodden with oxen and the stone. It is lifted in receptacles[171] but not buried (in storage-silos) for this burying would cause it to perish and rot.

Of this (category) also is the millet, the ear, from Wādī Buqlān in the district of the Banū Shihāb, i.e. the country of al-Qumlī[172] b. Sa'īd. So too they call seed-grain *ṣīb*[173] in the dialect of the Ḥaidān[174] people of the district of Kḫawlān in Ṣa'dah, and so too, in other districts, in places like the Ḥāzzah (area at the juncture of the mountains with the Tihāmah plain) of al-Qaḥmah and the places nearby they also call it so'.

5. The fifth variety is rice.

In *al-Ishārah* he said: 'The time for sowing it is the ten chosen [nights] of Nīsān/Naisān [8 May–17 May (?)][175] and it is harvested in Āb [14 August–]. If [it is grown] continuously on ground it ruins it, so when it is sown on any ground it is essential that it not be repeated there unless a year has passed. The places in the Yemen where it is sown are the mountains of Ḥaraz, Bura', and al-Liḥb,[176] it being sown at the same season [*matnam*][170] as millet in this afore-said country. I have sown it in al-Jahmalīyah and it sprouted and was harvested. The way in which it is cultivated is that the ground is ploughed for it, the ground being thoroughly cleansed (of weeds) by repeated ploughing, levelled with the scraper (*maḥarr*), and large bunds made also, to retain the water.'

In *Milḥ al-malāḥah*[177] he said: 'All bush or grass that springs up therein is cleaned away, and after ploughing it water is released onto it till it remains lying over it retained in accordance with the [height of] the bund. The more plentiful the water is the better. The water remains stagnant/retained in the parcel[178] of land until it clears, settling for an entire night. Each parcel of land is bunded off [*tu'qam*] by itself so that the water will not leave one parcel for another. When the water has cleared the rice is sown by scatter-sowing just like sesame, the rice [however] being in its husk. After being scatter-sown it stays seven or eight days then the rice-shoot[179] appears in it – thereupon the bund is opened and the surplus water flows away from it so that the root takes hold [in the ground]. When the root takes hold and some grass or bush comes up among it, once the rice has shot up about half a cubit, the grass and bush growing among it is removed from it. If the water dries up from it then it is given a second watering so that the ground where it is will not become dry. It stands six or seven months till it is harvested. The time for harvesting it shows by its turning yellow and the grain hardening like barley and *'alas*-wheat. Its head is then gently shaken into something that will contain it. It is protected from bird or beast intruding on it until it ripens and yellows, and at the time of its autumn cropping [*ṣurāb*] it is taken gently so that the grain will not scatter out of it. If one wants to detach it from its husk he should pound it with a pestle till the grain is detached from its husk.'

My father, God rest him, said: 'I sowed it by way of trial in my land at Muraizafān (?)[180] in the garden (*al-Bustān*) and it sprouted and bore. Then I harvested it, sowed it again as seed and it stood six months. That was in the year 731 H./1330–31 A.D.' He said, God rest him: 'Al-Mihtār[181] Aḥmad also sowed it in the year 737 H./1336–37 A.D. and harvested it, and I, al-Shihāb al-Ḥalabī and a number (of others) ate of it.'[182]

Ibn Baṣṣāl[183] said also concerning rice cultivation: 'It is sown in warm sheltered eastern exposures at walls with a western aspect. A thorough digging is carried out, rich wet matured dung applied, and plots made in it, [these] plots running along the length of the wall; a single load of the said dung is put down in each plot, and with this the plots are brought into a fine state of tilth. Then the said seed is sown in them, four *raṭls* to every ten plots, the plots [measuring] twelve cubits in length by four cubits in width, and thus it is sown, be it much or little. The seed is then well stirred in the ground[184] with hoes until it sinks down into it. Water is now released onto it twice a week till it sprouts and its shoot comes up evenly. Then one goes in to thin it out and put it into good order, hoeing it lightly after the shoot has become firmly fixed and strong. In thinning out one leaves a hand-span between each root and the next when one wants it to remain in the place where it is, though if one wants to transplant it one leaves it as it is. Sowing of that [rice] which it is intended should remain where it is takes place in the month of March [Ādhār],[185] and that which is to be transplanted, as previously stated, in the month of January. When that which is to be transplanted is plucked up for planting out, four or five roots of it, about a handspan, are brought together so that the hoe is able to hoe it [the plot?] and enter among them. In the breadth of the plot laid out in rows there shall be five rows of them, and [the plot] will be irrigated with water twice [a week].[186] Once the transplant takes hold and sprouts the water is cut off from it, and, when the moist earth [about] it is in good condition, it is lightly hoed over, then left until it needs water. The indication of this is that it turns dark and a blackness comes over it. Thereupon it is irrigated and water applied to it twice a week as previously stated, up to the first of August. Then the water is cut off from it, and it is not (re-)watered unless it be seen to need water – whereupon it is watered once, but no more than once, since, when it is [over-]watered, it turns soft, becoming concentrated on this[187] to the detriment of seeding. Sometimes it turns treacly, becoming like date-syrup and not forming its seed. On this account ground that is cut off from water, containing no moisture suits it, and whenever it is sown in fertile soil it ripens[188] and turns soft as I have said. If, at the time of transplanting, the transplant be weak, five or six [plants] are put in instead of one, but if it be sturdy from three to four roots are put in, for it is a plant which puts out side-shoots.'

'Rice is used only after it is husked. The best practice in husking it is that the rice be placed in bags made for it from camel-skins turned into sacks [*mushakkarah*] – they are filled with it but left slack [not quite full] then beaten with cudgels of oak wood and the [rice] brayed bit by bit. Some coarse salt [*milḥ muḍarras*][189] may be added with it and brayed along with it so as thereby to speed the husking. [Then it is sieved][190] – what is husked comes out beneath the sieve, and what is not husked remains on top of it, and the operation is repeated till it is free [of husk], if God the Exalted will, and God is most knowing.'

6. *The sixth species is* kinab/kinib[191] (Eleusine coracana).

My father, God rest him, said in *al-Ishārah*: 'It is sown at two periods, one of them along with Sābiʻī [millet], the second at the cutting of the date-crop [*thamarat al-nakhl*].'

In *Milḥ al-malāḥah*,[192] he (the author's grandfather) said: 'The ground is brought into good condition for it by scraping off the stones twice or thrice, this being [carried out] before it is watered, and the ground is divided up into square plots with sides of equal length. The *kinib* is scatter-sown like sesame is, covered over using iron mattocks, and watered following the scatter-sowing, then left four days and [re-]watered. If water is not readily available it can be watered [after] up to eight days. After four days it sprouts, and after two waterings irrigation goes on [at the rate of] a watering every half month – from the [time of the] beginning of its scatter-sowing it stands in the ground for four months.'

In *al-Ishārah* he said: 'It is sown only in places watered by *gharb*-buckets, and its ears come up small and round with separate extremities shaped like finger-nails, its grains being very small and much harder than Ḥamrā' [Red] millet. Harvesting it comes four months after it has been sown, its heads being plucked by hand [with a twist of the wrist], and whenever any is plucked more appears after it – one comes and goes to [pluck] it four times in the space of a month like bulrush-millet [*dukhn*]. Moreover, in Tihāmah, when it is watered repeatedly, on a single stalk there will be up to fifty or sixty heads, taken bit by bit. It is exposed to the sun on the threshing-floor until it dries, and beaten out with flails. Know this and understand.'

7. *The seventh species is lucerne* (qaḍb).

In *al-Ishārah*[193] he said, God rest him: 'The ground is prepared well as it is prepared for madder,[194] and in it is great fortune [*barakah*], for with watering and manure it lasts up to ten years. As often as it puts forth [new growth] whatever is above [the surface of] the ground is cut off, and it is irrigated, then the best there is is cut. What most harms it is that the ground and places where it grows should be pastured over. The time for sowing it is throughout the year except the days of autumn [*kharīf*] and much rain – for this weakens it and makes it muddy.'

In *Milḥ al-malāḥah*[195] he said: 'Through preparation of the land by ploughing, dunging, and flattening with the scraper-board[196] it is cleared of scrub, *wabal*-weed,[196] and grass. When the ground is right for it and one wants to sow it (the ground) one takes the lucerne-seed – with which a like quantity of soil or fine sand is mixed – and scatter-sows it by hand in uninterrupted sequence[197] in such a way that there will be ten grains under one's foot. There are some people who scatter-sow barley and *'alas*-wheat over it. When scatter-sowing it is completed it is ploughed over gently till it is covered with earth. The ground is divided up into plots, and the bund (enclosing) the ground is raised using an iron mattock so as to retain water. When the laying out of plots is finished the ground is irrigated with water,[198] plot by plot, without breaching a hole in (any) plot (i.e. so that water does not flow from plot to plot). After the watering one leaves it until it sprouts, then, after eight days, repeats the watering over it, continuing to water it when it needs watering, until the ears of the barley or *'alas*-wheat appear. It is cut along with the lucerne by cropping it close to the ground with the (fretted)[199] knife. Then one applies dung to it again if it be weak, watering it once every eight days, cutting it every forty days, watering it after cutting, and looking after it to remove any stones or pebbles that appear among it. One waters it as often as it needs water, that being when its lower leaf turns yellow. It is best not cut until it flowers and seed formation shows in it. Its leaf grows no less, but it goes on for ten years provided only that the watering and dunging be (constantly) repeated, and it be cut at its proper time only. No pest such as locusts[200] and the like touches it. What is most harmful to it is the pasturing of animals over the ground in which it is, or that it be cut above ground-level and some of its roots remain exposed, and it be cut before its (proper) time, i.e. at less than forty days.

If one wants to take seed from it, one lets it remain uncut till its seed forms and dries. Thereupon one cuts the seed off, exposes it to the sun, beats it out, cleans it of its husk and straw, and removes it (for storing). It is best not to take seed from it except when the lucerne is in a way to being exhausted. Then another piece of new ground should be (chosen) for it, brought into a good state of tilth, and sown with it.

8. The eighth species is Ṭahaf-*millet*[201] (Eragrostis abyssinica).

My father, God rest him, said in *al-Ishārah*: 'It is sown as *kinab/kinib* is sown and does not require a great deal of water, standing sixty nights and then being harvested. It does well and bears at whatever time it is sown. It is not eaten until it is well pounded so that nothing but the heart of its grain remains – it has more grains than *kinab*. Anyone who wants to eat it prepares it like rice and anyone who wants to make bread out of it does so.'

He (the author's grandfather) said in *Milḥ al-malāḥah*:[202] 'The way it is cultivated is that good places are ploughed up for it, and cleaned of grass by two ploughings, one lengthwise, the other breadthwise, and, in places where there is grass, by four ploughings. Well made bunds for it are then set up, and the places from which earth has been shifted by the scraper-board are ploughed, then the ground is watered thoroughly. The *ṭahaf*-millet is scatter-sown in the ground when it first drinks at the first watering. Then it is watered after scatter-sowing until the place is filled with water. Hand scatter-sowing of it is carried out in the same way as sesame is scatter-sown, and it sprouts the next day after its scatter-sowing. When it is half a month old it is given one irrigation with water, then left half a month and a third watering applied to it. It is harvested with the [fretted] knife [*sharīm*] and the sickle [*manjal*] which is the *sharīm*, and the *maḥashsh* [sickle] and the *miqṭa*ʿ [cutter] in the Yemen dialect, being harvested like wheat. Then it is borne off to the threshing-floor [*baidar*] which is the *jurn/jirn,*[203] after being left on the ground as one does with sesame. The leaving on the ground [*ta'ṭīn*] consists in one's leaving it after harvesting in place on its ground for a day and night. The threshing-floor ground will already have been rendered sound[204] with cow-dung and earth. It is then exposed to the sun till it dries, then beaten out with flails, i.e., sticks with curved ends used in beating. It is sown at any time and does well.'

9. *The ninth species is sesame* (simsim) *i.e.,* gulgulān[205] *in the dialect of Yemen folk.*

My father, God rest him, said in *al-Ishārah*: 'It is sown during the middle ten days [*lit.*, nights] of Tishrīn I [24 October–3 November] and flowers in forty-three days. It stands a hundred days and is plucked up by its roots, carried to hard ground, and made into sheaves [and stacked ?][206] load by load, and [by] two and three loads, and set up until it dries through [the action of] the sun and wind and its seed-pods open. Then it is turned to the left, and so on, so that the sesame [seed] it contains runs out. It is sown in Tihāmah and in those mountains near the heat, rough ground containing no sand being selected for it. In the mountains stony land with many pebbles is selected for it. In Tihāmah it does not need much watering – on the contrary it is enough to water the land once only before sowing, but in the mountains it must be watered a number of times. The ground is prepared for it by vigorous ploughing, four and more times. It is sown in the mountains – and if it be land watered by rain[207] alone this will be mentioned, if God wills; and if it be irrigated land it [is sown] in Tishrīn II [14 November–]. When its seed forms and becomes firm it is watered a second time and left till it ripens. Some is scatter-sown, and some sown in the furrowing of the ground behind the [plough-] oxen – this the cultivator calls *tanam,*[208] though some say *talam* [? vocalisation], both technical usage, not [classical] Arabic.'

'Sesame is of two varieties, one local [*baladī*], i.e., that with a white seed, of higher quality and better [both] for oil and eating. The second is "Chinese [*Ṣīnī*]" which is black with a bitterness to it, oil (extracted from) it being neither clear nor nice to taste, and its price below that of local [*baladī*] oil. "Chinese" [sesame] is mostly sown in summer [*ṣaif*] at the time when bulrush-millet [*dukhn*] is sown. If the ground be rain-land it is watered by rainwater only. It is scatter-sown at the first of Ḥazīrān [14 June] during the days of bulrush-millet sowing, and watering with rain is quite sufficient for it. After scatter-sowing it is gently ploughed over until covered with earth but not so that it is choked[?].[209] If it be in irrigated ground drinking from running streams the ground is brought to a good state of tilth for it, as above, and it is watered from the stream and left [unwatered] until it dries, although moisture to make the seed grow [still] remains in it, and it is ploughed. The sesame is scatter-sown as it is in rainland. In ground which drinks from running streams the time for scatter-sowing is two occasions – one in Tishrīn I [14 October–], as previously mentioned in connection with what is sown in the Tihāmahs, and another at the beginning of Shubāṭ [14 March–].[210] When it sprouts and two months pass over it, it is irrigated with water from the running streams and left unwatered until the seed forms. Once its seed grows firm and forms it is re-watered and left until it ripens. The period it stands, from the time it is scatter-sown till it is plucked up, is five months. Anyone who wants it in a hurry will take it after four and a half months, plucking it out by hand. It is bound into sheaves and stacked tentwise, each tent [stook] a load at a time, or more. If the tent [stook] consists of a load at a time it stands half a month stacked together. After that it is pulled apart with a stick – whereupon all the sesame[-seed] it contains comes down. If the tent [stook] be more than a load at a time it stands over twenty days and is [then] pulled apart just as previously mentioned. The way it is stacked tent-wise is that it is arranged into sheaves, two-handfuls or more in size [i.e. in circumference], with the heads meeting together at the top and its roots at the bottom – just as it grew. Then these sheaves are stacked tent-wise in a row, side by side with one-another, the roots next to the ground and the heads to the top. Once the said period from the time of stacking tent-wise has elapsed, it [the sesame] is pulled apart by undoing the tie, the heads of it are turned downwards and its roots upwards, and it is shaken, as I have said, until no more sesame remains in them[211] – until [the work] comes to its end.'

* * *

NOTE: You must know that the threshing-floor, *baidhar* (i.e. *mijrān* in the Yemen dialect), should be on an elevated place, removed from dwellings, vegetable-plots, cucumber-beds, vines and trees. The elevation is for the winds (to winnow it), and water, if it rains, flows away and none remains. Its being removed from dwellings is lest it (i.e. the fine chaff dust) harm people's eyes, their living and cooking. The distance from cucumber-beds is so that it should not harm the fruit, although dust from the threshing floor when it reaches the roots and branches of plants is beneficial to them in the same way as manure, but it is harmful to the leaves and fruit.

NOTES

* 1. I am indebted to Professor Maḥmūd al-Ghūl for a set of photographs of this MS. which contains a much abridged version of the *Bughyat al-fallāḥīn.*

2. Cf. Max Meyerhof, 'Sur un traité d' agriculture composé par un Sultan Yéménite du xivè siècle', *Bulletin de l' Institut d' Egypte*, Cairo, 1943, xxv, 55–63, 1944, xxvi, 51–65, and my 'Agriculture and horticulture: some cultural interchanges of the medieval Arabs and Europe', *Atti dei Convegni, 13*, Academia Nazionale dei Lincei, Convegno Internazionale 9–15 Aprile 1969, Oriente e Occidente nel Medioevo: Filosofia e scienze, Roma, 1971, 535–41. These articles will not be repeated here.

3. Cf. Ibn al-Daiba', *Qurrat al-'uyūn*, SOAS (London) copy, 82b, 'Ikhtaṣara K. al-Jamharah fi 'l-filāḥah', but, just possibly, for *filāḥah* one should read *baizarah.*

4. Meyerhof points out that 'Umar b. Yūsuf (al-Malik al-Ashraf) is actually the brother of 'Abbās's grandfather Dāwūd. He must be using *jadd* in the broad sense of ancestor.

5. In the Glaser Collection at Vienna. I am indebted to Professor Walter Dostal for advising me of the existence of this collection.

6. E. Griffini, 'I manoscritti sudarabici di Milano', *Rivista d. studi orientali*, Roma, 1916–18, vii, 55, no. 282(c), fols. 139, 140.

7. Dār al-Kutub al-Miṣrīyah, no. 155. I have re-foliated my Tarīm transcription after the Cairo copy, though it appears to be complete whereas the latter is not. The section translated *infra* is fols. 24b–35b.

8. J. M. Millás Vallicrosa y Mohamed Aziman ('Azīmān), *Ibn Baṣṣāl: Libro de agriculture*, Tetuan, 1955, an abbreviation of the original treatise.

9. Al-Tihāmī al-Nāṣirī al-Ja'farī, *K. fi 'l-filāḥah li-Abi 'l-Khair al-Andalusī*, Fez, 1357 H.

* 10. The author also of the almanac I translated in my 'Star calendars and an almanac from South-West Arabia', *Anthropos*, Posieux (Fribourg), 1954, xlix, 433–59. In this where 'winnowing' is mentioned 'sowing' should be read – the words are identical.

11. *Water rights and irrigation practice in Laḥj*, Cambridge, 1971, dealing with customary and *sharī'ah* law. It has a good bibliography. A. Ben Shemesh, *Taxation in Islam*, I, Leiden, 1958 and 1967; II, Leiden, 1965, presents interesting data on early land tenure and agriculture, some being Yemenite, but there are errors in rendering the Arabic.

12. An Egyptian (UAR) publication, *The Revolution in three years*, n.d., but probably Cairo, 1965, claims to give an account of agricultural and irrigational development during the occupation. Yemenis aver that development commenced before that date and little was done after 1962 up to 1967.

13. Cf. Maktari, 51. Mohamed Said El Attar, *Le sous-développement économique et social du Yémen*, Alger, 1964, is utterly unreliable on land tenure and social questions, but Muḥ. An'am Ghālib, *al-Yaman*, Beirut, 1966, is far more objective and informative.

14. Cf. my 'Materials for South Arabian History' II, *BSOAS*, London, 1950, xiii, iii, 591. A MS. of Bā Kathīr's *Īḍāḥ al-'uhdah* is now available to me.

15. Cf. my 'Forms of plea, a Šāfi'ī manual from al-Šiḥr', *Rivista d. studi orientali*, Roma, 1955, xxx, 1–15, Maktari, passim; Muḥsin b. Ja'far . . . Bū Numaiy, *Tashīl al-da'āwī fī raf' al-shakāwī*, Mukallā, 1954, 20–1; my 'Some irrigation systems in Ḥaḍramawt', *BSOAS*, London, 1964, xxvii, i, 61.

16. AGL:SF/YEM I, Technical Reports no. 9. Yemen Arab Republic, *Land tenure and water rights*, Tesco-Viziteru Vituki, Budapest, 1971. This was prepared by 'Abd al-Karīm al-Iryānī, co-manager of the project. I am indebted to Mr McCulloch of UNDP for a copy.

17. His biography in al-Sharjī, *Ṭabaqāt al-khawāṣṣ*, Cairo, 1321 H./ 1903 A.D., 37–9, makes no reference to this. The water-law of Zabīd valley must be very ancient and al-Jabartī's part probably that of an arbiter. The document shown in facsimile may be largely a re-statement of water rights.

18. F.A.O. Report, no. 80. Expanded Technical Assistance Program, *Report to the Government of Saudi Arabia on the state of agriculture in the south Tihama area*, by C. O. van der Plas, Rome, January 1953. Its descriptions of harvesting and water-rights are specially relevant to this study. The Yemen Ministry of Agriculture's *Irshādāt 'an zirā'at ahamm al-maḥāṣīl al-ḥaqlīyah wa-'l-khuḍar*, Yaḥyā 'Alī Zabārah and 'Alī Aḥmad al-Lāḥijī, Sirs al-Liyān, 1970, has little of interest, dealing mainly with chemical pesticides and artificial manures. For the Yemeni the instructions in our mediaeval *Bughyat al-fallāḥīn* would generally be more useful! I have been unable to consult A. J. Ivanov, 'The sorghums of north-eastern Africa and south-western Arabia', *Bull. Appl. Bot. and Plant Breed.*, 1929–30, xxix, ii, 273–300; A. A. Orlov, 'Barley of Arabia', *Appl. Bot. Gen. Plant Breeding*, Leningrad, 1929, xx, 283–346, 1933, xxv, ii, cited by R. Portères, 'Les appelations des ceréales en Afrique', *Journal d'agriculture tropicale et de botanique appliquée,* Paris, 1958, v, i and xi, and 1959, vi, i and vii.

19. Cf. note 10. Ḥaidarah's choice of materials varies considerably as between the 1946 and 1971 almanacs. Sometimes the same entries figure in both but under different dates.

20. As the 28-star calendar appears in R. Dozy and C. Pellat, *Le calendrier de Cordoue*, Leiden, 1961, it was presumably in use among the Spanish Arabs of Andalusia.

21. 'Abdullāh b. Aḥmad . . . Makhramah, *Fatāwā Bā Makhramah*, photocopy of Mukallā MS. in SOAS Library London, 293b.

22. Kharīf commences with the star al-Nathrah, Shitā' with the star al-Ghafar on 12 September, Rabī' with Sa'd al-Dhābiḥ on 12 December, and Ṣaif with al-Naṭḥ on March 13th.

23. His collection of short pieces attributed to Zāyid b. 'Alī, published in Arabic and Russian, p. 140. These 170 pieces range from aphorisms to short poems, and seem mainly to have been obtained from Qāḍī Muḥ. al-Ḥajarī. Yemenis are critical of this edition, and aver that far larger collections exist, but it is a useful book. Had he consulted E. Glaser, 'Die Sternkunde der südarabischen Kabylen', *Sitzb. der Kais. Akad. der Wissenschaften*, Wien, 1885, ii, 1–11, republished with corrections in my 'Star calendars . . .' 437, col. 2., he would have found a more accurate version, though 'Alān should be corrected to

'Allān. Ḥaidarah's almanac calls this calendar *nujūm zirā'at al-Mashriq*, the agricultural stars of Eastern Yemen, also followed in the neighbouring *qaḍā*'s (administrative districts). Agaryshev appears to accept the attribution of these aphorisms, many of which relate to agriculture, to Zāyid b. 'Alī without question. Yet some verses of Zāyid closely resemble lines attributed to the pre-Islamic poet and saint Qudum b. Qādim buried on the top of J. Dīn (which we visited in August 1972) north of Ṣan'ā', as quoted in Griffini, op. cit., 1914–15, vi, 1293 seq. These also resemble verses attributed to Ḥusain b. Manṣūr which I collected some twenty years ago, and this latter poet is mentioned at least once in Agaryshev's collection. Cf. 'Umar al-Jāwī, 'Ma'a 'l-Ḥumaid b. Manṣūr', *al-Kalimah*, al-Ḥudaidah, 1971, i, i, 46–51. Both poets are such shadowy figures as to be virtually legendary. Quite different is the case of the Ḥaḍramī poet in the same genre, Sa'd al-Suwainī, whose verses I published in *Prose and poetry from Ḥaḍramawt*, London, 1951, Ar. text 161 seq. He was an historical personage, and biographical particulars are known of him. All these three poets however have undoubtedly had large numbers of traditional verses fathered on them, some remounting possibly even to the pre-Islamic age. Cf. al-Qurashī 'Abd al-Raḥīm, 'al-Ahāzīj al-rīfīyah fi 'l-Ḥujarīyah', *al-Ḥikmah*, Aden, 10 Dec., 1971, i, viii, 61–9. The fashionable new preoccupation with the peasant and agriculture are represented by Muḥ. b. Muḥ al-Dhahbānī, *Dīwān Wādī Banā*, Ta'izz, 1970, 59, with a poem commencing '*Yā ṣabūḥ al-'inab*.' For proverbial material see Ismā'īl al-Akwa', *al-Amthāl al-Yamānīyah*, Cairo, 1968, i–.

24. Cf. al-'Izzī Ṣāliḥ b. Sinaidār, *Amthāl Yamanīyah wa-fukāhāt Ṣan'ānīyah*, Ṣan'ā' (?), 1379 H., and Aden, 1961, ii, 27, *Man bakar bi-'l-madhrā amla 'l-sūq wa-'l-mikhzān*. He who sows early (or-in Mabkar i.e. Nīsān?) will fill the market and granary (also *makhzan*). Sinaidār means *qaiyim al-masjid*, the intendant of the mosque. Al-Ṣulb was described to me by a Ṣan'ānī as three stars in a line. In the harvest days (*aiyām al-khair*) every Kharīf season it is to be seen, he said.

25. At Mabyan I came across a boy called 'Allān because he was born at harvest time (*sā'at al-ṣarīb*). A less happy name was that of a girl, Fitnah, because she was born at a time of trouble or war! 'Allān is also discussed by E. Rossi, 'Note sull' irrigazione et le stagione nel Yemen,' *Oriente Moderno*, Roma, 1953, xxxiii, 360.

26. In Jīzān the *ḥāmī* is a person scaring the birds from the fields, and his shelter is *sahwah*.

27. *Zawm* is boiled curdled milk (*al-laban al-rā'ib al-mufawwar*) over which a little boiling water is poured, and flour mixed (*yukhḍab*) in with it. This is relish (*idām*) for the tribes who cannot easily afford meat.

28. E. Rossi, *L'Arabo parlato a Ṣan'ā'*, Roma, 1939, 159, *malūjah pane . . . di frumento o di orzo.*

29. *Khair* can also mean, rain, harvest. In Jīzān people would say: '*Al-khair bi-yad Allah*, Rain is in the hand of God.' and '*Al-suyūl min 'ind Allah*, Floods are from God.'

30. Cf. my *Prose and poetry*, Ar. text, 164, no. 57, for a closely similar verse.

31. Al-Hamdānī, *Ṣifat Jazīrat al-'Arab*, edit. D. H. Müller, Leiden, 1884–9, 199. Cf. 'Star calendars . . .', 441, Maktari, op. cit., 177; Admiralty, Geographical Handbook Series, B.R. 527, Naval Intelligence Division, *Western Arabia and*

the Red Sea, Oxford, June, 1946, 606 seq. I have found it used in Jahānah of Khawlān which has two crop seasons, Ṣaif and Kharīf – in which latter rains fall.

32. 'Alī b. al-Ḥasan ... al-Khazrajī, *The Pearl Strings* ... *al-'Uqūd al-lu'lu'īyah*, edit. Muḥ. 'Asal, Leyden-London, 1914–18, iv–v, ii, 243, 313.

33. In Jīzān one says *madhrāt*, plur., *madhārī*, explained as *waqt badhr al-ḥabb*.

34. In the Jabal a calf of 1½ to 2 years old is called *ṭabī'* before being used for ploughing (*ḥirāthah*). One trains it to work (*yi'assif-uh*) and only then is the male called *thawr*.

35. *'Ajmah* means *qurāsh*, beasts.

36. Op. cit., 103a.

37. Muḥyi 'l-Dīn al-Nawawī, *K. al-Adhkār*, Cairo, 1356 H., 399, on *al-bākūrah min al-thamar*.

38. *Ṣifah*, op. cit., 198.

39. Ibid, 199.

40. Op. cit., 104a. For bread in Ḥaḍramawt of various cereals cf. *Prose and poetry* ..., 117, and also the first *Maqāmah* on the contest between the cereals and fruits.

41. *Jamma*, to become abundant and collect, of water.

42. Text, 'and sesame', but I have read it without the 'and'; perhaps this should be retained, and the 'and' before 'cow-pea' in the text omitted.

43. Usually the term *dijr* is used in South Arabia.

44. *Ṣubḥ al-a'shā*, Cairo, 1913ff., v, 8–9, where the readings are to be corrected.

45. Cf. my 'Two sixteenth-century Arabian geographical works', *BSOAS*, London, 1958, xxi, ii, 262. Here it is asserted that only when Tūrānshāh took Ta'izz and his brother Saladin sent him a quantity of various kinds of fruits from Syria (Shām) which he planted in Ta'izz, did the latter become a veritable Damascus for fruits and flowers – it seems implied that these were planted in Tha'abāt – this about 570 H./1174–5 A.D.

46. So corrected from Qalqashandī's Mahlah.

47. Khazrajī, op. cit., iv/ii, 377–82. For water led into Tha'abāt village and the building of a wall around it, ibid., v/ii, 60–1, and for the father of the author of the *Bughyah*, Mujāhid, developing its gardens, ibid, v/ii, 125. Cf. 'Abd al-Bāqī b. 'Abd al-Majīd al-Yamānī, *Bahjat al-zaman*, Cairo (?), 1965, 122, for the castles and gardens of al-Ma'qilī, and *Qurrat al-'uyūn*, SOAS copy, 93a.

48. Khazrajī, v/ii, 59, 75.

49. Ibid., v/ii, 40–1.

50. Ibid., v/ii, 90, *passim.*

51. Ibid, v/ii, 142.

52. Cf. *Qurrat al-'uyūn*, SOAS, 85b.

53. Op. cit., v/ii, 283 passim; *Qurrat al-'uyūn*, Vienna, 72b, 89a, this latter in the year 908 H./1502–03 A.D. Saryāqūs, possibly Siryāqūs, is a strange name in South Arabia; it might be taken from a village near Cairo of the same name, mentioned by Yāqūt.

54. Khazrajī, v/ii, 13.

55. *Qurrat al-'uyūn*, Vienna, 83b. Mawshaḥ in Dathīnah is a large town according to Hamdānī, *Ṣifah*, 91, but this may be another place.
56. Perhaps Hamdānī's *jammat* (note 41) should be read with a *ḥā* also.
57. Cf. note 199 infra.
58. Cairo writes *birr* which is the dialect of the north, Ṣan'ā' and Dhamār. In allusion to the difficulties in growing wheat, the Yemeni says: *'Lā taqul birr illā wahū fi 'l-ṣurr* – Don't say "wheat" till it is in the lap.' i.e. until it is in the grain-store, (Ismā'īl al-Akwa'). Al-Shāfi'ī, *al-Umm*, Cairo, 1321–5 H., II, 30 has some interesting remarks on types of wheat and other cereals when discussing the *ṣadaqah* payable on them.
59. The *Bughyah*, 19b, states that on the 4th Tishrīn II (17 Nov.) wheat is sown in the cool parts of the Yemen. It is called al-Qiyāḍ there and in it falls the *wasmī*-rain. Lane, *Lexicon*, describes *qiyāẓ* as seed-produce sown in autumn and the beginning of winter so as to be reaped in *summer.* For the *Bughyah qaiẓ* and *ṣaif* are apparently the same – in Jīzān it would end in late September. The Ta'izz almanac places the harvest of *qiyāḍ* from the 31st Kānūn II until the end of Shubāṭ (13 Feb.–13 March), which is Shitā'. Rossi, *L'Arabo parlato*, 151, calls Qiyāḍ *raccolto del tardo inverno* and Dithā' *raccolto de primavera-estate.*
60. F. Sezgin, *Geschichte des arabischen Schrifttums*, Leiden, 1971, iv, 310 seq. discusses what is attributed to Democritus. Pauly-Wissowa says of him: 'Als Schrifsteller über Ackerbau erwähnen den D. Varro und Columella, auch in den Geoponika wird manches auf ihn zuruck-geführt.' The reference of the *Bughyah* is probably to the *Georgica* attributed to Democritus transmitted by Ibn Waḥshīyah.
61. This word is uncertain in the text and might perhaps mean 'fertile'.
62. *Koran*, xii, 47.
63. Baiḍāwī, *Commentarius in Coranum*, ed. H. O. Fleischer, Leipzig, 1847–8, i, 462, *Li'allā ya'kula-hu 'l-sūsu*, lest the worm eat it.
64. Also *jāwarsh.*
65. Two texts read Wasnī, but Qāḍī Ismā'īl pronounces it Wisnī.
66. *Mudhannab* refers to the little tail at the end of each grain – the awm or hairs.
67. *Ṣurāb* is *ḥāṣil al-ṣaif*, cropping in late August to September. Cf. Bertram Thomas, *Arabia Felix*, London. 1938, 18, Surub, October to December.
68. The Ta'izz 1971 almanac, under 22nd Tishrīn I (4 Nov.) has the entry – most of the Ta'izzīyah – the *qiyāḍ al-ḥinṭah*; and under the 23rd of Tishrīn I, the last of the sowing (*dhary*) of al-Quṣaibah.
69. Or al-Qiyāẓ, cf. note 59.
70. 'Arabī is a small-eared variety.
71. Mayerhof reads *Milḥ*. Vienna MS., 3a.
72. i.e. in the mountains.
73. MM., 3a adds 'and in the Mashriq'.
74. Mikhlāf Ja'far (as above) is probably meant. It is so called after Ja'far al-Manākhī, and includes 'Udain, Ḥais, etc. A *mikhlāf* is a province or district; it is composed of a number of groups (*'uzlah* pl. *'uzal*) of villages.
75. E. Rossi, 'Note sull' irrigazione', op. cit.; 359, *'aqar, terre asciutte, non irrigate, bagnate solo dall' acqua piovana.* Cf. Hamdānī, op. cit., 199.

76. The history known as *al-Jawhar al-munīr* states that most of the grain of the Ṣa'dah district is Samrā', a kind of *burr*.

77. MM., 5b–7b.

78. Qāḍī Ismā'īl explained *th l ṭah* as *rāwiyah*, moist, but not *waḥal*, mud.

79. *Safakh, yasfakh*, to sow seed by scattering it broadcast by hand from a *masabb* a sort of leather sack, or along the furrow by hand.

80. Though in practice simple enough, this procedure is difficult to describe. The sectional diagram (Fig. 6) may help.

1st furrow with earth thrown up at its sides to form ridges A and B.

2nd furrow with the plough run at an angle so as to throw B on to A and form a single bank out of the two.

3rd furrow ploughed to the left of a space for a water-channel, throwing up ridges C and D.

4th furrow with the plough run at an angle so as to throw D on to C, and form a single bank out of the two.

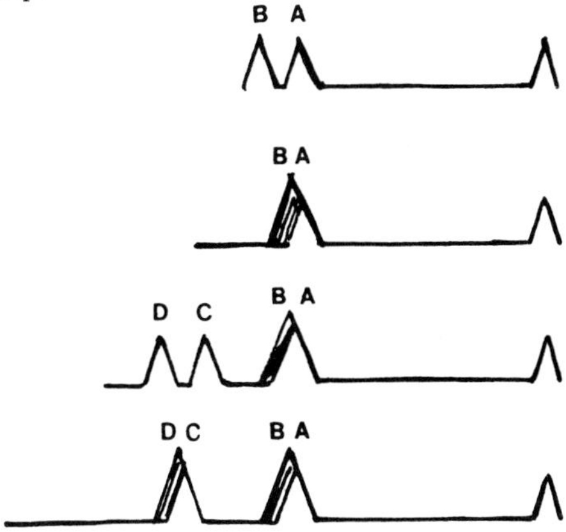

Fig. 6.

81. Parched or toasted grain, still a little green, is often put before casual visitors at these times, usually on small palm-leaf trays. I have eaten it in Ḥaḍramawt.

82. A *shamlah*, a bag either of camel-hair (*wabar*) or goat-hair, black or black and white, to carry wood or lucerne (*qaḍab*), etc., slung over the shoulder.

83. i.e. for foddering.

84. The ears are laid to the middle of the paved threshing-floor (*mijrān*) and the cut ends to the outside, to form a circular heap. The straw at the top will be so arranged that the heads of grain in the centre of the heap will be covered. It is left to permit the grain to suck in the goodness from the straw-stalks.

85. It is beaten out by treading (*daws*) into chaff small enough for the beasts to eat.

86. One treads out the crop and winnows (*yidūs al-zar' wa-yidhrī*), but those ears still remaining in the husk one collects together (*qashsh-ah, yaqushsh-ah*) to put aside and tread out again. After the treading (dawīm al-*birr* with a *midwam*) or beating of other grain (*labīj, khabīṭ* – one says *bā nakhbuṭ*) the grain is put into heaps.

87. *R k b al-tibn* – translation conjectural.

88. Qāḍī Ismā'īl says the measuring takes place in the early morning (*al-kail al-ṣabāḥ bākir aw ākhir al-nahār*) or at the end of the day – otherwise the white ant (*irḍah*) will attack it.

89. To this day an opening on the north side is put in a grain-store for ventilation (Q. Ismā'īl).

90. From Qā' al-Bawn, the famous wheat-growing district north of Ṣan'ā'.
91. *'Aqar* was here explained as ground watered with the *mā' al-Kharīf,* i.e. rain of the last part of the month Āb which remains in the soil. Cf. note 75.
92. Ḥadūr of the Banī Shihāb is west of Ṣan'ā'.
93. Probably thin bread of the type known as *ruqāq* (Qāḍī Ismā'īl). This sort of bread is mentioned by Hamdānī, supra, p. 33.
94. Maisānī is a Dhamār type of wheat. Cf. Hamdānī, *Ṣifah*, 199. The *Muntakhab min al-fiqh* of the first Zaidī Imām of the Yemen, (London Ph.D. thesis edit. A. K. Kazi), fol. 118, mentions *burr Maisānī, b. abyaḍ, b. 'Arabī,* and *b. Ṭishārī* (?). It is further mentioned by Jāḥiẓ, *K. al-Bukhalā'*, Cairo, 1948, 53. I found it growing in Ḍāli' (Dhala) in 1958. Maisān is said to be a star, and this is confirmed by the *Calendrier de Cordoue*, 176, but it is said to be really 6 days of the star al-Haq'ah.
95. Cf. al-Shāfi'ī, *al-Umm*, loc. cit., and iii, 91; Yaḥyā b. Ādam, trans. Ben Shemesh, op. cit., i, 82. Ibn Baṭūṭah, trans H. A. R. Gibb, *The Travels if Ibn Battuta* Hakluyt Soc., Cambridge, 1958–, ii, 383, mentions it at Ẓafār/Ẓufār, and says it is a sort of *sult* (barley). The Yemen Republic paper *al-Thawrah*, March 1966, calls it *farīd al-naw' wa-'l-nakhah*, unique of its sort and in its smell.
96. The *mijashshah* is said to be like the *raḥā*-quern, but simpler and it crushes more coarsely. It is the subject of a well-known Iraqi colloquial poem by 'Abbūd al-Karkhī, *Dīwān*, 1st edn., n.d., 2 al-*Majrashah*, 2nd edn., Baghdad, 1956.
97. Hamdānī, *Ṣifah*, 199 has Nusūl, as also Ḥusain al-'Arshī, *Bulūgh al-marām fī sharḥ misk al-khitām*, Cairo, 1939, 439.
98. This corresponds closely to the Ta'izz almanac of Ḥaidarah which records for the 30th of Kānūn I (12 January) the *qiyāḍ 'alas baiḍā'*.
99. A *mijrāsh* was described as a type of *maṭḥan*, a flat stone with a roller for crushing grain etc.
100. Cf. Yaḥyā b. Ādam, trans. Ben Shemesh, op. cit., I, 82. Muḥ. al-Khaṭīb al-Iskāfī, *Mabādi' al-lughah*, Cairo, 1325 H., 184, says *sult* is a sort of husked (*muqashshar*) barley, i.e. barley with no husk.
101. It is called Ḥabīb in Dhamār. A. Grohmann, *Südarabien als Wirtschafts-gebiet*, Brünn-Prague, 1930, 33, i, 215 gives types of barley as Ḥabīb, Shilb, Saqlah, Samrā, Bakūr. Grohmann's section on cereals, 207–17 is most informative on distribution of types of cereal by districts, times of sowing, etc. Ripe barley ears moistened and parched are called *qarī al-sha'īr*. In the north *qalī*, parched grains of *'atar*, peas, are eaten as in Europe one eats *fūl Sūdānī*, groundnuts.
102. Abū Muḥ. al-Ḥusain . . . al-Baghawī – cf. Brock., *Gal*, i, 363, *Sup.*, i. 620. al-Qāḍī Ḥusain b. Muḥ. al-Marw al-Rūdhī was his teacher in jurisprudence.
103. Or 'temperament, humour'.
104. *Al-Umm*, iii, 90.
105. Al-Buwaiṭī, Brock, *Gal.*, i, 179, *Sup.*, i, 304; al-Fūrānī, Brock., *Gal.*, i. 387, *Sup.*, i, 669. Al-Ṣaidalānī may be the person mentioned by al-Dhahabī, al-*'Ibar*, edit. Fu'ād Saiyid, al-Kuwait, 1961, iii, 69.
106. Doubtless the celebrated Shāfi'ī jurisprudent Abū Zakarīyā' Yaḥyā . . . al-Nawawī, Brock., *Gal.*, i, 394, *Sup.* i, 680.

107. *Al-Aḥkām* . . . , Cairo, 1960, 118. Abū Muḥammad may be al-Baghawī; I am uncertain about Abū 'Alī.

108. Cf. *al-Umm*, iii, 90; A. Grohmann, op. cit., i, 212–4. Bā Makhramah, *Fatāwā*, 95a–b, alludes to Baiḍā' and Ḥamrā' millet in Ḥaḍramawt. Ibn Baṭūṭah, op. cit., ii, 383 has an account of the well irrigation (*sināwah*) of millet at Ẓafār.

109. Raisī is called 'Umarī in Dhamār. Cf. E. V. Stace, *English-Arabic vocabulary*, London, 1893, 91. Muḥsin . . . Bū Numaiy, *Tashīl*, op. cit., 23, alludes to *Raisī dhurah al-shawl* as irrigated from the well-known *qanāt* near Ghail Bā Wazīr called Mi'yān at Ḥarth in the coastal region of Ḥaḍramawt. *Shawl* millet is of the Ṣaif season.

110. Lit., 'chosen nights'.

111. Ḥaidarah's Ta'izz almanac gives the following dates – 1st Nīsān (14 April) millet sown in some parts of Ḥujarīyah; 4th Nīsān (17 April) sowing in Dhī Sufāl and al-'Udain before the *ghurūb* of the Pleiades (i.e. up to 11th Nīsān (24 April)); 15th Nīsān (28 April) the ten chosen nights/days for sowing in Dhubḥān and al-Ṣalū; 18th Nīsān (1 May) sowing (*zar'*) of millet in Qadas, Ṣuhbān, and districts of the Ibb *liwā'*; 25th Nīsān (8 May) the ten chosen days/nights for the sowing in Tihāmah; 27th Nīsān (10 April) the close of the ten chosen days/nights for sowing in Dhubḥān.

112. Cf. discussion supra, p. 2.

113. Unidentified.

114. In 1958 I found Ja'aidī grows in Ḍāli' – it is Ṣaif millet. In 1972 it was suggested to me that the name might be derived from the Uj'ūd tribe of Ḍāli'.

115. Aiyār is also called Mabkar. Cf. my 'Notes on Ṣubaiḥī territory', *Le Muséon*, Louvain, 1953, lxvi, 129. The Ta'izz almanac says there is sowing on the 1st Aiyār in Shar'ab and J. Ḥabashī, and on the second sowing of grain in the running streams (*ghuyūl*).

116. The Ta'izz almanac for the 8th Aiyār (21 May) notes the *matlam* of Gharibah in the Shar'ab mountains. I noted Gharbah in Ḍāli'.

117. The Ta'izz almanac has identical information entered at this date.

118. A Kibsī informant said it is called Dahainī in Dhamār.

119. This variety, which takes so long to ripen, would be found in such places as Ba'dān and the Ibb mountains.

120. It grows in Ḍāli'. Cf. my 'Star calendars' . . . , 23.

121. The Ta'izz almanac enters Sābi'ī sowing from the 1st Tishrīn I (14 October) to the 11th Tishrīn I (24 October). The almanac enters the rising of the various stars of Ursa (Major) about twelve days after the dates given by the *Bughyah*.

122. The word *inṭarah*, so rendered, is not in the lexicons.

123. The verb is *kaḥaf, yikḥaf.* On 27 June 1972, a little north of Ta'izz, I saw the *kaḥīf* going on in the fields. There it is called *shaṣar, yashṣur, shaṣīr*, i.e. ploughing between the rows of millet. In the northern Zaidī districts in the mountains I was told that *kaḥīf* means to plough first along the left side of the space between two rows of millet, then along the right side, and finally, when the millet is breast-high, down the middle, so as to eliminate weeds. The first two ploughings, called *al-shāhik al-awwal wa-'l-thānī*, are to break up and soften the ground which has turned hard after watering. Qāḍī Ismā'īl quoted me a

proverb, '*Mā da'as-ah thawr al-kaḥīf fī 'l-wasaṭ akal-ah fī 'l-ṭaraf.*) This he explained as meaning that when the ox is ploughing over the field where the millet has grown high and breaks some millet stalk, it is only given the beast to eat at the edge of the field, not while working in the middle of it.

124. This phrase might also be understood as 'the greater or lesser size of the irrigation channel (*misqā*)'. This section in general differs as between the Cairo and Tarīm texts but I have made a composite version which appears to be accurate.

125. I.e. *kaḥīf* is ploughed in a wavy not a straight line.

126. Parched millet grain is called *tunfāsh* in some places. It is burned on a girdle (*yuḥraq fī sāj/ṭāwah*).

127. For 'intensely hot' MM., 18b, has *mustadiqqah*, fine, thin ?. C. de Landberg, *Glossaire datînois*, Leiden, 1920–42, 2712, *mallah – cendre ou terre chauffée sous le charbon incandescent*, but it also means a *tannūr* or jar-shaped flap-jack oven, and, in Ṣan'ā', the space under the floor of a Turkish bath through which the hot air passes.

128. This is probably to be read *maqāṭi'* with MM., 19a, instead of *maqāmi'*.

129. Sic, but probably *baidar* should be read.

130. Hamdānī, *Ṣifah*, 107 seq., speaking of the Khawlān b. 'Amr and Dhū Jurah districts, says they are called the Khizānat al-Yaman (Store of the Yemen), while Dhamār, Ru'ain and al-Saḥūl are dubbed the Egypt of the Yemen 'because millet, wheat and barley last a long time in these places. I have seen in J. Maswar wheat over which thirty years had passed without its stinking and changing (going bad). As for millet it is only in a hot district, and it is not stored in houses on account of the rotten state that soon overcomes it, but excavations are made for it in the ground and it is buried in silos (*madāfin*), a single one holding 5,000 *qafīz* or less. It is then closed over until perhaps even thorn bushes ('*urā*) grow on the cover, and it lasts a lifetime without being lost (? *infakhash*, a word not in the lexicons) except that its smell and taste alter. When a silo (*madfan*) containing it is opened up it is left for days until it cools and its fumes abate. Should anyone enter if when it is [just] opened up he would perish from the heat of it.' The *qafīz* is stated by the *Encyl. Islam* to consist of 48 *mudds*, and to vary from 25–55 litres, the lower figure in the early Islamic centuries being usual. A report on grain storage was made in 1970 to F.A.O. by R. C. O'Neill and D. J. Greig, containing a diagram of a *madfan* with a stone cover set in a stone surround fixed with a mud sealing plug. They estimate the *qadaḥ* which I imagine must be approximately equivalent to the ancient *qafīz* at about 40 litres, but there were, and indeed still are many different *qadaḥs* in the Yemen although the Ḥamīd al-Dīn Imāms attempted to establish an official *qadaḥ*. The silos, they say, are often lined with millet stalk before filling with grain. In 1954 I was told of two types – one constructed in the corner of a room in a house and called *dawbalī* (pl., *dawābilah* is filled from the top and the grain withdrawn from the bottom, the other the *madfan*; but I have since heard of a type called *maqṣūrah*. In the old Imāmic Government stores outside the capacities of the *madfans* were stated to vary from 500–1,000–2,000 *qadaḥs* capacity. In *madfans* are stored millet, barley, and *dukhn*, but not *burr, dijr* (chickpeas), *'atir/'atar*, *'adas* and *bilsin* (lentils) or *fūl* – these last being stored inside houses or in granaries (*shūnah* pl., *shuwan*) divided into compartments

called *kaid* (pl., *kuyūd*). Grohmann, *Südarabien* ... ii, 15, gives as names for silos or stores, *ḥaqb*, in Khawlān *ṭann*, and in Dhamār *kejt*, probably *kaid* supra. Cf. C. de Landberg, *Études sur les dialectes* ..., I, *Ḥaḍramoût*, Leiden, 1901, 87, for a description of the construction of a *madfan*.

131. Large heads of grain kept for seed are laid on cold earth or ash (*ramād*), but if grain is stored hot this brings the white ant.

132. This word is uncertain and the readings various. From Ḥajjah Robert Wilson however reports a well known millet grown in the Tihāmah called *bujaidah*, and this might be the correct form of the name.

133. For the different reckoning of the rising of these stars in the Ta'izz almanac see note 123.

134. In Wādī Jīzān *shabb* is sown, according to the local star calendar, in al-Dhirā' (from May 30th) in the mountains, and from the first star of Kharīf, al-Nathrah (12 June), which is the beginning of the agricultural year there.

135. MM., 21a.

136. For *bukr* see 'Star calendars', 450, Maktari, 156, *Gloss. dat.*, 191. F. M. Hunter, C. W. Sealy, J. W. B. Merewether, *Account of the Arab tribes in the vicinity of Aden*, Bombay, 1909, 4, say of Lahej that it has two agricultural seasons in the year '*ṣaif*' or summer and '*bukur*' or winter. Millet is the principal cereal cultivated, the white species being sown in *ṣaif* and the red in *bukur*; it is usual to have two harvests of each and occasionally three. The first crop which is called *ab* is the finest, the second is called '*aqb* (he writes *akb*), and the third '*aqb al-'aqb*. The '*aqb al-'aqb* of the *ṣaif* is frequently seen growing alongside the *ab* of the *bukur*. But however the season may be a land tax of 1½ dollars per *ḍimd/ḍamd* (acre) is collected in four instalments, and in addition tithes are levied on produce under the name of *firqah* (written *firka*). There are also dues on water. Cf. Landberg, *Études* ... II, *Datīnah*, Leiden, 1905, 232–3. *Fatāt al-Jazīrah*, 16 Sept. 1945 reports recent floods in Lahej and its Amīr as saying 'we'll plant *bukr* of millet, then sesame (*gilgil/juljulān*) and *ḥabḥab*-melons and Bainī millet, cropping after three months.'

137. Cf. Sharjī, op. cit., 179, 'a name for that part of Tihāmah near the mountain', pl., *ḥawāzz*. In Rasūlid times the two Ḥāzzahs of Zabīd had a special official appointed to them with supporting staff.

138. The abbreviated version of the *Bughyah* (supra) says, 'It is called al–Ḥaḍḍār (?) and it is called al-Tamrī because it is planted at the days of the palms.' He may mean at the time of the date-crop. The reading al-Ḥaḍḍār is not entirely certain.

139. MM., 21a. I noted Zi'ir in Ṣubaiḥī territory in 1940. It is also known in Jīzān. Zi'ir is planted especially in Ṣaif (13 March–) and called Ṣaifī, but it may be sown at any suitable time.

140. The Cairo Ms. inserts: 'and a third follows it' called *khilf*. In Jīzān this is pronounced *khalf*. For '*aqb* see note 138. Bā Makhramah, *Fatāwā*, 258a, is set a question about '*aqb* of Ṣaif millet and *bukr* which happen to be growing side by side. Should the owner of the '*aqb* be compelled to remove his crop since it harms the *bukr* millet? The answer was no!

141. In W. Jīzān in 1971 I learned the term al-Jinnīyah is still in use – because the crop is strange and unexpected – like a Jinnī!

142. MM. 21b.

143. Reading *shaṭṭ* for Cairo and MM., 22a, *saṭṭ*, a term which Qāḍīs Ismā'īl and Muḥammad al-Akwa' explained as the lowest part of the furrow. 'One opposite another' is best explained by the diagram in Fig. 7.

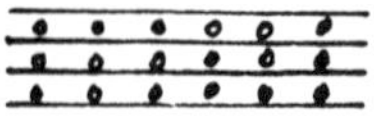

Fig. 7.

144. This is lest some grain be bad and should not sprout. If crowded together (*mitḥārish*) there is a fixed time for removing surplus sprouts so as to leave one sprout to grow by itself. The separating and removing of these sprouts, which are used for animal (*qurāsh*) fodder is *faqaḥ, yifqaḥ, fuqūḥ*.

145. I.e. the furrows.

146. Qāḍī Ismā'īl was not sure of the sense here, but one says *rās al-tilm wa-baṭn-ah*, the top and bottom of a furrow.

147. Reading *dharrat* with MM., 22a.

148. Vocalisation dubious.

149. The millet head is of course bent over, hanging from a stalk curved at the top. Ḥujainā is mentioned by the abbreviated version of the *Bughyah*, in mediaeval almanacs, and 'Star Calendars', 454, has *ḥijnah*. Ḥujainā is known in Zabīd today.

150. This dating, based on Glaser, is supplied from 'Star calendars', 437, but it is at variance with the date given in the latter part of the sentence here.

151. Lane, *Lexicon*, places the rising of al-Nasr al-Wāqi' about 25 November which would indicate that al-Wāqi' is meant here.

152. For al-Jaḥr cf. supra, p. 9. Qāḍī Muḥ. al-Akwa' defined it as *'al-qaiẓ 'inda-nā'*.

153. Lit. 'nights'.

154. See note 138.

155. A correct if unlikely reading might be Jaḥrī or Juraḥī. It is just possible that it might be named after a place, not well known, al-Ḥ r jīyah mentioned by Hamdānī, *Ṣifah*, 75.

156. See note 132. A further variant reading is *b dh n hā*.

157. The Tarīm and Cairo Mss. diverge so both are given. I am uncertain what to propose, but perhaps the original text said its husk was black like egg-plant, and its head crooked like Budaijā/Bujaidah.

158. Maktari, 166, calls a *sharj* a sub-channel. Cf. my 'Some irrigation systems in Ḥaḍramawt', *BSOAS*, London, 1964, xxvii, i, 54.

159. Cf. note 154.

160. For *dukhn* cf. al-Iskāfī, op. cit., 184, which he says means *jāwars*, cf. fn. 64. The Admiralty *Western Arabia . . .*, 595, has names of three types. In Dathīnah *dukhn baladī* and *dukhn sāḥil* are distinguished.

161. Rūmī today refers to maize, but in Rasūlid times it ought to mean 'Greek/Byzantine'.

162. *Ṭal'*, explained by Q. Ismā'īl as *sunbulah thamar.*

163. For *'aṣīd* cf. *Prose and poetry*, op. cit., Ar. text, 32.

164. The Tarīm text reads, probably incorrectly, 'from Abyssinia and Barbarah (Berbera)', but the Cairo Ms. has '*wa-bizru-hu kabīr*, and its seed is large.'

165. I have so far not traced this word, but it could also conceivably be from a dialect of the Tihāmah people of mixed African blood.

166. MM., 26a.

167. As I have seen on several occasions, it is actually tossed from behind to underneath the plough. I was told that, whereas for wheat and barley the sower can walk behind and scatter the seed, millet must be sown using the *qaṣabah*, a pipe with a funnel at the top of it attached to the plough, so that the seed will go down into the ground.

168. On the 21st September the Ta'izz almanac predicts rain in most districts.

169. *Kaḥīf* renders the ground suitably moist (*rāwiyah*) for the crops (Qāḍī Ismā'īl). The Qāḍī did not understand exactly what is meant by 'thorough-going application (*bi-'l-jumlah al-kāfiyah*)', but the text may be slightly corrupt here.

170. Cairo Ms. margin notes that *al-mantam* in Tihāmah dialect is *al-matlam*. There may also be a form *matmam*. The Qāḍī says *yatlimū* means *yabdhurū*, to sow, and *Gloss. dat.*, 235, states that *talam* is *faire des sillons, labourer*. Cf. note 116.

171. Receptacles are probably *gharā'ir*, woollen bags, but *dukhn* is covered in *dawbalī* (*ṭīn* – dust but I seem to have understood it as a sort of *madfan* – cf. note 132) and stored in pottery jars (*azyār*) or barrels (*barāmīl*) (Q. Ismā'īl).

172. The vocalisation of this name which I have so far found in no Arabic source is doubtful.

173. Cf. al-'Arshī, *Bulūgh al-marām* . . . , 429, *al-ṣīb al-badhr li-'l-zar'*; *Gloss. dat.*, 2159; Landberg, *Arabica IV* (Leiden, 1897), 136.

174. Reading so for the text's Ḥaddān. Nashwān b. Sa'īd is buried in Ḥaidān (Q. Ismā'īl). Cf. Muḥ. b. Muḥ. Zabārah, *al-Anbā' 'an dawlat Bilqīs wa-Saba'*, Cairo, 1372 H., 121.

175. The Ta'izz almanac confirms these are the chosen dates for sowing in Tihāmah but does not mention rice.

176. Sharjī, op. cit., 133, the mountains of al-Liḥb.

177. MM., 24a.

178. *Qiṭ'ah*, a term also used in Ahnūm and the western mountains of the north of the Yemen. Synonyms are *sabbah*, pl. *sibab, sibāb, tubbah*, pl. *tubab, salaq*, pl. *aslāq* defined as a division of the *jirbah* (field), also *qasam*, pl. *aqsām*, a small piece of ground similar to a *ḥawḍ* basin or plot.

179. Qāḍī Ismā'īl gave a valuable definition of *nabāt* as employed in the Yemen – *kull mā kharaj min al-turāb yusammā nabāt*, anything that comes up out of the earth is called *nabāt*. It is applied then to a shoot of any of these crops which has not grown above the level of the ridges of the furrow (*tilm*). See figure below.

Fig. 8. Furrow with a *nabāt* shoot in it.

180. This name appears entirely corrupt and is unknown to Khazrajī and other sources. Perhaps one should read al-Muzaiḥifah a well known village of the W. Zabīd (cf. Sharjī, op. cit., 58). Al-Bustān might be al-Bustān al-Sharqī discussed supra in the section on Rasūlid gardens.

181. Cairo Ms., margin, states that al-Mihtār Aḥmad was the maternal uncle of the Rasūlid monarch al-Malik al-Mujāhid. Generally *mihtār* in the Yemen today means *jins ḍa‘īf*, which may loosely be translated as 'low class' though this is not its exact meaning. Shihāb al-Dīn is probably Aḥ. b. 'Alī b. Ismā‘īl called al-Naqqāsh (ob. 750 H./1349–50 A.D.) known to Khazrajī, op. cit., ii, 82.

182. Under the annals for 801 H./1398–9 A.D. Khazrajī, op. cit., ii, 300, says: 'In the month of Rabī‘ I, rice was harvested (*ṣuriba*) from the properties (*amlāk*) of the Sultan in the W. Zabīd.' Again, ii, 318, he says, incorrectly, that the Sultan who died in 803 H./1400–01 A.D. was the first who planted rice in W. Zabīd. This monarch also introduced (*aḥdath*) the garden of Siryāqūs al-A‘lā of the same Wādī. The Mukallā proverb which I think I have also heard elsewhere, '*Mā kān fi 'l-ruzz quwwah, wa-lā fi 'l-Hindī muruwwah.* There is no strength (good) in rice and no manliness in the Indian,' expresses both the recognition of the poor nutritive values of (polished) rice (as contrasted with local cereals) and the Arab sentiment towards Indians!

183. *Libro de agricultura*, Ar. text 110 seq. This is a badly edited text, and from line 2 from the foot of p. 110 the subject discussed is rice – though this is not mentioned there, the text being confused at this point. It looks however as if the actual text of Ibn Baṣṣāl available to the author of the *Bughyah* was also confused since walls could hardly be necessary for rice-growing. Rice was already growing in Spain by the mid-tenth century A.D. according to S. M. Imamuddin, *Economic history of Spain*, Dacca, 1963, 85.

184. Ibn Baṣṣāl reads 'in the plots', and supplies additional data.

185. March and January, O.S. naturally.

186. Restored from Ibn Baṣṣāl.

187. Ibn Baṣṣāl, 111, reads for 'this' the word *tana‘‘um*; which would imply that the rice is diverted to turning soft rather than seeding.

188. Ibn Baṣṣāl omits 'ripens'.

189. Translation conjectural.

190. It seems essential to restore this phrase from Ibn Baṣṣāl.

191. *Kinib* is Ḥaḍramī pronunciation.

192. MM., 28b.

193. Cf. MM., 39b, to which *al-Ishārah* here and in many other places is closely similar.

194. *Bughyah*, 126b, states that madder has the ground ploughed up for it up to twenty times, and it must be specially free of weed.

195. MM., 37a.

196. A man stands on the board to keep it flat and it is drawn by a bull (*thawr*). This fills in the holes in the field by the sideways drag to the left and right of it – normally this operation is carried out in a straight line (Q. Ismā‘īl).

196. At Jahānah Qāḍī Ismā‘īl pointed out this weed to me, but *wabal* grass is common everywhere. It has long stringy roots like our Cambridgeshire twitch grass. In fields it weakens the crop even if no nearer to it than a yard or more,

possibly because of its deep roots. Though it is a great nuisance it is good fodder (*'alaf*) for beasts. Dr Said H. Mawly thinks *wabal* looks like *Cynodon dactylon*, and another weed *zibrah* he thought must be *Cyperus rotundus*, but as the samples I sent him were without infloresence he could not be certain. *Zibrah* he thought is a sedge not a grass, and both weeds serious and difficult to control. The action of getting rid of *wabal* is *yuwabbil-uh*. MM., 40a, says of it, 'It is the firmly rooted (*muta'aṣṣil*) grass which spreads (*yisraḥ*) in the ground and remains there without its roots leaving, and it is the greatest pest (*āfah*) to all crops/cereals (*zar'*).'

197. I.e. without a space between the first scatter-sowing and the next.

198. I.e. from a single channel (*sāqiyah*).

199. The *sharīm* knife (infra p. 57) generally used in all south Arabia and even found in pre-Islamic excavations at Ḥuqqah. If the lucerne be not cut at ground level it rots.

200. Reading thus from MM., 29b.

201. *Ṭahaf* is known to Aḥ b. Yaḥyā b. al-Murtaḍā (*ob*. first half of the 9th/15th cent.), *al-Baḥr al-zakhkhār*, Cairo, 1949, iii, 402; Landberg, *Arabica V*, Leiden, 1898, 213; Grohmann, op. cit., i, 215 for dates of sowing etc.

202. MM., 29b, reads for 'good places', 'good ground'.

203. *Jirn* is local pronunciation; *mijrān* is also used.

204. There are three alternative readings here – *m l j t, m l ḥ t, m ḥ l t*, the last from MM., 29b, means 'smeared, coated'.

205. The following section resembles MM., 31a. A 1954 note from Haitham b. Sa'īd of Dathīnah runs, 'Sesame has two seasons (*mawsim*), the first, sown from the end of Rabī' (probably March–early April) up to Ṣaif, being called Qaiḍī sesame (*gulgul*), and remaining in the ground seventy-five days. The second is Kharīfī sesame (*simsim*), remaining in the ground ninety days. To return to Qaiḍī sesame – you will see its taste after being pressed is delicious, and its colour white. The second variety however, the Kharīfī, has a bitter taste predominating and is red in colour. The reason for that is because the Qaiḍ (variety) comes in the Ṣaif season when the sun's heat is intense so that it ripens in seventy-five days, whereas the Kharīfī comes in the days of Kharīf which have much rain and cloud so that its proportion of fruit is small.'

206. The addition is made from the parallel passage in MM., 32a, which reads, 'And bound into sheaves and stacked (*yukhaiyam*), each stack consisting of a load and more than that.'

207. *Ḍāḥī* (pl., *ḍawāḥī*) is what is irrigated by rain (*mā yusqā bi-'l-maṭar*), while *sāqī* is what is irrigated by running streams (*alladhī yashrab bi-'l-ghail*). *'Aqar* (cf. fn. 91), synonym *al-ba'l*, is land sown in Shitā' with the rain of Kharīf (*yuzra'fī 'l-shitā' bi-maṭar al-kharīf*). (Qāḍī Ismā'īl).

208. Cf. note 170. A possible variant seems to be *matlam* (pl., *matālim*) which appears also in Ḥaidarah's Ta'izz almanac, and of course *tilm* is in common use.

209. This word is difficult, but the Cairo reading *yurtadd* is not supported by MM. *Irtadam* is 'to fall down together on top of one-another'. *Irtadam al-saqf/al-qabr*, the roof/sides of the grave, fell in. (Qāḍī Ismā'īl). Dozy, *Supplément*, says *irtadam* means heaped up.

210. The Ta'izz almanac says that white *gulgulān* is sown on the 3rd Nīsān (16 May), *simsim* is sown on the 30th of Aiyār/Mabkar (12 June), *gulgulān* is plucked on the 26th Ailūl (9 October). In my 'Star calendars . . .' 446, the earlier version of the almanac says *julul̄an* is sown in Tihāmah at the running springs on 17 February. Possibly the apparent discrepancy in dates is because the practice varies from district to district.

211. Cf. para. 1 of the section on sesame, supra.

ADDENDA

p. 60, n. 1: the manuscript will be published in facsimile in the Gibb Memorial Series, edited by G. Rex Smith and Daniel Varisco.

p. 60, n. 10: see article III above.

VIII

SOME IRRIGATION SYSTEMS IN ḤAḌRAMAWT

How deeply questions of irrigation affect the daily life of the peoples of southern Arabia was first impressed upon me when the summer floods came down to am-Fajarah in Ṣubaiḥī country lying west of Aden, where I happened to be stationed at the time. The villagers turned out to argue, with some violence, over the distribution of the flood-waters. It was, however, as Shaikh Durain,[1] a Lahej official with us in the village, pointed out, without weapons that the villagers had come to the fray, and though the women stood behind, disputing the issue no less fiercely than the men, egging them on, so that all would doubtless have come to blows, men and women alike, there would have been no stabbings or shootings. Shaikh Durain cynically implied that there was more of sound than fury in the rencontre, but of course this is not always so, and disputes arising over rights to the use of water can lead to blood-feuds. Since those days I have made some investigation into irrigation and its wealth of technical vocabulary in various parts of the Aden Protectorate, studying some systems in more or less detail, especially those near Mūdiyah village in Dathīnah.

The territory has a very wide variety of irrigation systems, ranging from the terraced hill-sides of Yāfi' and the Yemen, or other places, to the *qanāts* near Ghail Bā Wazīr, but my inquiries have been mostly concerned with flood irrigation in the *wādīs*, and not into the perennial streams (*ghail*), and only a little into terraced systems. Even where the systems themselves are nearly identical, the technical vocabulary differs from district to district, an important consideration when applying the said vocabulary of present-day southern Arabia to the interpretation of the 'Ḥimyaritic' inscriptions relating to irrigation, since each individual inscription is most likely to be interpreted correctly in accordance with the dialect of the district in which it was set up, given the fact that the population has changed little in composition from the earliest period of Islam.

Water-lore figures quite frequently in folk wisdom, especially that relating to wells. In Ḥaḍramawt one says, *Saiwūn bi-'l-mā' wa-lā saman Shibām* 'Rather water in Saiwūn than ghee in Shibām'. The allusion is to the sulphurous taste of Shibām water so that it is preferable then to live in poverty in Saiwūn than in affluence in Shibām! Of the salty brackish wells of Bōr, 'Abbūd of al-Ghuraf says,[2] *lā yisqūn-ak al-ḥālī*. The type of well found in a village might give rise

[1] *Durain* means 'fox'.

[2] R. B. Serjeant, *Prose and poetry from Ḥaḍramawt*, London, 1951, Ar. text, p. 62, verse 10.

to a nickname for it. Yet again, the author of *al-Ḥayāt al-saʿīdah* [3] quotes a saying,

إن مطرت والّا سَنَيْنا

(ونحن) إن مطرت والّا بكينا[4]

' If it rains (good) ! Else we water by well.
If it rains (good) ! If not then we weep '.

The allusion is to the *sināwah* when men, women, and animals draw water to irrigate the fields, and the hardness and tragedy of life if rains fail. Snouck Hurgronje has already published an excellent study on the Ḥaḍramī well [5] and it is not with this aspect of irrigation that we are here concerned, but let it be said that water diviners are naturally well known in southern Arabia and that a certain class of miracle in the hagiologies is concerned with discovering water or predicting the arrival of a *sail.*

Irrigation problems constitute an important section of the great *fatāwā* collections in the Yemen, be they Zaidī or S͟hāfiʿī, and this from the earliest period at which we possess legal works. In the *Kitāb al-Muntak͟hab* of the first Zaidī Imām [6] there is a section, *Bāb qismat al-māʾ baina 'l-ḍiyāʿ*, on the division of water among (various) holdings/properties. To the Zaidī sources cited by Rossi [7] should also be added *al-Baḥr al-zak͟hk͟hār* of Ibn al-Murtaḍā,[8] and of course all the MS Zaidite legal works in the British Museum, the Vatican, the Ambrosiana, and other large collections.

S͟hāfiʿī legal treatises and *fatāwā* collections are especially numerous and voluminous in Ḥaḍramawt. In the Āl al-Kāf Library at Tarīm I should cite *al-Bayān* of al-ʿImrānī [9] and al-Nawawī's *al-Rawḍah* [10] ; al-ʿImrānī is a well-known medieval S͟hāfiʿī authority from the Lower Yemen. There are also the

[3] *Al-Ḥayāt al-saʿīdah bi-Ḥaḍramawt*, Singapore, 1953, 4. This was sent me by Saiyid ʿAlawī b. Ṭāhir of Johore and might be his own composition. It is reviewed in the Aden paper *al-Nahḍah.* The anonymous author is critical of government activities in irrigation and anti-erosion work on technical grounds, and makes certain proposals of his own.

[4] If the bracketed word which the text includes as a part of the quotation be omitted, the verses are *rajaz* type.

[5] C. Snouck Hurgronje, ' L'interdit séculier (*rifgèh*) en H'adhramôt ', *Revue Africaine* (Alger), XLIX, 256, 1905, 92–9.

[6] See ʿAbd al-Khaliq Kazi, *A critical edition of the Kitāb al-Muntak͟hab fi 'l-fiqh*—a collection of the answers of the Zaidī Imām Yaḥyā b. al-Ḥusain to questions by Abū Jaʿfar Muḥ. b. Sulaimān al-Kūfī, thesis submitted for London Ph.D., 1957, 171.

[7] E. Rossi, ' Note sull' irrigazione, l'agricoltura e le stagioni nel Yemen ', *Oriente Moderno,* XXXIII, 8–9, 1953, 349–61.

[8] Aḥmad . . . b. al-Murtaḍā, *al-Baḥr al-zak͟hk͟hār*, Cairo, 1949.

[9] Abu 'l-K͟hair, Yaḥyā b. Saʿd b. Yaḥyā al-ʿImrānī, *al-Bayān fi 'l-furūʿ*, cf. Brockelmann, *GAL,* I, 391, Suppl., I, 675. He died in 558/1163 at D͟hu 'l-Sufāl. A copy of his *Bayān* is also available in the British Museum.

[10] cf. *GAL,* I, 396.

fatāwā collections of Abū Makhramah [11] and my MS of Muḥ. b. Ibrāhīm b. Ju'mān [12] but beside these there are many others.

In the late fifteenth or early sixteenth century four well-known questions on water law seem to have been sent to *'ulamā'* abroad from the Shāfi'ī part of the Yemen, though why these should have been well known in legal circles I have not as yet discovered. They run as follows :

١. (سؤال) فيما يملك من المياه المباحة وما لا يملك.
٢. (سؤال) فيما يتعلق بحكم الأراضى التى تسقى تشرب دفعة واحدة.
٣. (سؤال) فيما يتعلق بحكم الأراضى التى تشرب على التعاقب.
٤. (سؤال) فيما يتعلق بحكم انقسام المآء بين الشركآء فى سواقٍ متعددة.

1. Concerning what is possessed of common waters, and what is not possessed.
2. What appertains to the law of lands which are irrigated/drink at one time.
3. What appertains to the law of lands which drink in succession after the other.
4. What appertains to the law of the division of water between partners in various *sāqiyah*-channels.

If my surmise be correct these questions were answered first [13] by Ibn Ḥajar al-Haithamī (*ob.* 973/1565) in a MS which I found in the Āl al-Kāf Library at Saiwūn and which has not yet been reported as existing elsewhere,[14] and by Abu 'l-Ḥasan Muḥammad Tāj al-'Ārifīn al-Bakrī al-Ṣiddīqī (tenth/sixteenth century) in a MS at Batavia [15] of which I have a transcript.

In the Rasūlid kingdom it was the duty of the official known as *mushidd* of a district to look after the *'imārat al-wādī*, i.e. the maintenance of its irrigation works, including barrages (*'uqmah*) of earth and deflectors, details of which

[11] 'Abdullāh b. 'Abdullāh b. Aḥmad . . . b. Ibrāhīm Makhramah al-Saibānī al-Jūhī (1501–2–1564–5) a Ḥaḍramī author ; photocopy in SOAS Library.

[12] He died in 1034/1624–5. Cf. Muḥ. b. Muḥ. . . . Zabārah, *Mulḥaq al-Badr al-ṭāli'*, Cairo, A.H. 1348, 9.

[13] A MS work by 'Alī b. 'Abdullāh al-Samhūdī (*ob.* 911/1505) entitled *al-Anwār al-sanīyah fī ajwibat al-as'ilat al-Yamanīyah* (E. Lévi-Provençal, *Les manuscrits arabes de Rabat*, Paris, 1921, p. 249), I had thought might be another in the same series, but through the courtesy of Muḥammad b. Tāwīt who sent me a microfilm of the MS I have discovered that it does not deal with irrigation but with religious questions.

[14] Numbered Fiqh no. 85, and entitled *al-Ajwibat al-ḥasanah wa-jawābāt Ibn Ḥajar 'an al-as'ilat al-Yamanīyah.*

[15] P. S. van Ronkel, *Supplement to the catalogues of the Arabic MSS preserved in the Museum of the Batavia Society of Arts and Sciences*, Batavia, 1913, p. 292. His son Muḥammad it probably is whose obituary notice appears in Ibn al-'Aidarūs, *al-Nūr al-sāfir*, Baghdad, 1934, 414, under the chronicles for the year 993/1585. He himself figures in a list of Meccan *'ulamā'* including Ibn Ḥajar and several South Arabians, in Ulughkhānī, *An Arabic history of Gujerat*, ed. E. D. Ross, London, 1910–28, II, 637.

are given, the upper land 'drinking' before the lands lying lower down the *wādī*.[16] One is not to interpret this, of course, as meaning that the *mushidd* carried out this work himself or with subsidies provided for this purpose by the government, but simply that he saw to it that the work was done. The object of the central government was to ensure that there should be no diminution in its revenues through the neglect by the proprietors to maintain their bunds. I have heard it said that the tribal overlords in Ḥaḍramawt in former times were equally insistent that maintenance be regularly effected.

South Arabian historical literature contains many scattered references to irrigation which if collected could supply us with valuable data. As an example there may be cited *al-Nūr al-sāfir* [17] discussing the Ma'rib Dam and the *Sail al-'Arim* :

فبنـوا بين الجبلَيْن سدًّا من الصخر والقـار ونزل المآء العظيم خـارج السد، وجعلت في السد مشاعب أعلى وأوسط وأسفل ليأخذوا من الماء كلما احتـاجوا إليه فحرث داخل السد.

'They constructed a dam of rock and pitch between the two mountains, and the bulk of the water (?) came down outside the dam. They placed in the dam channels, at the top, middle, and bottom, to run off the water whenever they required it, and there was ploughing inside the dam.' I have not read that pitch was used in the construction but lead dowels were certainly employed.

One of Ettore Rossi's last tasks undertaken was to publish his notes on irrigation in the Yemen [18] ; these are excellent observations, but a comparison of the technical vocabulary he reports shows that it has little in common with the Ḥaḍramī data below. *Ghail* irrigation, on the other hand, resembles in certain aspects what is known from the Hejaz,[19] and even North Africa where the *ṭāsah* or cup used to measure the flow of water is also in use, for instance, south of the Atlas mountains. Lastly, two valuable theses have also recently been compiled on the pre-Islamic material which should make it easier for comparisons to be made between the existing and the ancient irrigation systems.[20]

[16] *Mulakhkhaṣ al-fitan*, fols. 11 r. and 12 v. Cf. Claude Cahen and R. B. Serjeant, 'A fiscal survey of the mediaeval Yemen', *Arabica*, IV, 1, 1957, 22–33.

[17] Ibn al-'Aidarūs, *al-Nūr al-sāfir*, Baghdad, 1934, 69.

[18] See p. 34, n. 7.

[19] cf. my 'Two sixteenth-century Arabian geographical works', *BSOAS*, XXI, 2, 1958, 257.

[20] A. K. Irvine, *A survey of old South Arabian lexical materials connected with irrigation techniques*, thesis submitted for Oxford D. Phil., 1962 ; M. A. Ghul, *Early southern Arabian languages and classical Arabic*, thesis submitted for London Ph.D., 1962. For a survey of earlier materials, see A. Grohmann, *Südarabien als Wirtschaftsgebiet*, Brünn, 1933, II, 19–33, 'Bewässerung', but the whole section 'Landwirtschaft' is not irrelevant. A more recent and general survey, Dante Caponera, *Water laws in Moslem countries* (FAO Development Paper No. 43), Roma, 1954, has little specific information on South Arabia.

Irrigation at Ḥuraiḍah

The valley of Ḥuraiḍah is called the Wādī Tajrūb, but it appears more commonly perhaps to be known as 'Aibūn, the latter name alone figuring on von Wissmann's map. Its flood-water would run into the main *wādī*, 'Amd. It might be observed at this point that two kinds of *misyāl* or flood-bed (and no doubt many more) are distinguished. The first is a *raḥbah* (pl. *riḥāb*), described as a *majrā al-mā'* 'covered with stones', a Daw'anī word.[21] This is to be compared with the ancient place-name Urḥūb [22] which appears in the *Kitāb al-Muḥabbar*.[23] The second kind, a *ḥidbah*,[24] is a water-course without a stony bed. Both words give rise to place-names, and my impression is that they are not confined to the Wādī Daw'an. At Ḥuraiḍah irrigation is confined to flood-water, since the wells are far too deep to provide water for irrigating the fields. The flood is deflected from the *wādī* down a *sāqiyah*-channel to the fields. The 'Aṭṭās Saiyids of Ḥuraiḍah sketched out for me two patterns of field-irrigation (figs. 1–2), not, of course, scale drawings, nor do they attempt to portray any particular field lay-out, although I surmise my informants probably had certain fields in mind, for the composite aerial view of Wādī 'Amd published by Miss Caton-Thompson [25] does seem to show, just above and to the south of Ḥuraiḍah, something resembling the field-patterns here.

To consider first the more elaborate field-pattern (fig. 1), there is at the head of the *sāqiyah*-channel deflecting the flood-water from the *wādī*, known as *rās al-sāqiyah*, and before the water enters the field, a little hole known as *minkī*, usually filled up with loose stones, etc., which can be opened up to divert the water and stop it flowing into the field area should this have taken sufficient water. In front of the actual entrances into the fields a stone apron, in a sort of rough cobbling, called *raṣa'ah*, is built into the *sāqiyah*. In Saiwūn such an apron of stones is called *raḍīkhah*, be it in front of the *maqāsim al-mā'*,[26] the channels leading off the water to various gardens, or in front of a gate. In Saiwūn the *maqāsim al-mā'* are constructed in stone, the stone floor or step between being called, it was said, a *raḍīkh* (fig. 3).

The divisions at the openings to the fields are made with great precision so that the flow of water into them is very accurately controlled. They seem to have special technical names. A *bidd* (pl. *budūd*) [27] is an open channel, masonry-lined, into a field, the word being used in Tarīm and elsewhere. A plank (*sufrah*), normally lying on the bank by the place where it is to be

[21] *Al-Ḥayāt al-sa'īdah*, 8. Irvine, op. cit., 253, quotes *rḥbm* from C540 as a proper name, but it could be taken in the sense given here.

[22] cf. the form Ukhdūd, *BSOAS*, XXII, 3, 1959, 572–3.

[23] A. F. L. Beeston, 'The so-called harlots of Ḥaḍramaut', *Oriens*, V, 1, 1952, 17.

[24] A *ḥadbah* was said to be a *qārah* (prominence) *aw bilād ṣaghīrah*.

[25] G. Caton-Thompson, *The tombs and moon temple of Hureidha*, Oxford, 1944, plate I. See also pp. 9–16, 'Ancient irrigation in the Wādī 'Amd', and pl. LXXII.

[26] See *Bibliotheca geographorum Arabicorum: indices, glossarium*, Leiden, 1879, 325, for *maqāsim al-mā'*.

[27] In Sa'īd of Upper 'Awlaqī territory (Wādī Yashbum), *bidd* (pl. *bidād*) is a channel.

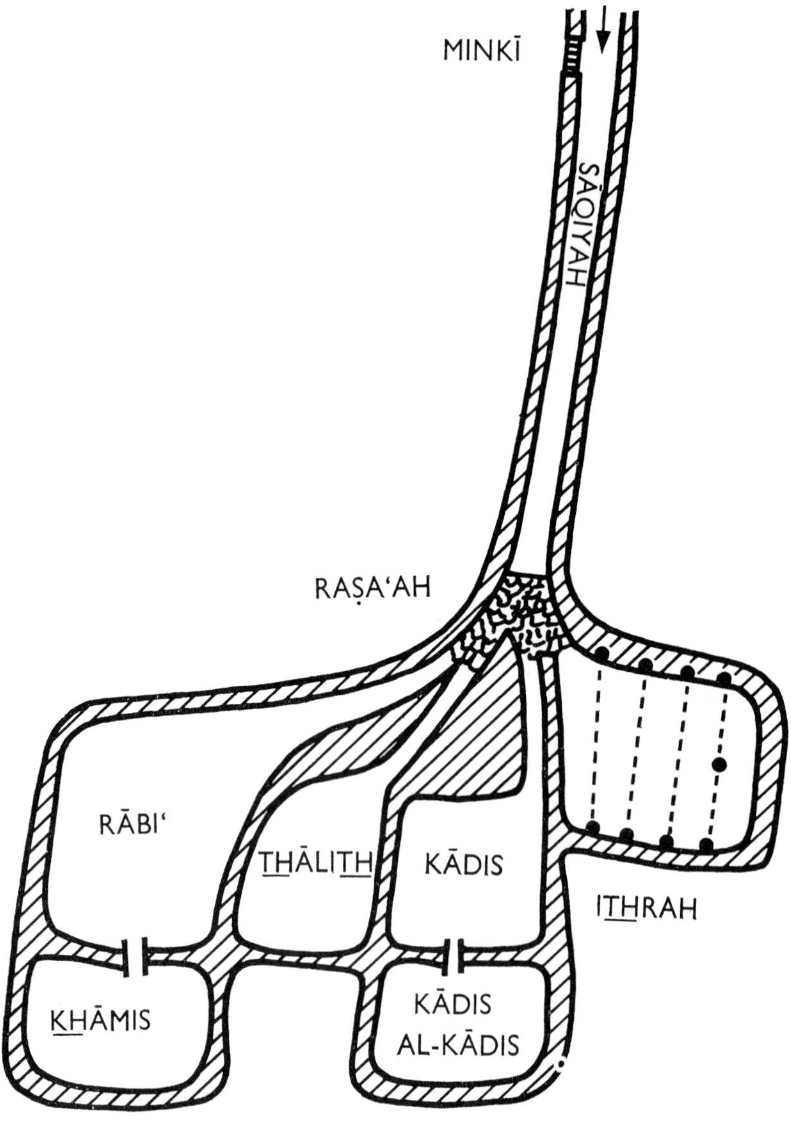

Fig. 1. A *sāqiyah* leads to the fields, with a *minkī* at the upper end which can be opened to allow surplus water to spill away. A stone-paved *raṣa'ah* lies at the openings on to the fields, each of which has its own name in the series. The *ithrah*-field is shown as divided into strips belonging to several different persons; these are marked by stones (*awthān*) set in the clay bund (*sawm*). These strips are called *khars*.

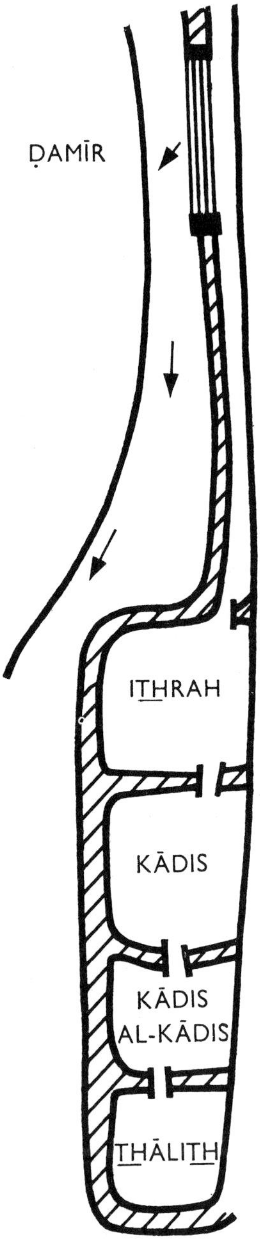

Fig. 2. A *sāqiyah* leads to the fields with a stone *ḍamīr* or barrage, possibly like that shown in Caton-Thompson, op. cit., pl. ii, over which water surplus to the fields' requirements will spill and be led to fields lower down the *wādī*. The fields from the *ithrah* to the *thālith* ' drink ' in sequence. A cutting is shown in the lower bank of each field, but its exact location would depend as to where it is most suitable to cut without causing damage or loss to either field.

employed, is inserted into the *bidd* when it is desired to stop the flow of water into the field, being fitted into grooved projections (*ṣadāwid*) (fig. 4) in the *bidd*. When in full spate this is quite a difficult, even a dangerous task. A *ḥarrah* (pl. *ḥirār*) is a channel with, as it were, a kind of stone bridge over it (fig. 5). At Ḥuraiḍah all works such as the *bidd* or *ḥarrah* appear to be plastered with *nūrah* or held together with some sort of cement ; they are *marṣū' bi-'l-ḥajar*,[28] revetted

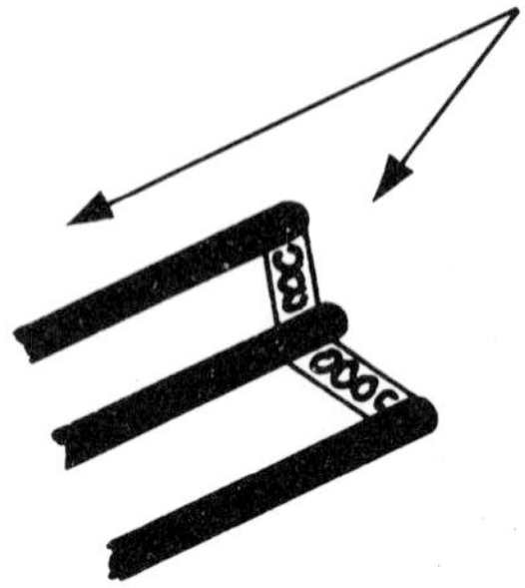

FIG. 3. The stone heads of the deflector channels (*sās*) leading from the flood-course, with a stone-paved *raḍīkh* running across the entrance to each channel to prevent erosion.

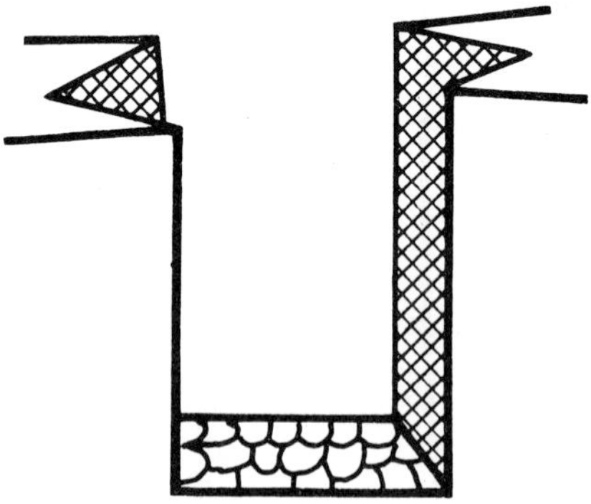

FIG. 4. Grooved projections (*ṣadāwid*) into which the sluice-plank fits, at the entrance of fields.

with stone, which is described as *al-ḥajar al-mugharraq bi-'l-nūrah*, stone overlaid with plaster.

The nature of the terrain conditions the number of fields which can be irrigated by any one *sāqiyah* ; in this case there are six, and the order in which they would ' drink ', as Arabs say, is as follows :

1. The *ithrah*.
2. The *kādis*, from which the flood flows into the *kādis al-kādis*.
3. The *thālith*, lit. ' third ', in order of ' drinking ' presumably.
4. The *rābi'* ' fourth '.

[28] cf. *marṣad* in p. 54, n. 73, and *ḥarrah* in *al-Shāmil*, 203.

5. The *khāmis* ' fifth '.
6. The *sādis* ' sixth ', also called *khadūd*.[29]

My notes unfortunately do not state precisely whether the *kādis al-kādis* is the same as the *sādis/khadūd*, though I think not, but it was affirmed that the *kādis al-kādis* drinks last of all the fields in this lay-out. The reason put forward is that the current is still strong when it comes to the second and third fields, but it is weak when the flood-water is turned to irrigating the fourth and fifth fields, and it is only when it has weakened a bit that the sixth field is allowed to ' drink '. I imagine this may be for one reason or another, or even both—namely that it is axiomatic that the higher ground ' drinks ' first in all systems known to me and the *kādis al-kādis* must lie below the level of the *kādis*, or else because a cut in the bank of the *kādis* while the flood is still violent might lead to the

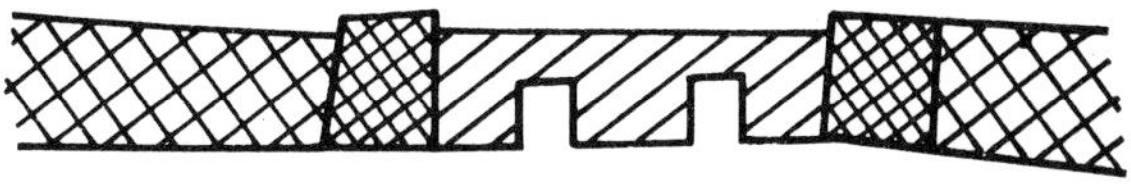

Fig. 5. A *ḥarrah* or stone bridge over the entrance for flood-water to a field.

erosion of the *kādis* into the *kādis al-kādis*. At Tetuan in Morocco I chanced to learn that the word *kaddūs* means a water-pipe, though this may only be a coincidence.

In Dathīnah where the banks or bunds of the fields are of clay in the systems that I studied, the owner of a field lying below another has the right to cut into the bund (*sawm*) of the upper field to water his own field. This right is of course subject to certain defined conditions which form a part of the customary law of the district. Some fields in Dathīnah, however, simply have a *takhrīj* made of stone, an exit over which the surplus water can spill and run away. In Ḥuraiḍah they have correspondingly a *mithwā* or *mansam*, the height of which was stated to be a *dhirā*' or a *dhirā*' and a half (approximately 18 or 27 inches). I have not recorded whether these are merely places in the clay bank, or whether they are made of stone. *Mansam* has, linked with it, a verb, *yinassim-uh* ' he opens it up ', so it may be merely of clay. I have also noted that the local name *faqīyah* seems to be more or less equivalent to the Dathīnah *maqṭa*' and that it is usually covered with clay (*marbūdah*) ; one speaks also of a *fatḥah marbūdah*, an opening covered with clay, from the verb *rabad*, *yarbad*.[30] Where there is a *minkī*

[29] cf. p. 37, n. 22. This word appears as a name on von Wissmann's map, near Ḥuraiḍah. In Dathīnah a *khudād* is a sort of furrowing in the ground made by a *sail*, always long-wise of course, a synonym of which is said to be *washar*, a ' sawing, saw-cut ' as it were.

[30] Landberg, quoting Hein, denies that the latter's assertion that *marbadah* means *siqāyah* is correct, but I think Hein is probably right, except that it may conceivably be so named because *siqāyahs* are covered with *nūrah* which makes them conspicuous. I think *rabada* should be taken as equivalent to *rabaṭa* ' to bind or tie ', and consider that the Ḥaḍramīs regard the *fatḥah* as bound together with clay. In Tarīm *rabad* means to close the breach of a clay bund (*sawm*) to make it hold water.

at the head of the *sāqiyah* channel, however, some systems at Ḥuraiḍah have no *mit͟hwā* since the surplus water can be diverted before it reaches the fields, and thus probably the risk of erosion will be lessened. A *maʿd͟har* is an emergency opening in a bank, filled with stones until it happens to be needed. The stone filling is then removed, if, for instance, the banks are in danger of being washed away, and the water is allowed to flow through it. It is a bound masonry channel.

A single field may belong, for instance, to ten persons, as in the case of the *it͟hrah* in fig. 1. Each of these may perhaps in turn divide his share into *maṭā'ir* (s. *maṭīrah*) [31] or plots, each separated from its fellows by a small bank (*mag͟hmas*). This *mag͟hmas* is covered by water at flood-time. The act of dividing the field among several owners is known as *tak͟hrīs*, and each strip as a *k͟hars*. Landmarks (*awt͟hān*),[32] stones or little heaps of small stones,[33] are set in the actual bund (*sawm*) to show how each individual strip of land runs. Sometimes a strip may be divided down the middle into two halves, called *yaks* (pl. *yukūs*), by a mark (*ʿalam*) known as a *qāriʿ*. I have recorded the phrase, *yarudd yiqraʿ ṣāḥib al-yaks*, probably meaning that the owner of the *yaks* may further subdivide the strip with a *qāriʿ* mark. In this last case the crop is sown, and at harvest time one man stands at each end of the field where the boundary stones are, and another walks between them—their three heads must be kept in line, and the middle man scores the ground to indicate where the dividing line should be made in the crop. The distance between bund and bund is known as *s͟haḥb*.

In order to allow the *it͟hrī/it͟hrah* to ' drink ' first, the other channels are stopped up with stones, this filling being known as *maʿs͟hā* (verb, *yaʿs͟hawna-hā* ' they stop up the channels with stones '), but these stones are removed from the *kādis* channel when it is time for it to ' drink '. The other fields drink in sequence. One single ' drinking ' of a field is known as *ziḥī*, but this term may be applied to the first and to the second ' drinking '.

If the first field to be irrigated with flood-water is arable soil (*ṭīn*) only, i.e. contains no palm trees, then you give it *mā' qaṣab-ah* only, lit. ' the water of its stalk ', and no more, for otherwise this would damage the *qaṣab*, i.e. millet-stalk which, of course, forms the staple animal fodder of the country.

31 The *maṭīrah* (in Tarīm 21 *d͟hirāʿ* square) is probably of an extent more or less fixed over all Ḥaḍramawt. ʿAlawī b. Ṭāhir, *al-S͟hāmil*, 173, discussing a certain *maṭarah*, apparently a large basin (*ḥawḍ*) which fills with water at flood-time, states its area in *maṭīrahs*, adding that the *maṭīrah* is 49 *d͟hirāʿ* square (nearly 79 feet); this measure relating presumably to the Wādī Dawʿan. The *Tarjī al-aṭyār* of ʿAbd al-Raḥmān b. Yaḥyā al-Ānisī al-Ṣanʿānī, ed. al-Yaḥṣubī and al-Fā'is͟hī (Cairo, A.H. 1369), 234, makes *al-maṭīr* equivalent to *al-mamṭūr*, quoting a verse *wa-lā zāl maṭīr*.

32 cf. Rossi, ' Note sull' irrigazione ', 359; Landberg, *Arabica*, v, Leiden, 1898, 146 seq., and *Gloss. daṯ*. A legal text I saw in Mūdiyah contains the phrase حول مقسوم بين شخصين متميّز بالأوثان. This perhaps may be read *ḥawl* with *Gloss. daṯ*., in the sense of a rain torrent, but Lane gives *ḥiwal* as a furrow or trench in the ground in which palm trees are planted. A Kibsī *saiyid* told me that a *ḥawl* simply means a *jirbah*, a field, and is still used in this sense in the Yemen.

33 On the Jōl west of the Wādī Dawʿan, Bedouin called a small heap of stones, approximately a foot high, where several paths branched off, a *kūt*; it had been made to indicate the correct path.

The *ḥazar*, explained as *qadr irtifāʻ al-māʼ*, the height of the water (at its maximum), is a *dhirāʻ*, approximately 18 inches for *qaṣab*. The water can then be diverted to the next field which might contain palms (*nakhl*) ; these would be given water in accordance with the height of the *sawm* or bund and its ability to retain the water.

A less complicated lay-out of fields is shown in fig. 2. At the top of the *sāqiyah* channel there is a *ḍamīr* [34] or barrage made of stone along one side of the channel. Clearly if the flood-water rises to the height of the *ḍamīr* it will spill over the top of it and be led away to another group of fields or go to waste. The flood-water required for the fields in the diagram would, however, pass down the *sāqiyah* and water the *ithrah*, then the *kādis*, and so on, each of these fields in sequence, the level of each field being somewhat lower than the field preceding it. To judge by Dathīnah, the point of exit of the flood-water from one field to the next would be selected in accordance with the terrain, and in Dathīnah if the flood-water were only sufficient to irrigate the top field, the others would not drink. This is likely to be the same in Ḥuraiḍah.

The ʻAṭṭās Saiyids told me that if people are afraid of excess water coming into a group of fields they take the stones out of the top course of the *ḍamīr* (*yunaiyiqūn al-ḍamīr*) which might be about 20 *dhirāʻ* (approximately 30 feet) across. The course of stones is called *ḥabl* in Ḥaḍramawt, and it is a word known also to ancient southern Arabia.[35] This causes the water level to be lowered and it does not rise to the fields they wish to protect. There is a saying, *al-maksar fi 'l-ḥajar mā yaḍurr* ʻ damage to the stone is of no account ʼ. The meaning of this maxim is that the stone barrage is easily repaired, but if the water attacks the soil (*ṭīn*) it scoops it out (*yajruf-uh*), i.e. erodes it, and this causes great damage, one might add in extreme cases, irreparable damage.

Where water is likely to scoop out or erode a clay bund (*sawm*) it is given

[34] Landberg, *Ḥaḍramoût*, Leiden, 1901, 184, gives a plural *ḍumur*, and describes it as a ʻ digue transversale pour faire entrer l'eau dans les champs ʼ.
يبْنوه بالحَجَر والنوره على عُرْض الوادى عل شان الماء يَعْشِى وُيطْلَع بذَبِر البلاد.
I have recorded a plural *ḍumūr*.

[35] cf. G. Ryckmans, ʻ Sabéen *ḥbl* = accadien *abullu* ? ʼ, *Archiv Orientální*, XVII, 2, 1949, 310–12. This article discusses a brief inscription, ʻ sur une grande pierre faisant partie des assises extérieures des grands temples de Ṣirwāḥ ʼ, from the Yemen. Ryckmans proposes that *ḥabl* is a term of construction equivalent to the Akkadian *abullu*, i.e. a large door, gate. Muḥ. b. Abī Bakr al-Shillī, *al-Mashraʻ al-rawī*, Cairo, A.H. 1319, I, 127, discusses a well known as al-Bīr al-ʻAlawīyah at Bōr dug by a celebrated Saiyid who طواها بحجارة كبار وكتب اسمه على كل حجارة من الحبل الأعلى وهو المدما ʻ lined it with large stones and inscribed his name on every stone of the top-most *abl*, which is the *midmāk* (vocalization dubious) ʼ. Lane gives *dumlūk* as applied to a stone in the sense of ʻ smooth, even ʼ. At Saʻīd of Yashbum it seemed to be the name of the two facing walls of a stone deflector, with a core of earth or rubble, this filling known locally as *wajlah*. The workmen called these two walls *ḥablain* (fig. 6). They used a sledge (*ʻishshah*) to carry the stone for the work, and the usual palm-leaf baskets for conveying the *wajlah*. Clearly this sense fits the phrase of *CIH*, 343, 5 *ʻdy ḥblthmw wʻbrthmw*. Cf. the discussion by G. Ryckmans, ʻ Epigraphical texts ʼ, in Ahmed Fakhry, *Archaeological journey to Yemen*, Cairo, 1951–2, II, 2. The Ḥaḍramī author

a stone revetment or facing called a *ḍil‘*; one says *yiḍla‘* ‘he makes a *ḍil‘* (revetment)’.[36]

A burst in the clay *sawm* or bund is known as a *hajīyah* (fig. 7); this is repaired by building a clay bank round it, the shaded portion of fig. 7. *Ṭawl* is the flowing of water over the top edge of the bund, thus damaging it, and *ghawl* is a sort of burrow in the bund through which the water enters the bank to destroy it. When this happens the owner of the field shouts out, *Yā fulān,*

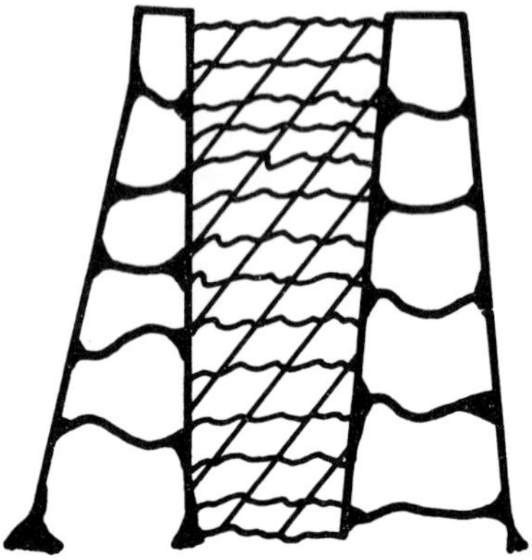

FIG. 6. Stone deflector (cross-section) at Sa‘īd, two external masonry walls (*ḥablain*) with a rubble core (*wajlah*).

yā fulān, maksar! maksar! ‘Oh, so-and-so, a break! a break!’ Those addressed all come and help him to repair it. *Ṭawl wa-ghawl* would form the sort of cliché in which Arabic delights and abounds. It is in Ḥaḍramawt part of share-cropping contracts that the cultivator should *yarudd al-makāsir* ‘repair the breaks’.

Silt is brought down by all the floods, known as *ṭafal, ṭamī, ghirain, gharyal,* according to the author of *al-Ḥayāt al-sa‘īdah*[37] though he does not specify

of *al-Mashra‘* evidently understands the word as a course of stones, but the sense in Yashbum is a little different. In the Tarīm area it was used of a course of stones in a barrage. *Al-Mashra‘*, I, 141, alludes to *shafīr al-bīr*, the ‘lip’ of the well. According to *al-Shāmil*, 177, the Ḥaḍramīs excavate for the foundations of a house until they come to the *nudūwah* or damp earth. They then lay the stone foundation, course (*ḥabl*) upon course, the lowest course being called *ḥabl al-nudūwah*.

[36] See p. 51.

[37] *Al-Ḥayāt al-sa‘īdah*, 1953. *Ghirain* is a word cited also in Muḥ. ‘Alī Luqmān, *Qiṣṣat al-dastūr al-Laḥjī*, Aden, 1952, 7. W. S. Blackman, *The fellāḥīn of Upper Egypt*, London, 1927, 146, records *ṭafl* as a yellow clay used as slip in the Nile valley. This silt is no doubt beneficial to the ground, since in Ḥuraiḍah the earth is cut up and *ṭifil* (colloquial pronunciation) is used instead of dung. Dung is not used in Ḥuraiḍah because of its heat except in places frequently watered. In the Yemenite *Bughyat al-fallāḥīn* (see p. 46, n. 44) there are many instructions for the application of dung to the soil, and human excrement is even sold for this purpose in some of the larger towns of Ḥaḍramawt.

in which districts these words are current. In reviewing [38] Bowen's article on 'Irrigation in ancient Qatabân', actually at Baiḥān, I have suggested that this silt which is piled up at the sides of fields to keep them at their existing level may, in many cases in South Arabia, have raised the area of the bunds to a point where it is no longer economic to continue cultivating the ever shrinking piece of land, hence abandoned irrigation systems may have been the result of silting rather than neglect or political disturbance. The action of piling the earth deposited on the surface of the field by the flood-water on the bunds is called *ṭarḥ*, and the bank where it is deposited is known as *ḥawmil.*

FIG. 7. A *hajīyah* or burst in the clay bund or *sawm* showing a repair bank (shaded).

Each man knows which is his own *ḥawmil*, but this is an issue upon which disputes sometimes occur. No earth may be deposited on a bank of a *sāqiyah*-channel. The reason for this rule seems to me obvious, namely that if it falls into the channel in any way the latter may become silted up and be raised to a level where it is too high for the water to irrigate the fields, thus causing expense to clean it out which would of course fall on all proprietors of fields watered by the *sāqiyah.*

On Pemba Island I was shown several Ḥaḍramī MSS by Saiyid Hādī b. Aḥmad al-Haddār of Āl al-Shaikh Bū Bakr b. Sālim, one of which mentioned *ḥaithat al-sāqiyah*, and the verb *ḥaiyath* which was explained to me as *akhadh al-turāb/al-ṭīn* 'to remove the new earth', and equivalent to the word *al-ṭarḥ*, would be the phrase that occurred in the same MS, *ilqā' ṭīn al-ḥaithah bi-'l-arḍ.* One says *ḥaiyath al-nakhlah* 'he removed the new earth from (the roots of) the palm'. I do not know if these terms are also current in Ḥuraiḍah.

In Ḥuraiḍah they mentioned two kinds of soil to me, *ṭīn*, the word always used for cultivated land or the soil of that land from early times,[39] *zibr* [40]

[38] R. LeBaron Bowen, Jr., and F. P. Albright, *Archaeological discoveries in South Arabia*, Baltimore, 1958, 43–88, reviewed in *BSOAS*, XXIII, 3, 1960, pp. 582–5. Cf. the article '*Ukhdūd*' cited in p. 37, n. 22. An Aden paper, *al-Janūb al-'Arabī*, no. 164, p. 9, speaking of Shibām, refers to طرح التطيور على السوافح الواطئة which might mean placing the spoil at the side of the field.

[39] e.g. in the *Sīrah* of Ibn Hishām, cf. *BSOAS*, XXI, 1, 1958, 3.

[40] *Zibr* is pronounced *zabr* in Dathīnah where it was explained to me as *turāb*, but there also a *zabīr* is a mass of earth like a *sawm* or bund.

which is *al-ṭīn al-rakhū ila 'l-bayāḍ aqrab*, soft soil nearest white (in colour) with no sand in it, and *al-jidfir*,[41] which is *al-ṣulb* [42] *al-māyil ila 'l-ḥumrah*, hard inclining to redness. *Shaḥmah* is the word employed in Ḥaḍramawt, certainly in Tarīm and Saiwūn, to mean ' good soil ' or perhaps the goodness in the soil, and in *al-Ḥayāt al-sa'īdah* [43] occurs the phrase *al-suyūl . . . qad jarafat shaḥmata-hā* ' the floods have eroded its good soil '. The phrase *ishtallat al-shaḥmah ḥaqqat-hā* ' the good soil of it has been carried away (by the flood) ' is commonly employed. Certain information is supplied by the medieval Rasūlid monarch al-'Abbās b. 'Alī b. Dā'ūd al-Ghassānī in the *Bughyat al-fallāḥīn*,[44] on Yemenite soils, though much of what he says is repetition of Ibn Baṣṣāl, and it is only his Yemenite observations that would concern us.

When the flood-water comes down the *wādī* at night all the village turns out with lamps to work—a lively scene, especially if it be a *sail ḥamīm*,[45] a powerful, strong, and noisy flood. Saiyid 'Abd al-Raḥmān laughed at Saiyid 'Alī b. Sālim al-'Aṭṭās because of his tremendous and admirable activity on such occasions, and his fearlessness of the floods. If a man cannot get at his fields, or if he is cut off by the flood from them, or for any other reason, then those already on the spot look after the control of his water, without payment. At Ḥuraiḍah co-operation in such labours is almost a duty, especially when it comes to inserting the sluice-plank (*sufrah*) in the openings to the fields or to removing it. This compares with the customary law of the fisher-folk of the coast where a range of emergency duties must be executed without recompense. Similarly in the event of fire (*ḥarīq*) at Ḥuraiḍah—which often breaks out about the time the *dhurah*-cane (*qaṣab*) is stored in the houses, all turn out to help extinguish it,[46] fetching from their houses whatever water is to hand, in pots, *zīrs*, etc., and the *jābiyah* (cistern) [47] of the mosque is opened to supply water

---

[41] Many places in Ḥaḍramawt are called al-Jidfirah/Jidfarah. Cf. H. v. Wissmann and R. B. Serjeant, ' A new map of southern Arabia ', *Geographical Journal*, cxxiv, 2, 1958, 168, ' a stratum of good hard clay without sand '. *Al-Ḥayāt al-sa'īdah*, 14, says that *jadāfir* (pl.) are wide open places like desert tracts (*ṣaḥārā*), they being cultivated places from which the course of the *wādī* has descended, so they have become dried up and hard. Cf. *al-Shāmil*, 170. Raḥaiyam explained *jidfarah* as *al-nukhr/nukhur* (cf. p. 71, n. 158), and as الطين الصلبه الذى فجّرها الماء من السيل.

[42] cf. Rossi, ' Note sull' irrigazione ', 359, *ṣulbī* ' terra non coltivata '.

[43] op. cit., 12.

[44] In addition to my transcript from Tarīm two copies are available in the Dār al-Kutub though not in the printed catalogues. Cf. Max Meyerhof, ' Sur un traité d'agriculture composé par un sultan yéménite du xve siècle ', *Bulletin de l'Institut d'Égypte*, xxv, 1943, 55–63, and xxvi, 1944, 51–65. Many passages in the text can correct or improve *Ibn Baṣṣāl, Libro de agricultura*, Tetuan, 1955, ed. and trans. by J. M. Millás Vallicrosa and Mohamed Aziman, a source upon which the *Bughyah* draws extensively. I am indebted to Fu'ād Saiyid for information on the existence of the two Cairo copies.

[45] *Al-Ḥayāt al-sa'īdah*, 14. *Ḥamīm* was stated to mean a flood that *yahumm* ' makes a noise ', cf. *Gl. daṭ.*, ' gronder ', of a torrent. It is described in *al-Ḥayāt* as being destructive. In Aden one speaks of *al-dardashah ḥaqq al-sail*, the flowing noise of a flood, or a perennial stream (*ghail*).

[46] For corporate organization to deal with fires, see *Prose and poetry*, preface, 27.

[47] Aḥmad b. al-Ḥasan . . . al-Ḥaddād, *al-Fawā'id al-saniyah*, fol. 104v., cites a word مباضى as meaning *jawābī* (pl. of *jābiyah*). *Mayāḍī* (s. *mīḍā*) should probably be read.

to put out the fire—all this because the wells are so deep that the delay in obtaining water in sufficient quantity would be too great for it to be of any avail against the spread of the fire. In cases where an old man has an irrigation channel choked by silt at the flood-time, all come and help to clear it for him.

People may select a *rā'iḍ* [48] *al-mā' alladhī yir'aḍ al-mā'*, a man who distributes the water. They may not perhaps choose this water-man from among themselves, but take an outsider. He keeps the *sawms* or bunds from breaking, and attends to everything at the time of the watering of the fields. Thus one person only, may look after the watering of the *jirbah* or field, even if it be divided among several. In return for this work he has the right to the *rumām*, grass [49] and other growth of the *sāqiyah*-channel.

With the term *rā'iḍ* one may compare the word *mir'āḍ* which I heard in Tarīm, with the synonyms *mizyāb* [50] and *khārūr*, a baked clay irrigation spout or drain, described to me in Arabic as *majrā ḥaqq al-mā'*, and used to carry water from one *maṭīrah*-plot to another.[51] Other informants seemed to imply *mir'āḍ* was used for draining the rain off buildings only, as indeed *mizyāb* would be applied. My MS copy of the *Manāqib* of Bā Hārūn [52] alludes to the *khuluṣū al-mir'āḍ* [*sic*] ' hole of the water-spout ', leading out of an upper room. In Saiwūn *khuluṣ* (pl. *akhlāṣ*) are holes in a garden wall for the flood-water to enter and irrigate it, to be compared with Rossi's *manfas*.[53] A *ṣalīf* is a little opening in a clay bank of a *maṭīrah* to allow of the passage of water from one *maṭīrah* to another, this word being recorded from Tarīm though it is probably widely used. A small pottery spout to enable water to enter the *ḥufrah* or circular runnel round the roots of a palm tree is called *qarā* (pl. *qiryah*) [54] in Tarīm.

When the dam or deflector (*ḍamīr*) has been broken by recent floods at Ḥuraiḍah, a drum is beaten each night to announce what the peasants are to bring next day in order to work upon the repairs of it. This is called the *taṭrūbah*, and has the sense of ' proclamation ', but it is used, in my experience, especially amongst the tribes in South Arabia for summoning the tribesmen in time of

[48] cf. Landberg, *Ḥaḍramoût*, 163, *ra'aḍ*, ' faire dévier l'eau du *sêl* dans les champs '.

[49] *Gloss. daṯ.*, ' foin '. The word is known also to Lane.

[50] cf. also ' Building and builders in Ḥaḍramawt ', *Le Muséon*, LXII, 3–4, 1949, 283. Cf. *al-Nūr al-sāfir*, 252, *'imārat mīzāb al-raḥmah min al-bait al-sharīf*, i.e. the repair of the rain-water spout of the Ka'bah.

[51] The Yemenite plot called *qaṣabah* (pl. *qaṣab*) is apparently the same as *libnah* (pl. *liban*), a plot of land 12 *dhirā'* square. One says, *Bā nasīr niḥbil buq'ah* ' We're going to go and measure out a piece of land '. In the Lower Yemen *libnah* is also known as *shaklah*. Rossi, op. cit., 359, reports much the same. Cf. the section on *masāḥah* in the Ambrosiana MS no. 112, *Rivista degli Studi Orientali*, III, 1910, 908. My ' Star-calendars and an almanac from south-west Arabia ', *Anthropos*, XLIX, 3–4, 1954, 450, also has some names for fields and irrigation-works.

[52] My acephalous MS, fol. 62r., perhaps to be identified as *Uns al-sālikīn*.

[53] op. cit., 358.

[54] *Al-Ḥayāt al-sa'īdah*, 11, alludes to *tanānīr qi'ār al-nakhl* at Baḥrān. *Qa'r* (pl. *qi'ār*) seems to mean a dead palm trunk, and a *tannūr* is a large earthenware jar in the bottom of which a fire is lighted and flapjacks cooked on the warm interior sides. In this context it might mean some sort of pot in which a young palm had been planted, but this is surmise. This would be different from the *qarā*. Cf. *Lisān al-'Arab*, *qarīy*, *majrā al-mā' fi 'l-rawḍ*, *majrā al-mā' fi 'l-ḥawḍ*.

war, etc.[55] Everywhere in southern Arabia, to the best of my knowledge, a channel or deflector bringing water to a number of proprietors must be maintained at the expense of them all. At Ḥuraiḍah each peasant works on the *ḍamīr*, but should he not for any reason, he will pay instead his contribution towards the cost of the operation, at the rate of 10 East African shillings per 100 *maṭīrah*-plots to be watered. The *ujrat ṣalāḥ al-wādī*, cost of repairing the *wādī*, also known as *sawq* [56] was in 1954 as follows. Field no. 1 (fig. 1) pays a contribution of 10*s*. per 100 *maṭīrah*-plots, no. 2 pays 10*s*. per 200, no. 3 10*s*. per 300, and so on up to the seventh field in a series in some areas, where the proprietor pays 10*s*. on 700 *maṭīrah*-plots. I suppose the justice in this arrangement is that the proprietor of the first field can be certain that if there is any flood-water his field will get what is available, but the chances of the other fields being flooded diminish in proportion, and from my casual observations in various parts of Ḥaḍramawt it is evident that the lower-lying fields in some places ' drink ' only in good rain years, for one often sees them lying uncultivated and unwatered while the crop is doing well on the upper fields. It appears that the *hijj* (pl. *hawāyij*) or *ḍamd*,[57] the acre, or area that can be ploughed by a yoke of oxen in one day, is not used here as it is elsewhere. In Ḥuraiḍah town there exist agreements on the rates of payment for each category of agricultural work and repairs,[58] no doubt written documents which should be interesting, but time did not permit of my collecting samples and having them explained—not an easy matter. At Hainin in the main Wādī Ḥaḍramawt, and at one time a sort of capital, I was told in 1948 that there was what they called a *k͟haiyil*, a Nahdī tribesman, an old man whose function it was to divide out the work on dams, state what the wages of workers as individuals in relation to the work should be, and in general deal with costing all agricultural work.

When I last visited Ḥuraiḍah in August 1954 I was surprised to see both *s͟harā'if* and *sādah* working alongside the peasants in the fields. In other parts of the country one would not be likely to see this; especially unusual would it be to see *s͟harīfahs* working in the open, but I did have friends among the younger *saiyids* in Tarīm in 1947 who told me that as a voluntary effort they had taken to working on well irrigation (*sināwah*) during the famine in the latter half of the war, and they had found it excellent exercise, but this is quite outside what has long been regarded as the norm in that part of Ḥaḍramawt, that the *saiyid* does not engage in agriculture. I should not like to say whether the agricultural activities of the *sādah* in Ḥuraiḍah have continued from the stage in the development of Ḥaḍramī society when *saiyids* and *mas͟hāyik͟h* did engage in manual work, as I think I have seen hinted or noted in the volumes of *Manāqib* of the saints, or whether the recent hard conditions in the *Qiblī*

[55] cf. the section on corvée, *infra*, p. 59–60.

[56] The word also occurs in the *Mulak͟hk͟haṣ al-fiṭan*, fol. 12v., *sawq al-mā' ilā ummahātihi min maqāni' al-ra'īyah fi 'l-s͟hu'ūb wa-'l-wādī*.

[57] I heard a Yemenite use *ḍumaid* as a pl. to this word.

[58] Where food forms a part of the payment it is known in Ḥuraiḍah as *ḍuḥā*, explained as the *g͟hadā* of other districts.

or western part of Ḥaḍramawt, commencing perhaps with the war-time famine, have broken down this convention.

In this connexion it is worth recording that the *jihāz* which I have rendered as ' bride-price ' [59] in Ḥuraiḍah was at the time of my visit, as follows :

Saiyids (who call it *ṣadāq*)	50*s*. for a virgin
Mashāyikh	80*s*. to 100*s*. for a virgin
Peasants	300*s*. for a virgin
	150*s*. for a non-virgin

I have not recorded payments for non-virgins in the other cases, but it is probably half also. In all classes the *mahr* is 9*s*. It was stated that the *ṣadāq* of the *sādah* was formerly 5 ocques of silver. It will be seen that marriage in Ḥuraiḍah is not expensive, but as elsewhere in Ḥaḍramawt of the interior because of the massive emigration, women far outnumber men. As I have earlier remarked [60] the bride-price is indicative to some extent of the economic value of a married woman's services.

A pleasant ceremony, which I witnessed from the Āl al-'Aṭṭās house, is called *al-Būb*. At the season of the date-harvest village people come round the houses of Ḥuraiḍah, accompanied by a couple of drums and a conch-shell upon which, from time to time, they blow a note, dancing and playing to the tune under the houses. Women and even unmarried girls are to be found in the party, dancing and clapping their hands. When they stop before a house they sing to those watching them from the first and upper floors,

دار مِن يَا مَحْياها   الله يِسِلِّم مَوْلاها

' Whose house is this ? May it prosper,[61]
May the good God save its master '.

They ask the house for dates, and each house will scatter to them a *khubrah* or two, i.e. a basket-cover [62] full of dates. By way of thanking the people of the house when the latter sprinkle the dates down on them (*yarushshūn al-tamr 'alai-hum*) they sing,

يــا الله على عَيْبُون   من كلِّ بارق سَيْل
وسيــل بــالبُكْــرَه   وسيل تالِى الليــــل

' Be there, O God, upon 'Aibūn,
From each lightning, a flood that flows,
At early morning time, a flood,
At night a new flood that follows '.

[59] Ibn Ḥajar, *al-Fatāwā al-kubrā*, Cairo, 1938, IV, 111, alludes to *jihāz* as the present of the bride before marriage. See ' Recent marriage legislation from al-Mukallā ', *BSOAS*, XXV, 3, 1962, 482 *passim*.

[60] ' Two tribal law cases, 2 ', *JRAS*, 1951, Pts. 3–4, 156. The Ḥuraiḍah prices for 1954 are much lower than those for Tarīm in 1947.

[61] More strictly literally the sense appears to be ' how flourishing it is with *karam* (generosity) '.

[62] The *khubrah* is placed over the ripening dates to keep off birds.

Irrigation at Tarīm

Although the irrigated palm gardens or plantations in fig. 8 were sketched for me at Tarīm, and the nomenclature of the fields and structures is consequently that of this district, and might not be current elsewhere, the actual method of dealing with flood irrigation is probably practised over many parts of Ḥaḍramawt.

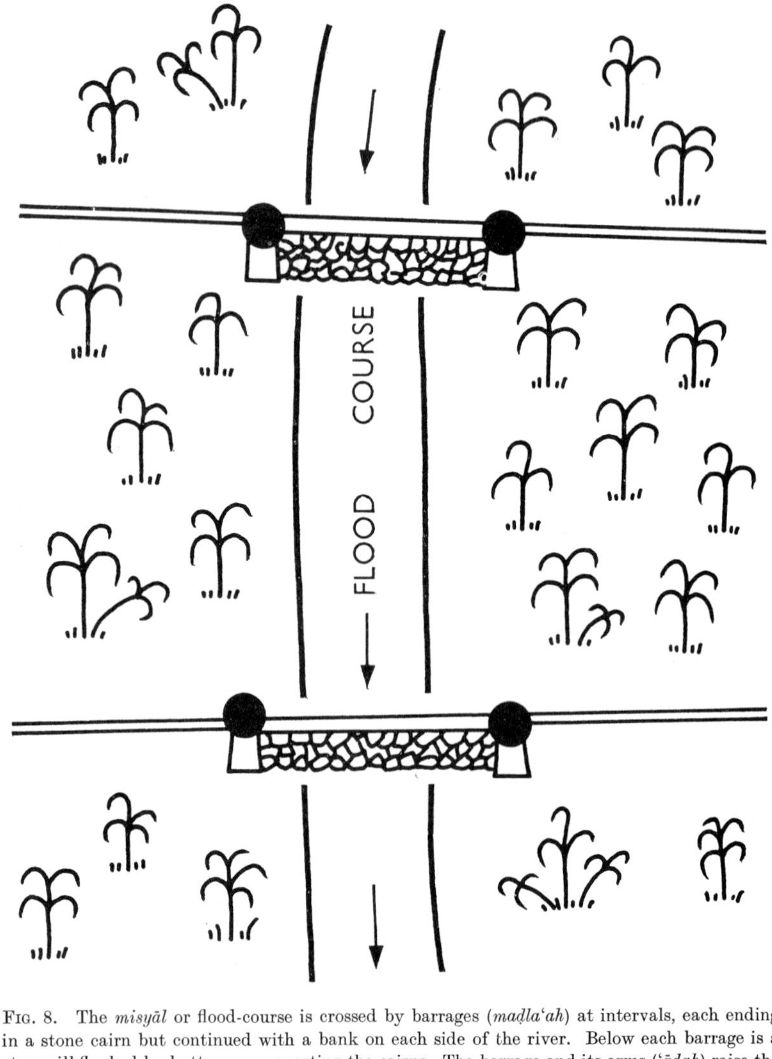

FIG. 8. The *misyāl* or flood-course is crossed by barrages (*maḍla'ah*) at intervals, each ending in a stone cairn but continued with a bank on each side of the river. Below each barrage is a stone sill flanked by buttresses supporting the cairns. The barrage and its arms (*'ādah*) raise the level of the water to irrigate the palm trees in the gardens on each side of the flood-course, and the surplus spills over the barrage to water areas below.

The flood-course down which the torrent pours after rains in the mountains is called *misyāl* or *qaṣabah*. At appropriate places a *maḍlaʿah* or barrage is constructed across the flood-course, so that in effect one might find a series of barrages descending the *wādī*. Each barrage will deflect a portion of the flood-water to irrigate its own sector while the surplus spills over the barrage to irrigate sectors below it. There is some reason to believe, though this awaits detailed survey, that the Wādī Ḥaḍramawt and its tributary *wādīs* down to Qabr Hūd where it narrows markedly, had, in ancient times, a more or less complete succession of barrages. The force of the floods was thereby broken, or perhaps the barrages never allowed the floods to gather great force to destroy and cause deep erosion, while utilizing the waters to the full for irrigation purposes, so that, as legend has it, one could walk most of the way under the shade of palm trees from Ḥuraiḍah to Hūd.

Since those days, however, floods have scored deep ruts in the *misyāl*, and at ʿĪnāt, for instance, have eaten back into the *misyāl* up-stream. At the famous *Nuqrah* [63] below ʿĪnāt the flood-waters have carved a small canyon, perhaps 50–60 feet deep, in the alluvial clay bottom of the Wādī Ḥaḍramawt. In the lower Wādī Ḥaḍramawt especially there are large areas where Shanbal in the early tenth/sixteenth century, and even later historians, knew prosperous villages and little towns where now there is hard dry plain, but there has been erosion in the upper Wādī as well. At Ḥaid Qāsim, just above Tarīm, one can see dying palms, but Tarīmīs tell me they can remember flourishing plantations there in their own lifetimes.

To return, however, to fig. 8, the *maḍlaʿah* or barrage consists of a stone wall constructed in courses (*ḥabl* [64]) running from one side of the *misyāl* to the other, the top course being known as the *mardam* [65] or sill. At each end of the barrage is a *bikrah* (pl. *bikār*),[66] a stone mound or cairn; these *bikār* have buttresses on the lower side known as *zāḥimah* (pl. *zawāḥim*), a term also applied to a buttress to a house.[67] The barrage will generally be sloping or inclined (*maḍlaʿah mukharraṭah* [68]), and in front of it lies a stone apron (*salqah*, pl. *salaq*; the word *salqah* in other districts known to me as meaning a palm-leaf mat). The flood-water pours on to the apron but does not, in consequence,

---

[63] cf. *Gloss. dat.*, *nuqrah* and *naqar*. Cf. D. v. d. Meulen and H. v. Wissmann, *Ḥaḍramaut*, Leiden, 1932, 148–9. Raḥaiyam describes a *nuqrah* as *majrā al-māʾ*—*al-masīlah, min al-ṭīn al-mutawassiṭ al-rakhwah*. W. Popper, *Extracts from Abū ʾl-Maḥāsin ibn Tughrī Birdī's chronicle* (*Ḥawādith al-duhūr*), Berkeley, 1930–42, glossary, cites *makān n q r*, a tract of land, once a *birkah*.

[64] cf. *supra*, p. 43, n. 35. *Al-Shāmil*, 186, defines a *maḍlaʿah* as a barrier (*ḥājiz*) of stone, constructed at a field (*jarb*), or basin containing palms (*ḥawḍ al-nakhl*), or the side of a *sāqiyah* to revet (*ḥajaza*) it against collapse and retain its water and soil.

[65] 'Al-Lughat al-dārijah bi-Ḥaḍramawt', *al-Rābiṭat al-ʿAlawīyah* (Batavia), I, 3, A.H. 1347, 169–71, citing *mirdam* as used in the area.

[66] *Gloss. dat.*, 192, *bikār* 'colonne'.

[67] To buttress is *zaḥḥam*. Raḥaiyam said, *Al-zāḥimah tarfud-uh* 'the buttress supports it'. *Rifād* means 'support'.

[68] I have also noted a form *mukhraṭah*.

form a pool below the barrage, though I have seen just such a pool below a barrage in the *masīlah* below Tarīm. Beyond the cairns on each side lies a clay bank, usually distinguished here as *baḥrī* (southern) and *najdī* (northern), the clay banks themselves being known as *'ādah* (pl. *'īd*).[69] These act as extensions to the barrage to hold the flood-water, and deflect it back up-stream to irrigate the *ḥuyūd* (s. *ḥaid*) or palm plantations lying on each side of the *misyāl*.

Usages of which I have noted examples include the following: *Zād 'ala 'l-'ādah ilā nuṣṣ al-bikār* 'It (the flood-water) rose to the *'ādah* to half-way up the cairns'[70]; *Ṭalli' ḥabl fi 'l-'īd al-yumnā* 'Add a course to (heighten) the right-hand clay banks'. One speaks of a *sail 'affāsh*[71] *farash fi 'l-ḥaid*, rendered as a flood which *istamadd aw 'amm al-nakhl bi-'l-saqy* 'covered all the palms with water'. *'Ammat al-madla'ah* means that a flood covers the barrage and the palm plantations.

In the side of the Masīlah below Tarīm, I noticed, a good few years ago now, a short passage either cut or tunnelled through the hard clay bank at an angle

Fig. 9. Dam or barrage at Tin'ah, with extensions (*illīd/al-yad*) terminating in cairns (*'arūs*).

to the flood-bed, leading the water off the Masīlah to palms. It was called an *'ambarah*. At this distance in time I cannot recollect whether it was for the purpose of filling a pit from which the water would be used to irrigate the palms until it was exhausted, as indeed I think, or whether some other use was made of it. However, the *ḥasī*, a shallow well dug to catch the *sail*-flow which is then used for irrigation, is well known, so that there is even a saying, *Man bagha 'l-mā' yaḥsī luh* 'He who wishes water must make a well to catch it'.

The irrigation system in fig. 10 was described to me as *jarb fawq jarb*[72]

[69] At the village of Tin'ah our Tamīmī camel-man called the part corresponding to the *'ādah* by the name *illīd* (i.e. *al-yad*), and the cairn, *'arūs* (fig. 9). According to *al-Rābiṭat al-'Alawīyah* (Batavia), I, 3, A.H. 1347, 169, it is a wall or column (*ḥā'iṭ aw usṭuwānah*) in a house. In Saiwūn *al-yad* is called *al-ḥāmil*.

[70] My rendering made without further reference to the informant is slightly dubious here, for it might mean that the flood-water was higher than usual, half-way up the cairns.

[71] *'Affāsh* does not figure in the lexicons consulted.

[72] For various forms of this word cf. *Gloss. daṯ.*, 275 seq. The hagiology known as *al-Jawhar al-shaffāf*, story 379, speaks of *sāqiyat jarb Abī 'l-Ḥasan* 'the channel of the field of Abī al-Ḥasan'. *Al-Shāmil*, 210, speaks of *jurūb fi ḥijl Ṣīf*, i.e. 'fields in the cultivated land of Ṣīf', the latter a village of Wādī Daw'an, known as *Jurūb al-Darasah*, or 'the Students' Fields', because they were a *waqf* to maintain students. A part of these being destroyed by flood, it was put out on a *munāsharah* or *mufākhadhah* contract (cf. p. 62, n. 110) by which the cultivator (*fakhīdh*) who restored and replanted it obtained half the fields and half remained with the *waqf*. I do not know whether there is any difference between *jurūb* and *dhabr*. Landberg describes *dhabr* as 'terrain arrosé par la noria ou la pluie, champ'. He states that it is often pronounced *dabr*. I have seen it written in MSS as *ḍabr*, with *ḍād*. This brings it near to the root *ẓbr*, rendered by A. Jamme, 'South Arabian', in J. B. Pritchard, *Ancient Near Eastern texts*, Princeton, 1955, with a query, as 'land-measure or weight'. Bā Hārūn, op. cit., fol. 45r., has an example of its usage, *Jā Wādī Dammūn bi-sail 'aẓīm wa-sharibat jamī' dhubūr al-bilād*.

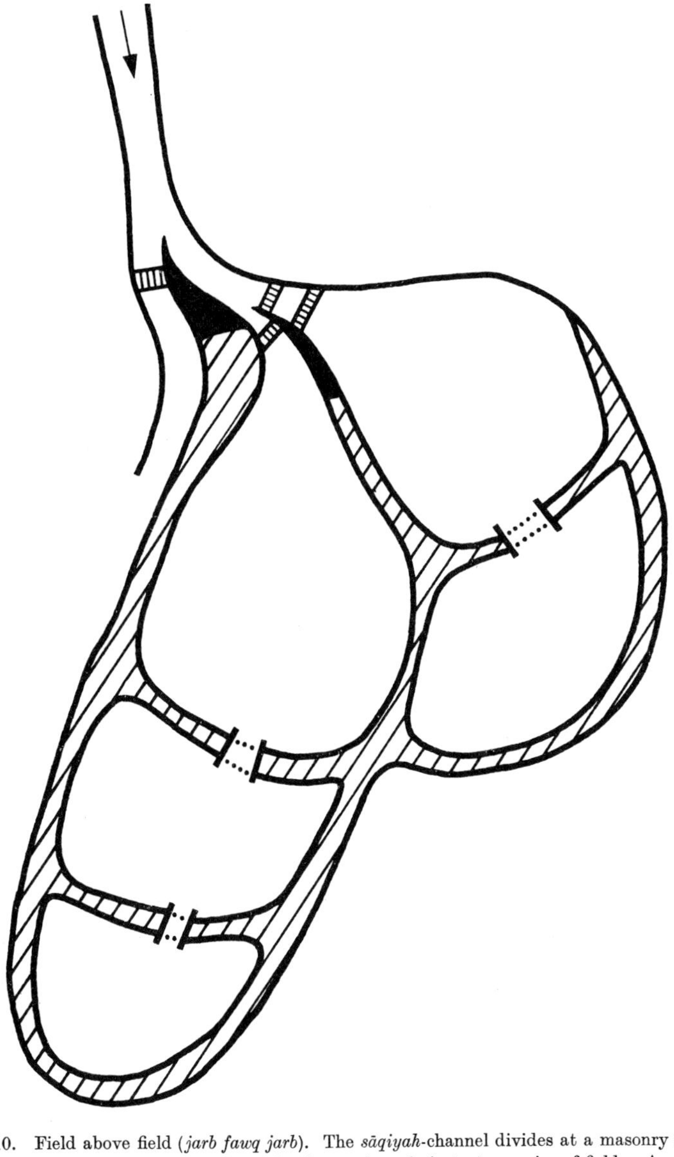

Fig. 10. Field above field (*jarb fawq jarb*). The *sāqiyah*-channel divides at a masonry tongue called a *ḍark*. One of the channels subdivides again to irrigate two series of fields. Across the entrances to these are low stone *marṣads* to prevent erosion, but they do not impede the flow of water. The higher set of fields ' drink ' first. Cuttings are made in the banks to allow the lower fields to be irrigated in sequence.

'field above field', and this is probably the commonest type of irrigation system in the Aden Protectorate.

The main water-channel is divided, and at the right-hand smaller channel as it branches off from the other, is a *marṣad*, this being described to me as a low barrage (*maḍla'ah*), about 6 inches in height, really in fact a stone line across the bed of a channel or *wādī*. *Marṣad* was stated to be a term applied to anything smaller than a *maḍla'ah*.[73] At each side of the *marṣad* would be a stone cairn (*bakrah*). Further down, the left-hand channel is once more subdivided into two small channels leading to the fields and separated by a masonry tongue called a *ḍark* (pl. *ḍurūk*). The fields (*jurūb*) 'drink' in succession, the lower from the higher, but of course where the channel is divided the higher ground would 'drink' before the lower ground.

In Tarīm an *'uqdah* (pl. *'uqad*) seems to be a sunken piece of land lower than *jurūb*; it is said to be synonymous with *fijrah* (pl. *fijār*).[74] You could find *'uqdah ba'd 'uqdah*, one piece of sunken ground lower than another, and one speaks also of *shurūy wa-'uqad*.[75] Another common word for an irrigation channel is *'atim*[76]; I have noted a phrase, *makhdūm al-'atim 'alaih*, described as *maṭrūq bi-nūrah* 'covered with lime-plaster'.

The Āl Bā Ḥuraish[77] are master masons (*ma'ālimah ṣanā'ah al-aḥjār*) at Tarīm who construct barrages (*sudūd*) but there are many *fakhāyidh* of masons there, the Bā Sumbul, the Āl Bā 'Adail, and the Āl Bā 'Awḍān. At Dammūn next to Tarīm the Āl Nīnū carry out the same type of work. From another *wādī* in the district, the Āl 'Abūdān have experience in the management (?)[78] of water (*'inda-hum khibrah bi-ta'dīl al-mā'*), they are estimators for water-channels (*mukhamminīn li-'l-masāqī*), for the management of water-channels

[73] Ibn Ḥajar, op. cit., III, 64, discussing the *manāfidh* 'exits' of a *sāqiyah*-channel, says, بعض منافذها مرصد بوضع أحجار فيها ولا يسد من المنافذ لشرب آخر. In Tarīm *marṣad* (pl. *marāṣid*) was said to mean *maqāsim al-mā'*, a place where the stream is divided (cf. p. 40, n. 28), but I was also told that a *marṣad* is a low *maḍla'ah*, about 6 inches high, really a stone line across a *wādī* or a channel-bed. Cf. *raṣa'ah*, *marṣū'*, etc. Lane and Dozy are not very helpful on either root.

[74] Raḥaiyam defined a *fijrah* as *al-arḍ al-ghawīṭah al-nāzilah min al-mustawiyah al-mutaḍā-yiqah*.

[75] *Shurūy*, *shurūj*; cf. *Gloss. daṯ.*, 2034; 'Star-calendars', 450. Vatican MS no. 1362 in *Bāb al-suyūl wa-'l-miyāh*, though disappointing in that the usages all seem to be purely classical, records that *al-shirāj* are *madāfi' al-mā' min al-ḥizzān ila 'l-suhūl*. The Yemenite *Tarjī' al-aṭyār*, 308, states, الشرج جمع شريج وهو القطعة من الأرض وهو فى الأصل مدخل الماء من السائلة الكبرى. My own notes from Tarīm describe a *sharj* as a part of the *wādī* in which there is arable soil (*ṭīn*) and *'ilb* trees, and where one has fields. It is near the *misyāl* or else it may be watered direct from the spill from the mountains. A *sharj* can be irrigated by a *sail* or by rain, but not by *sināwah*.

[76] An Aden newspaper defined the *'atim* as *majārī fī 'l-arḍ makshūfah li-tantafi' bi-'l-mā' al-ashjār wa-'l-mazrū'āt al-mujāwirah li-'l-majārī*.

[77] They are also the grave-diggers; cf. 'The cemeteries of Tarīm', *Le Muséon*, LXII, 1–2, 1949, 159.

[78] *Ta'dīl* might mean 'levelling, adjusting'. Perhaps the Āl Bā 'Adail are so named because of their specializing in this craft. The Bā Sumbul are stone-workers.

(*li-ta'dīl masāqī al-mā'*), i.e. water dividers and series of steps (*marāṣid wa-marāqī* [79]). Raḥaiyam added that, in ancient times, the Āl 'Abūdān used to construct the channels which take the water from higher up and go deep into the earth and appear in another area at the top of the earth (*al-budūd alladhī ya'khudh al-mā' min a'lā wa-yaghūr (yaghīṣ)* [80] *fī 'l-arḍ wa-yaẓhur min jānib thānī fī a'lā 'l-arḍ*). I do not think that he was alluding to *qanāts* but rather perhaps to some underground or covered channels such as one may see in the ruined town of al-Ḥāfah in Dathīnah.

Raḥaiyam also quoted to me the popular saw, *Kull arḍ razzim-hā bi-ḥajar-hā*, which might be rendered, ' Build up strongly each piece of land with its own stones '.[81] He stated that *razzam* meant *makkan-hā wa-sawwas-hā* but he also said that it meant to fill up a hole without much arrangement.

Legal questions arising from flood irrigation

Naturally disputes over the use of flood-waters give rise to many questions in law, be it customary law, or be it *sharī'ah*, and the three cases which follow are merely those which have come to my notice and appear to be illustrative of differences which may come about between the various parties.

The *Fatāwā* of Bā Makhramah [82] include one case where land is rented for agricultural purposes. The agreement was concluded before the land had ' drunk ', and was for the period of a year. The land in question usually ' drinks rain-water either in *Ṣaif* (April–June) or in *Kharīf* (July–September) ' (*tashrab min mā' al-maṭar immā ṣaif-an aw kharīf-an*). If there is no rainfall (*ḥuṣūl al-maṭar*) during this period, asks the questioner, does the contract still hold good ? Bā Makhramah, reasonably enough, holds that it does and that it cannot be rescinded.

Another problem in the same collection [83] is stated in the following terms :

مسألة رجل استاجر أرضاً ثم حرثها وعمرها ثم جاء السيل وأجبر عبرها فلم يمكنه سقيها إلّا بعمارة العبر وامتنع اهل الأرض من العمارة فهل له الفسخ واذا فسخ هل له الرجوع على الموجر بما اصرفه فى حرث الأرض ام لا؟

' The question of a man who rents ground which he then ploughs and cultivates. The flood then comes and cuts (?) [84] the bank of it, so that it is impossible to

[79] A *marqā* means a *darajah* or step, as in Lane.

[80] *Ghāṣ, yaghīṣ*, *Ḥadramoût*, 672.

[81] In the Yemen *razm* can mean a water-channel, and *Gloss. daṯ.* discusses this root at some length. The *Tarjī' al-aṭyār* says that a *marzam* is ' what is placed upon something to make it firm and prevent it from moving and shifting from its place '. The proverb resembles another I heard in Tarīm, *Dāwī 'l-ḥimār min rawth-ah* ' Cure/treat the ass by its dung '. The dung is to be mashed up and placed on the animal's sore.

[82] Photocopy in SOAS Library, fol. 291r.

[83] ibid., fol. 792v.

[84] *Ajbara* is probably to be so rendered ; cf. Dozy, *Supplément*, *jabr al-baḥr* ' cutting of a canal bank ', an Egyptian usage. *'Ubar* could be the bank leading to the field, which it probably is in this case.

irrigate it without reconstructing the bank, but the owners of the land refuse to repair (it). Can he renounce the contract, and if he renounces (it) may he come back to the party who has let it out (to him) for what he has expended on ploughing the land or not ? '
Bā Ma<u>kh</u>ramah decides that he may renounce the contract and claim from the party that let to him what he has lost.

The third case is taken from a MS work in my possession, of Bā Ṣabrain,[85] and may apply more particularly to the Wādī Daw'an or the western districts of Ḥaḍramawt.

إذا أخذ السيل حرثَ أو حجرَ جماعةٍ معلومين، أو جماعة فى حِجال معلومة، فهو مال مشترك بينهم بنسبة أموالهم، لا مُباحًا ولا ضائعًا، فلا يُملك بإحرازه فضلاً عن مجرّد وضع علامة عليه. فمن أعظم المنكرات التحجّر على شىء من ذلك لغير شريك فيه، أو على زائد على ما يخصه منه. أمّا ما يجيبه السيل من حجر وطين أو رَبْش فمباحة، [sic]، فلمن أحرزه وحازه ملكاً، ولمن علّم عليه علامةً أُولَويّةُ استحقاق. فلو علّمه رجل وربّخه الثانى ملك المربِّخ، لا المعلّم عليه علامة. أو مملوكة لمجهول فهو مال ضائع، لا مباحًا ولا مشتركاً. أو لمعلوم فهو باقى على ملكه ومثل ذالك الخشب والمَقَالِع.

'If the flood-water takes plough-land or stones of persons known, or of a group (with ownership in certain) known fields,[86] then this is property held in common among them in proportion to their properties (i.e. land), neither in a state of being open to acquisition, nor lost. Nor does it become possessed by being acquired, far less by merely placing a mark upon it. A major reprehensible action is the appropriation of any of it by someone who has no share in it, or the appropriation of more of it than that to which one is entitled.

Concerning what the flood brings in the way of stones and earth or mixed dung and earth,[87] this is open to acquisition. It then belongs to the person who acquires and obtains it as a possession ; the person who first marks it with a mark has priority of entitlement. Should one man mark it and a second loosen [88] it (the earth), the person who loosens (it) and not the marker of it with a mark has possession of it.

[85] 'Alī b. Muḥ. b. Sa'īd Abū Ṣabrain, *al-Muhimmāt al-dīnīyah fī ba'ḍ al-murtakab min al-manāhī al-rabbānīyah*, p. 24, no. 66. The author flourished in 1294/1877. Cf. 'Materials for South Arabian history, II', *BSOAS*, XIII, 3, 1950, 593.

[86] *Ḥijl* (pl. *ḥijāl*) is cultivated land of any area. 'Alawī b. Ṭāhir, *K. al-<u>Sh</u>āmil* (SOAS photocopy) calls it *al-ḥaql wa-'l-jurūb*.

[87] *Rab<u>sh</u>* is the dung brought down by the flood mixed with all kinds of debris. Cf. *marba<u>sh</u>ah*, the palm-leaf basket used by cultivators, presumably used for carrying *rab<u>sh</u>* as well as for other purposes.

[88] *Rabba<u>kh</u>* 'to loosen', probably here, 'to remove'. The text is not easy to interpret at this point, and seems almost to contradict itself.

PLATE I

A large *karīf* at Qaidūn of Wādī Daw'an, of stepped and plastered construction, for holding water probably for use in irrigation.

PLATE II

[*Courtesy Royal Air Force*

Flood irrigation at Shibām. Shibām from the north-west. Jabal al-Khibbah (also known as Jabal Shibām) is the name of the massif projecting from the Jōl plateau, and bounded on the east by Wādī Bin ʿAlī. The Masīlah runs across in front of J. al-Khibbah from west to east. Under J. al-Khibbah lies the suburb al-Saḥīl with cultivation and palm-gardens. On the north side of the Masīlah lies Shibām, surrounded on all sides except where it borders the Masīlah by palm-gardens. The large circular patch of lighter ground to the west of Shibām is the cemetery, Jarb Haiṣam ; it may stand a little higher than the fields which surround it. From the right of the illustration the *sāqiyah*-channels can be perceived, radiating out into the cultivated area, although the upper part of these *sāqiyahs*, lying further west, does not appear. The medieval *Manāqib Bā ʿAbbād* refers to the *sāqiyat al-balad Shibām*.

PLATE III

[*Courtesy Royal Air Force*

Flood irrigation at Shibām. Direct aerial view of Shibām. Jabal al-Khibbah occupies the upper part of the illustration with al-Saḥīl beneath it, its gardens receiving their irrigation from the overspill of the mountain, and at other times by well-watering. The tower known as Ḥuṣn al-Khibbah which protects Shibām from rifle-fire from the top of J. al-Khibbah can be seen as a dot in a white circle at the top of the ridge along the top of the picture. It is reached by a path running up from the right-hand tip of the mountain, i.e. west of Shibām town. On the Jōl at al-Khibbah there are many small firing sangars (*mirbāh*, pl. *marābī*) all over the mountain top, as also some areas for grazing and some small *sudds* to collect water. The Masīlah runs from west to east across the illustration, its broad western end being planted with palms in the actual bed of the *wādī*, but doubtless in an area where it is not anticipated that a *sail* would come with sufficient force to uproot them. Shibām town lies along the Masīlah. It is walled, and upon a fairly high mound which the *sail* would be unable to destroy—for this reason Shibām is sometimes called al-Dimnah. Drinking water, rather brackish to taste, is obtained from wells in the actual Masīlah in front of the town. This illustration shows very well the high banks round the fields which I believe to be the result of hundreds of years of piling up the surplus silt brought down by the *sail* on to the fields, on to the field-banks.

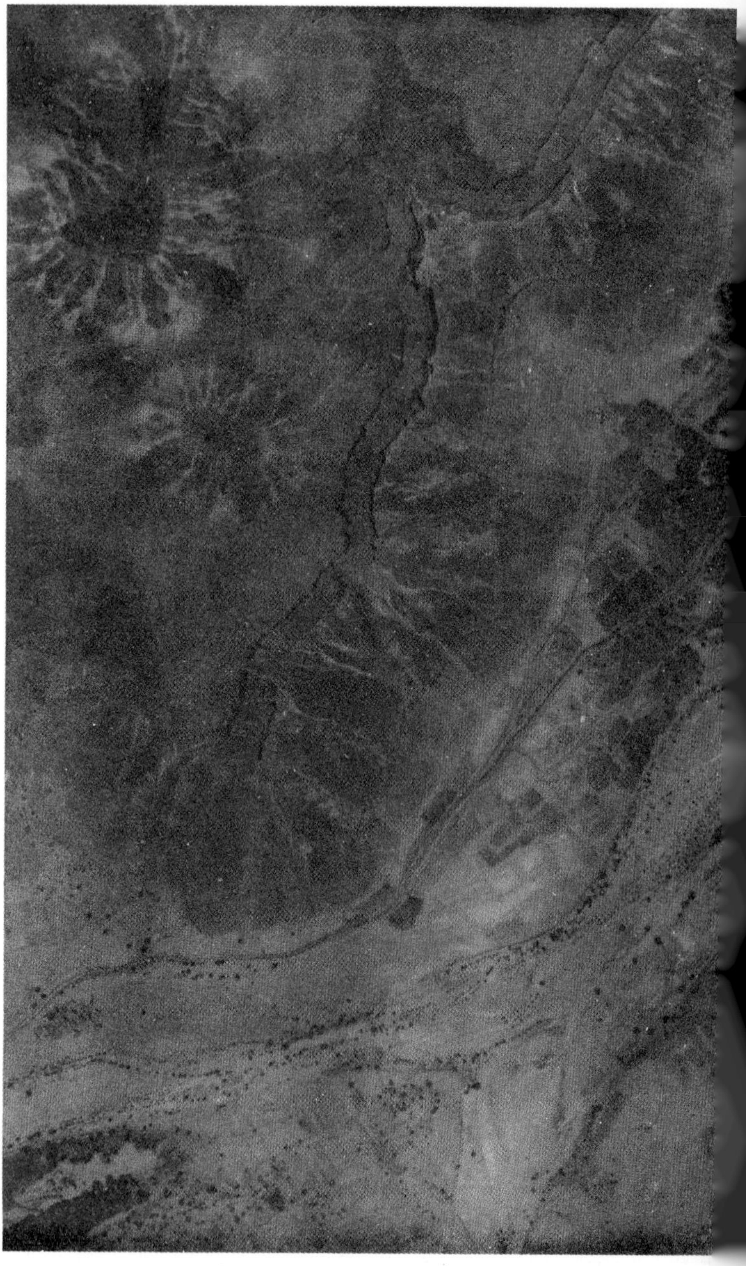

Wādī irrigation in Ḥaḍramawt. In the bottom left-hand corner two *wādīs* join to form
and lead it to the fields on both sides of the *wādī*. The villages stand on higher ground o
palms even perhaps in the *masīlah* itself though in parts where the current is not likely t
which meets the alluvial soil of the *wādī*, and it is this that I understand to be the *safḥ a*

PLATE IV

[*Courtesy Royal Air Force*

gle *masīlah* or flood-bed. At flood-time the deflectors and *sāqiyah*-channels catch the *sail* eminences in the *wādī* to which the flood will not mount, but there has been planting of strong; perhaps these can be described as *nukhr*. The mountains terminate in stony scree al to which the water rose and covered the lower section in the flood of 894/1488–9.

Or else it is possessed by a person unknown—then it is lost property, neither open to acquisition, nor held in common. Or else it belongs to a person known—in which case he continues in possession of it.

Wood and palm trees [89] are the same.'

Qanāts

The *qanāt* system is generally associated with Persia, though it is thought that under Arab influence *qanāts* were introduced into North Africa and even into such a place as Madrid.[90] They appear to be recorded as in existence in pre-Islamic Arabic from the evidence which will be cited, but this does not preclude their having been introduced there by the Persians who are believed by ancient authors to have constructed cisterns at Aden, Jeddah, and possibly other places. In default of established data, however, it would be idle at present to argue where and by whom the *qanāt* system was invented. The anonymous author of *al-Ḥayāt al-saʿīdah* [91] discusses them in the following passage.

'*Al-qinīy* is the plural of *qanā*, and *qanā* is the plural of *qanāt*, while *kiẓāmah* has a plural *kaẓā'im*.[92] Their essential meaning is that they are the wells/pits dug into the earth one after the other, in order that their water may be brought forth and flow upon the face of the earth. In the Tradition [93] there are tithes (*ʿushūr*) upon what the heavens and the *qanāts* irrigate. In the *Lisān al-ʿArab* [94] he says, " The *kiẓāmah* is a *qanāt* in the inside (*bāṭin*) of the ground in which water runs ". In the Tradition (it says) that the Prophet came to the *kiẓāmah* of a tribe and performed the ritual ablution from it and rubbed over his boots. The *kiẓāmah* is like the *qanāt*, its plural being *kaẓā'im*. Abū ʿUbaidah said, " I asked al-Aṣmaʿī about it, and expert people of the Hejaz, and they said, ' They are wells in a series, excavated and set at a distance from each other, then what lies between each two wells is pierced by a *qanāt* conducting the water from the first to that which lies next to it under the (surface of the) earth so that their waters come flowing together, then emerge at the end of them flowing on the surface of the earth ' ". According to the Tradition of ʿAbdullāh b. ʿAmr, " When you see that Mecca has been dug (full of) *kiẓāmahs* and its buildings are level with the tops of the mountains then know that the end has drawn near you " '.

To the east of Ghail Bā Wazīr is a rocky plateau in which there are large

[89] *Maqlaʿ* (pl. *maqāliʿ*), a palm tree.

[90] Jaime Oliver Asin, *Historia del nombre Madrid*, Madrid, 1959. Cf. Norman N. Lewis, ' Malaria, irrigation, and soil erosion in central Syria ', *Geographical Review*, XXXIX, 2, 1949, 287, for *qanāts* and '*foggaras* ' in Syria.

[91] op. cit., 26 seq.

[92] Rossi, ' Note sull' irrigazione . . .', 356, describes a ' *kadāmah* ' apparently similar to our *kiẓāmah*. Cf. G. van Vloten, *Liber Mafâtîh al-olûm* (al-Khwārizmī), Leiden, 1895, 71, *al-kaẓā'im al-miyāh al-jāriyah taḥt al-arḍ mithl al-qunīy* ' *kaẓā'im* are waters running under the ground like *qanāts* '.

[93] I have not discovered the Tradition in this form in A. J. Wensinck, *Concordance . . . de la Tradition musulmane*, Leiden, 1933– .

[94] Beirut edition, 1956, XII, 521.

quarry-like holes perhaps 30 feet from ground level to the surface of the water with which they are filled. These are of natural origin and their water is warm and has a sulphurous smell; they appear to be deep. These holes, in one of which we used to bathe, are known as *ḥawmah* (pl. *ḥuwam*), and from them an underground *sāqiyah*, punctuated at intervals by vertical shafts leading to the surface, around which the excavated debris lies, conducts the water from the *ḥawmah* to the lower ground to the south which it irrigates. I recall that in some places there seem to be small plots irrigated by bucket from wells or these shafts, and tobacco is grown there.

To Mr. A. M. Bā Ḥas͟hwān I am indebted for additional information on the *qanāts*. These he says are spoken of locally as *mi'yān*, which is the name of the channel itself, the vertical shafts being called *naqab*. At G͟hail Bā Wazīr the family which constructs these irrigation works is called Bar'īyah, and there would be a *muqaddam* or foreman to each three or four workers. There is a well-known *ḥawmah* called Ḥawmat al-'Arūs in which people avoid bathing since, it is said, a bride was being conducted on a camel to her husband when during the brief absence of her escort, the camel slipped into the *ḥawmah* and was drowned. At al-Qārah the largest *mi'yān* is called al-Ḥart͟h, but the water of it being brackish, is not drunk.

There may be *qanāts* of this type further to the east of al-Mukallā and al-S͟hiḥr, but I know of nothing to the west although I have travelled along most of the coast, apart from some curious underground channels ending in a sort of pond at Ruḍūm in the Wāḥidī sultanate.[95] *Al-Ḥayāt al-sa'īdah* [96] even alludes to a wooden *mīzān* (lit. ' balance ', perhaps a level ?) which he describes as *d͟hu 'l-s͟hāqūl* and *ālah d͟hāt al-s͟hāqūl*, i.e. with a plumb-line, which is used to ascertain the level of the bottom of the well and the slope of the hill to the cultivated ground below. He states that he saw Saiyid Muḥ. b. Ṭāhir al-Ḥaddād use this instrument when constructing a *sāqiyah* at Qaidūn and another at Jidfarat al-Zinjī.

Water supplies on the Jōl

Very few people live on the Jōl, but there are occasional small villages of some stone-built houses. Visiting al-Najadain in 1948, I saw a cistern (*naqabah*, pl. *niqāb*) being hollowed out in a slight depression near the fort. An iron bar was drilled into the stone by hand to break it up, the operation involving no small labour.[97] The rain-water runs off into the *naqabah* which is usually

[95] Other writers allude to *qanāts* which I have not seen, and there may be more perhaps also at Aḥwar which I have not yet visited. For Oman one may probably find information on *qanāts* or relevant thereto, in 'Abdullāh al-Sālimī, *Jawhar al-niẓām*, Cairo, A.H. 1345, especially II, 56 seq., *Bāb al-sawāqī*.

[96] op. cit., 27–9. According to the *Tāj al-'arūs*, the *s͟hāqūl* is an iron-shod pole in use among the cultivators at Basra, about 2 cubits (*d͟hirā'*) high, but from the rather summary description I cannot discern how it is used. Dozy, *Supplément*, describes *s͟hāqūl* as ' plomb ou fil à plomb '. Cf. *Liber Mafâtîh al-olûm*, 255, which describes it as a plumb-line used by carpenters and builders.

[97] 'Alawī b. Ṭāhir, *K. al-S͟hāmil*, 89, says of a *naqabah* that the ground is excavated to the depth of a fathom (*tunqab al-arḍ qadr qāmah*).

wide at the bottom, clearly to minimize the loss by evaporation. The *naqabah* I saw was not more than six feet deep, and two men could stand in it to work the iron bar.

A larger type of receptacle for water conservation is to be found in the *karīf* which is a sort of pond, generally formed by damming a convenient hollow and leading the rain-water to drain into it from the surrounding ground, by little runnels consisting of lines of stones leading to the *karīf*. There are many of these all over the Jōl. The people drink from the *naqabah* and keep the *karīf* water for animals. There are of course flood-courses on the Jōl and for a short time after rain they run merrily, but they are very soon empty once more. A larger and more complicated *karīf* is sometimes used in the Wādī Daw'an and elsewhere, such as the stepped *karīf* at Qaidūn (pl. I) with cemented sides. The *karīf* of Ḥabbān known as Sāqiyah Badr Bū Ṭuwairiq is famous; it lies beside the *wādī*-bed and the *sail* is led into it by a short tunnel.

The late Sulṭān Ṣāliḥ b. Ghālib al-Qu'aiṭī,[98] writing of the Jōl territory in the Daw'an province, states that pious Daw'anīs build small huts (*sirīn*) and storage cisterns for rain-water on the roads. These cisterns are either roofed (*musaqqafah*) and called *niqāb*, or without roofs and called *kuruf*. It is a trait of the Daw'anīs to devote *ṣadaqāt* to this purpose to help travellers, and many of the *Bādiyah* of Daw'an write in their bequests (*waṣāyā*) that a *sirīn* and *niqāb* should be built in their name, and this is why there are so many in Daw'an province.

Corvée

In connexion with the maintenance of irrigation works it is relevant to describe how corvée operates in Ḥaḍramawt. Corvée, unless the republican government has abolished it, which is hardly likely, was an institution in the Yemen up to the present day, known under the term *sukhrī*.[99] In Ḥaḍramawt the word *wakad* is apparently used to cover compulsory labour, but it has other senses. Should a sulṭān, for example, wish to build a house or a tower-fort (*kūt*) he compels labourers to come and work on it at a much lower rate than they are normally paid. His action is called a *nā'ibah/nāyibah*, paraphrased as *ish'ār wa-ṭalab*, i.e. notifying and request, and explained as *ilzām bi-'l-ḥuḍūr fi 'l-waqt al-mu'aiyan*, compulsion to turn up at a specified time. Raḥaiyam distinguished three categories—a *wakad dawlah*, a sulṭān's demand for compulsory labour, *wakad manṣab*, that of a *manṣab* or head of a sacred enclave, and a *wakad qabīlah*, the imposition by a tribe of compulsory labour, *scilicet* upon the labouring population under its protection. These categories of corvée would

[98] *Al-Riḥlat al-sulṭānīyah*, Cairo, A.H. 1370, 23. 'Alawī b. Ṭāhir, op. cit., calls the *karīf* a *ṣahrīj*, while Muḥ. b. Hāshim, *Riḥlah*, Cairo, A.H. 1360, 12, describes it as a *khazzān*. *Al-Shāmil*, 199 seq., has an instructive account of local disputes before the decision to build the Qaidūn *karīf*, with a poem on the event describing local wells as 100 *qāmahs* deep. Water was led into the *karīf* from a spring through pottery pipe sections (*jibḥ*, pl. *jubūḥ*) joined together with *nūrah* beaten up with cotton and (sesame ?) oil.

[99] E. Rossi, *L'arabo parlato a Ṣan'â'*, Roma, 1939, 202.

depend on which type of government was exercised over any given district, but it is interesting to see how the theocratic type of rule as exemplified in the case of the *manṣab* has also its corvée, though I expect the summons would be couched in different terms from that of the sulṭān or tribe. A case cited to me as typical would be when a *manṣab* announces a *wakad* for the repair of a large dam or deflector, and the labourers *yajurrūn madar*, lit. ' drag earth ', i.e. with oxen and a scraping board, or perhaps by other methods. Orders relating to corvée are normally given through the *Abū* or headman, and I have a document dated 1270/1853–4 detailing the procedure when a *nāyibat sulṭān* is sent to the Sūq quarter of Tarīm city. As seen already, the *wakad* is not confined to irrigation repairs, but might be for repairs to the town walls in time of danger or other purposes. This enables us to form some conception of the procedure probably followed in providing labour for the massive public works of antiquity in the ruins of which southern Arabia abounds.

The *nā'ibah* indeed is an institution which goes back to the earliest age of Islam and includes the repair of dykes/bridges (*qanāṭir*) and the stopping of breaches caused by water in a bank (*sadd al-buthūq*) ; it is a term known to the western Arab world and Spain, continuing in Morocco up to modern times.[100]

Wakkad was paraphrased as *ista'jar* ' to hire, engage ',[101] and ' to make an assignment (*wa'd*) with '. *Ana muwakkad 'indak* means ' I am employed by you '. One of my corpus of documents relating to the quarters of Tarīm alludes to persons as *muwakkad fī walīmah*, i.e. employed in unpaid service at marriage festivities, at the direction, of course, of the headmen of the quarter in the way which I have already described.[102] You can, however, have *wakad al-riḍā* free employment, employment by consent. In general, there appear to be two sorts of employment, *taṣrūḥ* under which you receive ' lunch ', i.e. *tamr, lukhạm, qahwah*, etc., dates, shark, coffee, and a *ja'ālah mu'aiyanah*, a fixed wage, the master (*mu'allim*) receiving more than an ordinary workman, and a boy less. The other type is termed *qiṭā'*, i.e. when a man estimates for a job, employs his own men to execute it, and gains or loses on the contract as he may.

Share-cropping

Share-cropping is discussed in the earliest legal works of Islam, and where South Arabia is concerned I possess ample documentation both in printed works and in MSS, as for example in the *Fatāwā* of Bā Makhramah,[103] for naturally

[100] cf. M. J. de Goeje, *Liber expugnationis regionum . . . al-Beládsorí*, Leiden, 1866, glossary, 106, النائبة النازلة ونوائب المسلمين ما ينوبهم من حوائج كإصلاح القناطر وسد البثوق وغير ذلك. Dozy's *Supplément* contains further examples.

[101] *Prose and poetry*, Ar. text, p. 22, l. 18 ; *Gloss. daṯ.*, *wakkad* ' préparer, apprêter, mettre en ordre '.

[102] ibid., pref., p. 27.

[103] See p. 35, n. 11 ; photocopy fol. 277 r.–v. Cf. Yaḥyā b. Ādam, *Kitāb al-kharāj*, trans. A. Ben Shemesh, *Taxation in Islam*, Leiden, 1958, 102.

in the case-law of an agricultural community this is a subject of frequent occurrence. I should also refer to my edition of Baḥraq,[104] a treatise composed about the beginning of the sixteenth century (between A.H. 894 and 920) which contains a *musāqāh* plea. However, it is appropriate first to cite three types of contract from a much more recent author, the s̲h̲aik̲h̲ Muḥammad b. ʻAbdullāh b. Aḥmad Bā Sawdān, who composed his *Taḥṣīl al-maqṣūd fi-mā ṭuliba min taʻrīf al-ʻuqūd* [105] in 1315/1897, though his local information refers more particularly to Dawʻan than Tarīm.

(١) صيغة المساقاة

فيكتب: الحمد لله، وبعد، فقد ساق [106] (كذا) فلانٌ فلانـًا على النخل المعروف بكذا، مساقاة شرعية، مدة سنة كاملة، أوّلها شهر كذا، بناصفة الثمرة؛ وعليه إصلاح النخل المذكور، وتلقيحه، وتنقية نَـهْرِه، وإصلاح الأجاجين، وتنحية الحشيش، وحفظ الثمر، وجَـذاذه وتجفّـفه؛ يفعل ذالك بنفسه، أو بنايِـبه.

(٢) صيغة المخابرة والمفاخذة

ويقال لها المناصبة؛ وهى دفع الأرض لمن يغرسها من عنده، والشجر بينهما بالسَوِيـَّة أو التفضيل. فيكتب: الحمد لله، وبعد، فقد اتّـفق فلان وفلان على أن يغرس فلان مكاناً، أو ارض الفلانى. ثم يحددها بما به حفرة [107] بما شآء من أنواع النخل، وعلى الفخيذ المـَقـَالـِع، والمـُؤَن [108]، والسقى، والتنمية الى التعتيق بعـُرف الجهة؛ وذالك فـَخـْذ النصف، أو على المناصفة [109] تخابرا على ذالك مخابرة ومفاخذة صحيحة شرعية. كان ذالك بتاريخ يوم كذا من عام كذا.

(٣) صيغة المزارعة والمخابرة فى الارض لمن يزرعها

فيكتب: الحمد لله، وبعد، فقد قال فلانٌ لفلان زارعتـُك على ارض كذا؛ وينعتها؛ لتزرعها، وعلىّ بذرها، وعليك نصف ثمرتها. ومثله: عاملتك على ارض كذا،

[104] ‘ Forms of plea, a Šāfiʻī manual from al-Šiḥr ’, *Rivista degli Studi Orientali*, xxx, 1–2, 1955, 11.

[105] In the Āl al-Kāf Library, Tarīm. The family of Bā Sawdān lived at al-K̲h̲uraibah of Dawʻan, and are described as of Kindah. Muḥammad and his famous father ʻAbdullāh have many works in MS as yet unpublished of which I have some record. According to ʻAlawī b. Ṭāhir, *al-S̲h̲āmil*, photocopy in SOAS Library, 141, Muḥammad died in 1281/1864–5. For his biography see ʻAbdullāh b. Muḥ. al-Saqqāf, *Tārīk̲h̲ al-S̲h̲uʻarā' al-Ḥaḍramīyīn* (Cairo, A.H. 1353–), III, 196–201, with a slightly different title for this *risālah*.

[106] Read *sāqā*.

[107] The *ḥufrah* is the trench or runnel, and appears to be used here much as one would say ‘ so many head of horse ’.

[108] To be read *mu'an* (?). Bā Faqīh al-S̲h̲iḥrī's history (my MS, fol. 53r.) has the phrase مونة السفر, baggage.

[109] This phrase appears to mean that they have equal shares in the palm plantations resulting from the efforts of the planter. The term *taʻtīq* which precedes, probably means until the land and plantations are in a proper state of cultivation.

وعليك بذرها ، ولك نصف ثمرتها ، مزارعة ، أو مخابرة ، صحيحة صريحة شرعية ، جامعة الشروط المقتضية للصحّة المرعية. كان ذالك بتاريخ كذا في عام كذا.

In the first of the aforegoing patterns of contract, the peasant engages to work on the palms for a year, in return for which he is to receive half of the date-crop. He undertakes to tend the palms, fertilize them, clean the trunks, maintain the runnels about their roots, keep them free of weed, to look after the dates, cut and dry them, either labouring himself or delegating another person to execute the work on his behalf.

The contract that follows is known as *mukhābarah* or *mufākhadhah*,[110] but Bā Sawdān adds a synonym *munāṣabah* not so far known to me from other writers or from informants ; possibly it is Daw'anī. The ground is given to a man to plant with date-palms, the trees to be shared in equal parts (*bi-'l-sawīyah*) or unequally, between the two parties to the agreement. The number and types of palm to be planted are specified. *Inter alia* the working partner has to deal with the young palms (*maqla'*), irrigate them, and rear them up to the *ta'tīq*, this last term not being precisely known to me. *Mu'an*, Bā Ḥashwān informs me, means ' gear ', and would include the date-basket cover (*khubrah* in Wādī Ḥaḍramawt, but perhaps more often *shāyah* (pl. *shawāy*) on the coast), the *simsim* oil used to oil the young date shoot, and the rags bound round it to shield it from the sun. He remarked that the *mawwāl* or capitalist in a share-cropping contract who owns the palms would give the *ḥarrāth* who looks after the palms, one sprig (*khīl*) of dates from each palm as his share of the contract, but this last item of information is in application to the first type of contract in this series. To this type of contract perhaps, a *rajaz* that I heard in Tarīm, sung at the ploughing (*'amālah*) and at the sowing (*ṭaiyar dharī*) alludes :

من ما مَعَهْ حرّاث يبيع المال وآلّا ينشُرُهْ.[111]

He who has no peasant-cultivators must sell his property or else give it to someone else to cultivate it (*ya'mur-uh*). The verse, I was told, alludes to that category of contract where the cultivator and owner each receive half of the crop.

The last form of contract (*supra*) appears to cover cases where the peasant

[110] The *Liber Mafâtîh al-olûm*, 16, defines *mukhābarah* as *al-muzāra'ah bi-'l-thulth aw al-rub'*, or the like. I am, however, inclined to link it with *khubrah* the basket-cover placed over the ripening dates, but the material quoted by Lane is interesting and not necessarily at variance with this suggestion. Particularly interesting is the reference to Khaibar in this connexion, where dates were an important crop. *Mufākhadhah* might be connected with *fukhtah* the fertilization of the female palm tree, but *fākhadh* means ' to divide ' as does *nāṣab* and one can say *fākhadh al-laḥm* ' he divided the meat ', according to A. M. Bā Ḥashwān. Al-shaikh 'Alī b. 'Abd al-Raḥīm Bā Kathīr is known to have composed a *Manẓūmah fī Aḥkām al-muzāra'ah wa-'l-mukhābarah* (*Tar. Sh. Ḥaḍ.*, ed. cit., II, 71) but no copies are yet known. 'Abd al-Raḥmān b. Muḥ. . . . al-Mashhūr, *Bughyat al-mustarshidīn*, Cairo, 1936, 162 seq., has much additional informative material on *mughārasah, mufākhadhah, munāṣabah*, and *mukhāla'ah*.

[111] *Nashar* ' to give one's land to someone else to cultivate '. *Munāsharah* is similar to *mufākhadhah*, and a *nashīr* is defined as *alladhī ya'mal al-arḍ 'ind istiqbāl al-suyūl*. *Nishr* was explained to me as ' sharing in the crop '.

is required both to cultivate field-crops and attend to palms planted on the property, as they undoubtedly would be planted along the irrigation channels leading from the wells, and probably elsewhere.

In Tarīm a *ḍaʿīf* or peasant of the category called *ṣālī* (pl. *ṣilāh*) always works under a *ṭabīn* or capitalist, with a fixed wage (*murattab*), plus dates and grain, and a share in the crop (*qism fi 'l-ḥarth*). It may be observed in passing that at al-Khashaʿah where there has been considerable agricultural development of latter years, one speaks of having a share in the district (*khushr fi 'l-Khashaʿah*), i.e. one is *mushtarik* 'sharing' in the cultivation of a piece of land.

Of late years the introduction of pump-irrigation on a large scale in Ḥaḍramawt has led to the modification of these very ancient and traditional forms of share-cropping contract to meet the new conditions imposed by the use and maintenance of these pumps. Another recent innovation in one or two places is the introduction of piped water from perennial streams, a type of project usually financed by Ḥaḍramīs who have made fortunes in Saʿūdī Arabia.

The type of dispute that can arise in share-cropping contracts is in evidence in the two types of plea which I quote below from Saiyid Muḥsin b. Jaʿfar b. ʿAlawī Bū Numaiy,[112] the first of these being a complaint preferred by the owner of the property to be irrigated and cultivated, against the peasant with whom he has contracted to execute the actual work. The second complaint is that wherein the peasant claims that the owner has not fulfilled his side of the contract, and where he demands immediate payment of what is due to him.

باب المساقاة

(١) صورة دَعوى المالك المساقاة

أن يقول زيد: أدّعى بأنى ساقيت عمرًا هذا على البستان الفلانى بجميع ما فيه من النخل على اختلاف انواعها، على أن عليه سقيها وتعهدها وتسوية سواقيها وإصلاح الحفر التى فى أثناء السواقى وتلقيحها وحفظ ثمرها وقطعها وغير ذلك مما فيه صلاحها، وله فى مقابلة عمله الثلث من الثمرة الحاصلة منها، وقد امتنع من ذلك، أو هرب، أو قصر فى عمله، وانا مطالبه بالوفاء بما التزمه من العمل، فمُرْه، أيها الحاكم، بوفاء ذلك.

(٢) صورة دعوى العامل بمقابلة عمله فى المساقاة

أن يقول عمرو: أدّعى ان زيدًا ساقانى على البستان الفلانى بجميع ما فيه من النخل على اختلاف انواعها، على أن علىّ سقيها وتعهدها وتسوية سواقيها وإصلاح حفرها وتلقيحها وحفظ ثمرها وقطعه وغير ذلك مما فيه صلاحها، ولى فى مقابلة عملى الثلث من الثمرة الحاصلة منها، وانا مطالب له بما شرط من الثمرة حالاً، وهو يلزمه وقد امتنع من ذلك، فمره، أيها الحاكم، بتسليمه الىّ.

[112] *Taḥṣīl al-daʿāwī fī rafʿ al-shakāwī*, al-Mukallā, n.d. (but received in 1958), 21–2.

At the fertilization of the palm trees I have already reported that certain songs are sung [113] though I had at that time no examples. Professor J. Robson has somewhere stated that it is believed that evil spirits attend the fecundation of the date-palm and the collection of the date-harvest in southern Arabia, so that the peasant going up the tree sings to scare away the Jinn.[114] In February 1954 in Saiwūn I heard a man in our garden singing as he climbed the tree to perform the *fukhtah* or fertilization, the charming ditty,

يـا طُيور الخـَلا والصَيْـد والنُـوب
يا خـَيْـر مطلوب

' O birds of the open country, game, and bees,
O what an excellent thing to ask'.

The allusion to birds has probably a significance that is at present unknown to me, but the reference to game at once brings the ritual Ḥaḍramī hunt to mind, and honey has properties next to magical.[115] Another ditty which I heard in Saiwūn seems, if it has been correctly given me, to be of rather unconventional form.[116]

وُٱزْرَعْ يا زارِعْ
عَسَى مِنْ حُضُورَكْ
وَلا ضَار يَضورَكْ
يا ٱلله بِصَلاحَكْ
يا مَوْلَا الصَلاح
والكَرَع فى جُذورَكْ
وَسِّعْ نُحُورَكْ
والزَكاه والمعروفْ
والخَيْر المَوْصوفْ
قِسْمِى وقِسْم الغُرابْ
وبا فَرارَهْ ما يِجيهْ
ولا يَقْرُبْ حَوَالِيهْ
وُٱزْرَعْ يا زَارِعْ.

[113] *Prose and poetry*, preface, 37–8, mostly an ' ooooooo ' sound, with words appearing here and there.

[114] While I cannot positively affirm that this is also the case in Ḥaḍramawt it seems to me very likely.

[115] Its curative properties were praised by the Prophet and to this day these are highly esteemed.

[116] Neither of these two pieces seems to have a consistent metre.

'Till, O husbandman.
May it be that you come [117]
With no harm to befall you.
O God, for your well-being,
O Lord of well-being,
The rain-water at your roots
Make wide your fronds.[118]
Charity and customary benefit,[119]
Outstanding prosperity
In which I and the crow have our shares.
May the worm Bā Farārah [120] not come to it,
Nor come anywhere near about it.
Till, O husbandman.'

It is to my regret that, at the time of taking down this ditty I thought the Arabic simple enough to dispense with detailed comment, if indeed that could have been given for it by the peasant. It now seems to me that this is almost a hymn to the palm tree, or an invocation to it to grant of its fruit, and to Allāh to protect it. This poem would then be further evidence in support of my contention that traces of palm-worship survive to this day in the folk memory of Ḥaḍramawt. Although I obtained this ditty in Saiwūn I was told at Ḥuraiḍah that it was also used there, so perhaps it is widely spread throughout the country.

Documents relating to property

Documents or deeds relating to the possession of property are known as *khuṭūṭ māl*. In pre-Protectorate days the Kathīrī sulṭāns used to take a fee of 10% or more of the price of houses and land for their signature to the transfer documents, but this is no longer charged. In general tribesmen with protection rights (*shirāḥah*) [121] over palm plantations used to exact a fee when one palm-owner disposed of his property to another party. In Tarīm when the owner of ground is not known it is called *arḍ majhūlah*, a term probably universal. I am inclined to consider it axiomatic that there is no inch of Arabia that is not owned by someone, however desert it may appear, and though we might be inclined to regard a piece of land which the Arab calls *arḍ majhūlah* as having no owners,

[117] I had thought the palm-cultivator to be addressing himself, but I now think he must be speaking to the palm tree. This has made me wonder if *ḥuḍūr-ak* could mean 'to ripen'. *Yā mawla 'l-ṣalāḥ* must, I think, be apostrophizing the palm tree, and it is reminiscent of the title given to saints, even unknown saints such as Mawlā Maṭar.

[118] *Gloss. dat.*, *kara'* 'eau de pluie'. *Naḥr* according to Landberg, *Ḥaḍramoût*, 720, is 'couronne du palmier', but *nuḥūr* was explained to me as *sa'f*.

[119] For the *zakāt* see *Prose and poetry*, Ar. text, p. 24. By *ma'rūf*, the *ṭarḥah*, or small gift of dates to a beggar at the date-harvest may be meant.

[120] Bā Farārah, described as a *dūd tawkul al-kharīf*, a worm which eats the ripe dates.

[121] For a *maqāmah* on this institution see *Prose and poetry*, Ar. text, p. 16 seq. A dispute over *shirāḥah* figures in Muḥ. b. Hāshim, *Tārīkh al-dawlat al-Kathīrīyah*, Cairo, 1948, 134. *Māl* is stated by *al-Shāmil*, 175, to be used in Ḥaḍramī dialect in the sense of cultivated fields (*jurūb*).

I feel that behind this term lies the Arabian conception that it is owned by someone even if the owners be unbeknown to us. I do not, however, propose to go into this subject though I think the sections on *iḥyā' al-mawāt* in Arabian law-books would probably enable us to form a true conception of how Arabians regard the holding of waste land.

During one of our sessions in Tarīm, Raḥaiyam dictated to me the following document which, though it relates to no particular holding, unless indeed it be he had some property of his own in mind, may be considered typical of property in palm trees. Though a holding of five palms may seem small, it must be recalled that some owners possess as little as one palm only or a share in a palm.

لفلان بن فلان خمس حُفَر نخل، حمراء وصَفَارِى، عند مَضْلَعَة مَتّاش، من ذلك دَوَّار حمراء أُنْهُر خمسه قِبْلى نخلة فلان ونَجْدِى نخلة فلان وبَحْرِى فوُصِح (خالى) وشرقى عَيْن المَسِيله. وله من الحَيْد القبلى ثلاث حُفَر حمراء وصفارى، الأُولَى سُرَيَّعِى مثلوث شرقى حمراء المذكور ونجدى قِرَيْن فلان بن فلان وبحرى مَنْقَر فلان بن فلان وقبلى المسيله فوق الحيد ومع زار حيد شرقى السريّعى المذكور ونجدى خَلْعَة فلان وبحرى مدينى فلان بن فلان.

'So-and-so (*A*) owns five circular runnels [122] of palms, red date type and yellow date type,[123] at the Barrage of Mattāsh,[124] consisting of a multi-trunk palm [125] of the red date type with five stems,[126] west of So-and-so's (*B*) palm, north of So-and-so's (*C*) palm, south of an open (empty) piece of ground,[127] and east of the channel of the flood-bed.[128] Of the western palm plantation [129] he owns three circular runnels of red date type and yellow date type, the first of which is a *Suraiya'ī* [130] with three stems [131] east of the said red date type,

[122] See p. 61, n. 107.

[123] For these types see *Prose and poetry*, Ar. text, p. 128, v. 15.

[124] Mattāsh would be so called after one of the nine *fakhīdhahs* of Yāfi' which used to control Tarīm.

[125] A *dawwārah* is a palm with three or more trunks stemming from the root.

[126] *Nahr* (pl. *unhur*), syn. *judhū'*, trunks or stems of the palm ; where a palm has several stems from one root these are known as *awlād*. It has occured to me that the word *anhār* in Baḥraq, ' Forms of plea ', 11, may be another plural in this sense.

[127] *Fūṣiḥ* was said to mean an open piece of ground without *'ilb* trees or palms. Perhaps it is to be linked with *fasaḥ* in the *Gloss. daṯ.*

[128] The Masīlah is the flood-bed of the Wādī Ḥaḍramawt, which is very wide in places, the *'ain* seems to mean the channel in it where floods normally run.

[129] *Ḥaid* is palm-grove at the side of the Masīlah, cf. fig. 8. *Al-Ḥayāt al-sa'īdah*, 5. mentions Ḥaid Qāsim which one passes coming into Tarīm. *Qāsim* is apparently *maqāsim al-mā'*, a place where the flood-waters are distributed, not a proper name. In coastal Ḥaḍramawt a *ḥaiṭ* is a *bustān*, and I think that *ḥaid* in the sense of palm-grove is to be linked with this word, not with *ḥaid* in the sense of a mountain.

[130] *Prose and poetry*, Ar. text, p. 128, v. 14. A red variety becoming black, called ' swift ' because it is the first date to ripen.

[131] I have not confirmed *mathlūth* in this meaning.

north of the double-stemmed [132] palm tree of So-and-so (*D*), south of the tall single palm [133] of So-and-so (*E*), and west of the flood-bed above the palm plantation, along with a *Zār* [134] palm of a grove east of the said *Suraiya'ī*, north of the young fruit-bearing palm [135] of So-and-so (*F*), and west of the *Madīnī* [136] palm of So-and-so (*G*).'

In Saiwūn I was informed that a small group of palms is called *ḥailah*, but do not know if this term is used in other places, and there may be other technical names for groups of palms also.

Floods in Ḥaḍramawt

The list of floods *infra* has, unless otherwise stated, been based upon notices in *Tārīkh Shanbal* up to the year 905/1499–1500, some of the other authors quoted being from annotations to my edition, in preparation, of this chronicle. For the remainder of the tenth/sixteenth century, unless otherwise stated, the notices are derived from the chronicle of Bā Faqīh al-Shiḥrī. The list is defective in several respects—obviously not all of the great floods of the period have been chronicled, and many other MS sources, as for example *al-Sanā' al-bāhir*, and the *Mukātabāt* al-Ḥabīb 'Alī b. Ḥasan al-'Aṭṭās cited by *al-Ḥayāt al-sa'īdah*,[137] could supply data. Shanbal sometimes draws upon sources which know nothing of Ḥaḍramawt of the interior, and vice-versa. The dating of rains occurring at certain stars has been established from the *Ḥisāb al-Shibāmī*.[138]

It is surprising to discover that after so many destructive floods there is no evidence quoted of cultivated areas being abandoned, and the same localities appear to continue in cultivation. Were the notices in this list to be compared with the evidence on the ground it is just conceivable that the effect of some of the major floods of olden times could be assessed. Daw'an valley, doubtless because it is so narrow and takes the spill of so much of the Jōl, suffers recurrently from violent floods, as indeed it did in September 1954 when I was in

[132] *Prose and poetry*, Ar. text, p. 130, v. 30, a *qirain* is a palm with two stems from one root.

[133] A *manqar* is a tall single palm.

[134] *Prose and poetry*, Ar. text, p. 131, v. 41, the *zār* is a yellow type of date.

[135] The *khal'ah* (pl. *khilā'ah*) is a small young palm which has reached the fruit-bearing stage. Syn. *naqīl*. Bā Hārūn, MS cit., fol. 8r., speaks of a *saiyid* who had *khali' fī 'l-Maqta' taḥt Rawghah . . . idhā waqa' al-kharīf mā yukhallī aḥad yashraḥ khali'-ah*, i.e. palms in al-Maqṭa'(?) near Rawghah . . . when the ripe dates came he would not let anyone act as watchman over his palms. Raḥaiyam commenting on *Prose and poetry*, Ar. text, p. 163, stated that *khali'* is pl. of *khal'ih* ' palm '. *Khilā'ah, Prose and poetry*, Ar. text, p. 130, is the act of planting palms. Bā Hārūn, fol. 9v, contains the phrase *yaghris khil'ān*, the latter word also a pl. of *khal'ih*. A celebrated *saiyid* ancestor is known as Khāli' Qasam, i.e. the man who planted palms at Qasam village. Raḥaiyam told me that at Rawghah Satan became a watering ass to one of the *saiyids* and used to work on his *khal'ah*. This field where Satan worked is still known at Rawghah, lying by the mosque there.

[136] *Prose and poetry*, p. 128, v. 13, the *Madīnī*, said to originate from Medina, is a type much esteemed for its quality.

[137] *Al-Ḥayāt al-sa'īdah*, 18.

[138] cf. my ' Star-calendars and an almanac from south-west Arabia ', *Anthropos*, XLIX, 3–4, 1954, 434–5. The almanac in question has entries relating to flooding, but these apply to the Yemen.

Ḥaḍramawt, but it never seems to be abandoned. There are many places where the fields have been turned into stony *wādī*-bottom, but perhaps the soil may be piled up in other places. Exceptionally heavy floods sometimes seem to drive up-stream as in 794/1391-2 when the waters of Wādī Kh̲ōn mounted some miles up-*wādī* to 'Īnāt. The flood of 894/1488-9 is remarkable for rising up the mountain-side, i.e. above the termination of the scree line, and washing away threshing-floors placed there so as to be clear of this danger. Locusts in great number often follow the floods, a phenomenon that may be well known to the Desert Locust Survey.

For reasons I have not explored Ḥaḍramīs consider that the south wind is beneficial at times of flood, for a Tamīmī *rajaz* which I heard runs,

شَارَهْ بِلَا بَحْرِى كَمَا   بَاب آرْتكب مَا لُهْ قَوَام

' Rains (and floods) without the *Baḥrī* wind,
Are like a door set up without supports '.[139]

Preliminary list of rains and floods in Ḥaḍramawt, coast and interior

658/1259-60. Many houses destroyed by rain. (This reference may perhaps apply to the coastal area only.)

685/1286-7. Violent rain, tempestuous wind, high seas at Ẓufār. Crops ruined, many wells destroyed, and many (coastal ?) districts ruined.

721/1321-2. Famine in Wādī Ḥaḍramawt causing much death to man and beast, implying thereby a failure of rains.

725/1324-5. Rain at Ẓufār destroys houses, mosques, and gardens.

758/1356-7. Heavy rain in Wādī Ḥaḍramawt followed by locust swarms.

775/1373-4. Great flood in Wādī Daw'an sweeps away the palms of Raḥbat Daw'an entirely, carrying some palms with their crop as far as al-Kasar, clearly a summer flood.[140] The *Tārīkh Bā Sharāḥīl* chronicles a severe famine this year, perhaps a result of the flood, remarking on the high prices in Wādī Daw'an.[141]

787/1385-6. Heavy rains, thunder and lightning at al-S̲h̲iḥr destroying houses and part of the town wall.

794/1391-2. General rains in the month of Rajab. From S̲h̲ibām and its districts come great floods. Wādī Kh̲ōn floods and (its waters) mount to 'Īnāt causing much destruction. Wādī T̲h̲ibī floods, carrying off palms at al-Naqar (between Tarīm and Ḥuṣn Fallūqah). In this year there was great ease.

810/1407-8. Rain in areas in Ḥaḍramawt (Wādī) and al-Kasar results in low prices for commodities, i.e. millet, dates, and ghee.

811/1408-9. Commodity prices still lower, implying perhaps a continuation of good rains.

---

[139] *Qawām*, the supports or posts, the lintel being *kafāh*.

[140] *Al-Ḥayāt al-sa'īdah*, 15.

[141] For this author cf. ' Historians and historiography of Ḥaḍramawt ', *BSOAS*, xxv, 2, 1962, 242, as also for Bā Sanjalah, *infra*.

812/1409–10. General floods in all districts. A great flood comes from Ju'aimah (north of Shibām). This is followed by locusts.

823/1420–1. A pestilence in Ḥaḍramawt with great scarcities and much death from hunger, even cannibalism being alleged. This may indicate a failure of rains.

832/1428–9. In the star al-Dhirā' (14–26 January), and again at al-Ṣarfah (20 March–1 April) rain and floods at Shibām. (*Tārīkh Bā Sanjalah.*)

838/1434–5 or 839/1435–6. Rains in nearly every district which, in Daw'an, carried off many palms and destroyed fields (*aḥjāl*). (Bā Sanjalah.)

848/1444–5. After a period of high prices for commodities, al-Kasar and other places have rain in Kharīf (July–September).

853/1449–50. The barrage (*ḍamīr*) of Wādī Thibī was built up at the cost of Sulṭān 'Āmir of the Yemen, but a *sail* from the Wādī Thibī destroys it.[142]

860/1455–6. Heavy rain in all districts.

866/1461–2. Heavy general rain in Ḥaḍramawt (interior ?).

891/1486. All Ḥaḍramawt had rain with two great floods in the star al-Ṣarfah (20 March–1 April).

894/1488–9. A great flood in the star al-Iklīl (24 May–5 June) from al-Khamīlah (south of 'Amd) which was swelled by the flow from 'Ain, 'Amd, Daw'an, Sar, 'Idim, Thibī, etc. It destroyed threshing-floors (*aṣārī*), many palms, barrages (*maḍāli'*). It destroyed Qasam, Ribāṭ Āl al-Zubaidī, part of Bōr, then people got it under control. It rose about 10 cubits (say 15 feet) on the foot part (*safḥ*) of the mountains.[143] Locusts in quantity followed and their young (*dabā*).[144] The locusts ate the crop but its roots, from which a new crop would spring, remained, and the people seem to have done even better because of this. Bā Sanjalah adds that some of Masīlat Āl Bā Dhīb,[145] Bait Maslamah, and other places, perhaps Wādī al-Mukhainīq (but the MS is badly wormed at this point) were destroyed. Four years seem to have been spent in the repair of some places.

896/1490–1. Rains in Daw'an, al-Kasar, Wādī 'Amd and most of (interior ?) Ḥaḍramawt. The barrage (*'aqm*) [146] of Wādī Shibām was broken (Bā Sanjalah).

900/1494–5. A great flood in Daw'an destroys many palms and much land (Bā Sanjalah).

901/1495–6. In Jumādā II in the middle ten days there was rain in Shibām

[142] *Al-Ḥayāt al-sa'īdah*, 15, but in the following year Bā Sanjalah records rains in Ḥaḍramawt after much dearth, the MS unfortunately being wormed at this point so that it is not clear that rains had been short previously.

[143] cf. *Prose and poetry*, Ar. text, p. 129, *Fī safḥ Daw'an luh awṣār*. The *safḥ* is the point at which the mountain foot touches the level plain.

[144] Young of locusts, with wings about one inch long.

[145] The letters are undotted in the text ; my reading is conjectural.

[146] Perhaps this should be read '*uqam* with Landberg, *Ḥaḍramoût*, 660, deflectors. These would be the deflectors which lead the water into the fields as in pl. IV. Cf. the Aden paper *Ṣawt al-Janūb*, II, 30, 1963, العقمة عبارة عن سد طينى يقـام لتحويل مياه السيل.

and ‘ the *wādīs* ’, the flood entering the Shibām cemetery (west of the town (pl. II)) (Bā Sanjalah).

903/1497–8. A flood at Shibām and elsewhere on the first of the star Sa‘d Bula‘ (25 July–6 August) (Bā Sanjalah).

905/1499–1500. In the star al-Haq‘ah (1–14 December) Ḥaḍramawt had good rains (Bā Sanjalah).

929/1522–3. Rain and a great powerful wind called in al-Shiḥr *al-Shillī* [147] which caused destruction in al-Mishqāṣ, Ṣarīḥ/Ṣuraiḥ (?), and the Wādī (Ḥaḍramawt ?).

930/1523–4. Rain and wind but less than the *Shillī* of the previous year.

935/1528–9. In al-Shiḥr district at the close of Rabī‘ II and beginning of Jumādā I heavy rain destroying buildings in Shuḥair, Rawkab, etc.

939/1532–3. On 28 Shawwāl heavy rain and contrary winds in al-Shiḥr district destroy many palms. News arrives from Daw‘an that a number of palms had been destroyed by floods the like of which had never been known, 4,000 being destroyed by Riḥāb alone. Their fields (*dhubūr*) had been so destroyed that only a few were left in Daw‘an and only a few palms. The flood had risen to places to which it had never occurred to anyone it might attain.

945/1538–9. A famine in Ḥaḍramawt, Daw‘an, and al-Shiḥr, from which one may imply a failure of rains.

948/1541–2. In Jumādā II prices became cheaper and rain (*ghaith wa-raḥmah*) was general, contrary to the high prices and famine of the preceding year (947).

968/1560–1. Famine in the Ḥaḍramawt and al-Shiḥr districts from Rajab to Ramaḍān (implying that the spring rains had failed).

970/1562–3. *Al-Nūr al-sāfir* [148] states that on 2 Shawwāl befell, ‘ the mighty great *sail* in Ḥaḍramawt, the like of which has never been heard, destroying many palms. The people of that province recall it to this day and date (events) by it. By them it is known as *Sail al-Iklīl* (24 May–5 June) ’. After quoting a chronogram he continues, ‘ It is said that in ancient time there was a *sail* or two like it or approaching it, but there is no strength or power but in God ’. Bā Faqīh al-Shiḥrī places the rain as falling on the last night of al-Iklīl, in al-Shiḥr district accompanied by a tempest lasting day and night, for some 48 hours. Then followed heavy rains, destroying buildings everywhere so that people fled from their houses into the streets. These disasters took place in Ḥaḍramawt (of the interior), al-Shiḥr and al-Mishqāṣ districts, even as far as Hurmūz in the Persian Gulf. It carried off people from Tarīm and al-Masfalah : some even appeared dead in the sea off Ḥairīj, but others escaped by clinging to palm-trunks. Countless palms were washed away and the cultivated valley land (*dhabr al-Masīlah*)

[147] T. A. Shumovsky, *Tri neizvestnye lotsii Aḥmada ibn Mādzhida, arabskogo lotsmana Vasko da Gamy*, Moscow, 1957, fol. 88r., l. 24, reads this word as *al-Shullī*, but in G. Ferrand, *Instructions nautiques*, Paris, 1921–8, I, 148 and elsewhere, *al-Shillī*.

[148] p. 274.

of Wādī Bal-Ḥāf and its gardens[149] were turned into a *raḥbah*.[150] It swept away all the palms of al-Ghail al-Asfal and al-Ghail al-A'lā, the *'ilb* trees of al-Raidah and about 170 palms (or *'ilbs* ?) in Maifa', as well as the palms of Wādī Dhahabān of Shibām.[151] Al-Shiḥrī gives an account also of the shipping destroyed along the coast, adding that much verse had been composed about this mighty flood.[152]

979/1571–2. On 11 Jumādā II, heavy rains on al-Shiḥr district.

998/1589–90. *Al-Nūr al-sāfir*[153] chronicles a great flood in Wādī 'Idim at the star al-Thuraiyā (22 November–4 December) greater than *Sail al-Iklīl*; it is known as *Sail al-Thuraiyā*.

1049/1639–40. A flood is mentioned by 'Abdullāh b. Ja'far b. 'Alawī in the Wādī 'Idim,[154] called *al-sail al-mahīl* ' the terrible flood '.

1056/1646–7. According to *al-Ḥayāt al-sa'īdah*[155] the second flood of al-Iklīl took place some time before this date. It was less serious than the flood of 1124 (*infra*).

1124/1712–13. *Al-Ḥayāt al-sa'īdah*[156] places in this year *al-Sail al-Hamīm* in the star al-Ḥūt (1–13 October) or 16 Ramaḍān, a flood which swept away palms and *dhubūr*. Most of the destruction took place in Bait Muslimah and the Masīlah of 'Idim[157] and the alluvial earth of the *wādī*-bed (*baṭḥā'*) of Shibām, taking away palms from the latter. It also swept away many *'ilb* trees. It passed the *ghiyāḍ* (gardens) of al-Mishqāṣ district, eroding its *dhubūr*. It appears to be the same flood that is described by 'Abdullāh al-Ḥaddād[158] as destroying palms and huts about Tarīm. It then entered the village of al-Nukhr[159] at 'Īnāt and its cemeteries. It also destroyed most of the houses in Qasam and its Jāmi' mosque.[160] The text as quoted in *al-Ḥayāt* is a little obscure

[149] My MS has *ghaiṭ* for which *al-Ḥayāt al-sa'īdah*, 16 (based on Bā Faqīh al-Shiḥrī) has *ghiyāḍ*.

[150] A stony-bottomed flood-course, see p. 37 *supra*.

[151] In the Wādī Dhahabān is the *mawza'* of Shibām, of which Muḥ. b. Hāshim, *Tārīkh al-dawlat al-Kathīrīyah*, Cairo, 1948, 165, says, that it is *li-ṣadd miyāh al-suyūl wa-ḍabṭ-hā*. He adds *wa-'alaihi madār al-raiy fī 'l-maghāris ḥawla-hā*.

[152] Bā Hārūn, MS cit., story 253, refers to *Sail al-Iklīl* which carried off the palms of Ḥaḍramawt, and story 274, to the same flood which carried off the palms of al-Maqṭa' (?), houses in Qasam, etc. *Al-Rābiṭat al-'Alawīyah* (Batavia), II, 7, A.H. 1348, 275, in a note on floods in Ḥaḍramawt, alludes to this flood and the large number of palms destroyed by it in the Tarīm district, as well as some 20 mosques in the area.

[153] p. 458.

[154] Author of *Tadhkirat al-mutadhakkir fī-mā jarā min al-sail al-mutabaḥḥir*, Dār al-Kutub MS, no. 1257, a verse *maqāmah* of about 20 pages with *akhbār al-suyūl al-qadīmah*, and references to the Iklīl flood and *al-Sail al-Hamīm*.

[155] p. 17, but this statement may be incorrect.

[156] p. 16.

[157] Bait Muslimah is *qidā 'Idim* ' in 'Idim direction '.

[158] A celebrated eighteenth-century Ḥaḍramī *ṣūfī 'ālim*.

[159] A nukhrah (pl. nikhār) was defined by Raḥaiyam as, أرض ليّنة ينخرها الماء – يعني يفجّرها.

[160] In the Ahl Sahl Library, Tarīm, Muḥ. b. Zain b. 'Alawī b. Sumaiṭ, *Ghāyat al-qaṣd wa-'l-murād fī manāqib Shaikh al-'Ibād, 'Abdullāh b. 'Alawī b. Muḥ. al-Ḥaddād*, story 279, has an account of the great flood that swept away the Jāmi' mosque of 'Īnāt. For B. Sumaiṭ cf. ' Materials for South Arabian history, II ', *BSOAS*, XIII, 3, 1950, 582.

at this point but it appears that a part of this district had the aspect of a bay of the sea. Al-Ḥaddād is quoted from another place as saying that this was a flood of a kind never previously known, worse even than the floods of al-Iklīl, affecting from Tarīm to 'Īnāt and Qasam, destroying 10 mosques built alongside the Masīlah. It ran into Nukhr of 'Īnāt into the cemetery and among the houses. It appears that the perennial streams (*ghail*) of Sāh were also heavily swollen.

1332/1913–14. 'In the star al-Fargh (18–30 September 1914), Upper Ḥaḍramawt drank, from Wādī 'Amd and Daw'an—to al-Ḥijail it reached, and Sadbah, Ḥawrah, 'Arḍ Bin Makhāshin the water reached, and may a good raining be hoped for . . . . Before that Rakhyah, 'Irmā, Duhur, and al-Qiblī (the western area of Ḥaḍramawt) drank in the star Sa'd (al-Akhbiyah ? commencing 23 August ?) and al-Hajarain and al-Ghabar. Also they mentioned the harvesting when there was already there a first *mūsim* (flood-millet) crop . . . . 'Awaḍ b. Sa'īd wrote this in al-Qa'dah, 1332 '.[161]

Undated. *Sail Bin Ramaidān* is a flood mentioned by *al-Ḥayāt al-sa'īdah* [162] at which Raḥaiyam says he was present and upon which he composed a poem. It could therefore be dated from local information, but it does not appear tc have been so important.

Appendix I

Irrigation terms in Oman

Oman is a territory so little known that it is interesting to note that irrigation there is undoubtedly as complex as it is in south-western Arabia, but the technical terms employed appear to be totally different from those to be met with in Ḥaḍramawt or the Yemen. The vocabulary *infra* is culled from al-Sālimī's *Jawhar al-niẓām* [163] but it is not to be assumed that it is more than a small portion of a vocabulary as extensive as that in any other part of Arabia.[164] Al-Sālimī defines a *wādī* in the following verse :

الوادى مجرى الماء فى السيول   من جملة الشعاب والفحول

[161] My rendering of the Arabic is tentative. I found this passage in a MS in the library of the late Sulṭān 'Alī b. Ṣalāḥ at al-Qaṭn. It would appear that when the first crop of millet was being harvested the floods caused the roots to sprout a second time.

فى نجم الفرغ عـام ١٣٣٢ شربة علوا من وادى عمد ودوعن الى الحجَيْل بلغ وسدبه وحوره وعَرْض بن
مخاشن بلغ الما فعسا قطر خير . . . وقد شربت قبل ذلك رخيه وعرما ودهر والقبلى فى نجم سعد والهجرين
والغَبَر كذلك ذكروا اسراب وقد فيه باكر موسم . . . كتبه عوض بن سعيد فى القعده ١٣٣٢.

More correct spellings would of course be شربت, وعَلْوَى, الصُراب. 'Awaḍ b. Sa'īd b. 'Alī Bā Faḍl was secretary to Sulṭān 'Alī.

[162] p. 18.

[163] II, 66, 293, 56, 50, 54, 50, 57, I, 293, II, 65, 58, 69, I, 43.

[164] Al-Sālimī demonstrates that an elaborate system of share-cropping obtains also in Oman, referring (II, 39) to the need for the hired man (*ajīr*) and partner (*sharīk*),

لاقتعاد الأرض والمياه
ليزرع الأرض ويسق الواهى
ويصلح الدروب والسواق.

73

The *shi'b*, he says, is *al-wādī al-ṣaghīr*, and *al-fahl* is *al-wādī al-kabīr*. The *falaj* is defined by him as *tur'at al-mā' wa-'l-nahr al-ṣaghīr aw kull mā shaqq fi 'l-arḍ li-taqsīm al-mā'*. *Māt al-falaj* means that the water of the *falaj* has ceased to flow. The *sāqiyah*-channel he describes in the following lines :

وهى جوائز وحملان ترى　　فالجائز الذى يكون اكبرا
خمس اجايل حوى من أسفل　　وقيل اربعا حوى لا من على

والحملان مسلك تشعبا　　من جائز ليسقى ما قد قربا
مسقاه دون المسقى للجوائز　　هذا الذى يقال غير حائز
والقائد الساقية الكبيره　　وهى التى بها اعتماد الديره
وكل ما يمنع جرى الماء　　يخرجه الشاحب والافتاء

فالجدول الوعب على السواقى　　او غيرها من كل وعب باقى
وذلك الوعب يسمى دكّا　　فى عرف بعضنا لننفى الشكّا

Not all of these words are explained but definitions are given in certain cases only.

الجائز　— نوع من السواقى وهو ما كان منها مشتركاً بين متعدد فى توزيع وسقى ومنها الاجازة والحملان والقائد.

شحب الفلج — قشره بالمسحاة.

الوجين　— شاطىء الساقية . . . وشاطئ الوادى. الوجين أيضًا الارض الغليظة الصلبة والعارض من الارض.

ظهر النهر　Perhaps the bank of a water-channel ?

والسيل إن كسر ظهر النهر يجوز أن يسقى بذاك الكسر.

السبية　— الماء الجارى فى الساقية بعد رد النهر ويسمى أيضاً المجرى.

الاثر　A quantity of water remaining in a stream.

الآد　— نوبة السقى دوران الماء فى الفلج.

اجاله (ج. اجائل)　A sort of *sāqiyah*-channel.

تصريج السواقى — طلاؤها بالصاروج وهو شىء يخلط بالنورة ويطلى بها الحياض ونحوها.

الثجّ/ثجّ — السَيَلاَن.

Appendix II

Additional irrigation terms in Ḥaḍramawt[165]

(1) Classical terms still in current use:

جُرْف, defined as موضع اثر السيل فى الوادى.

ريد (coll. رَيْدَه) defined as نواحى الوادى لحدده.

شيّد داره, to plaster (with lime, *jiṣṣ*) his house.

صنبور holes of a tank (*ḥawḍ*).

قلت defined as رَصَفَة, and as,[166] الماء العِدّ الدائم يكون فى رؤس الأودية غالباً.

كَبَسَ, of a channel (*qanāt*), to fill it with earth. In Oman also, كبس السيل فى الأفلاج means الطمّ والامتلآء بالتراب.[167]

مجهورة, a well when its water is exhausted.

(2) Modern terms:

قَيْد (Wādī ʻAmd), a name given to a pillar, possibly the end of a ruined deflector, for *qaid* in Dathīnah means ' digue '.[168]

مَجْفَا (Saiwūn), a *sawm* or bund.

مَدْمَار (Saiwūn) supporting bank at foot of wall (fig. 11).

مَوْطَا (Saiwūn), a wide low *sawm*.

Fig. 11. The *midmār* or supporting piece at foot of a clay wall.

The following are from Saʻīd of Upper ʻAwlaqī territory:

بَدّ (pl. بداد), a channel. Cf. *al-Shāmil*,[169] الجدول والساقية المتفرعة من الكبيرة.

حَبَط piece of uncultivated land near a field.

شَبَب piece of uncultivated land in a field.

غَبَرَه spring with little water.[170]

مَرِيرَه واوثان low dividing *sawm* between two shares in a single *widin* (field) with land boundary marks.

الحامل حق الودن a *sawm* or bund.

[165] *Al-Lughat al-dārijah*, 169 seq.
[166] *Al-Shāmil*, 165.
[167] *Jawhar al-niẓām*, I, 302.
[168] *Gloss. daṯ.*, 1416.
[169] *Al-Shāmil*, 175.
[170] *Al-Shāmil*, 186.

مِسْمَار means a furrow (*tilm*). I have noted, but without an explanation, the phrase مسار بطون.

مَشْصَن (pl. مشاصن), ' espèce de digue '.[171]

In Sa'īd the local experts who go out to decide in land and water cases are known as *'urrāf*.

Appendix III

While in Johore in August 1963, through Saiyid Ṭāhir b. 'Alawī's courtesy I consulted more of his late father's *Kitāb al-Shāmil* than was hitherto available to me. There (pp. 214 and 224) a most curious account is given as to how the Bedouin of western Ḥaḍramawt seek rain from the celebrated 'Amūdī saint, Shaikh Sa'īd b. 'Īsā (a photograph of whose tomb appears in *BSOAS*, XIII, 2, 1950, opp. 281), a ceremony so essentially pagan that it might almost serve as commentary to certain pre-Islamic inscriptions. The following summary version may throw some light on ancient rites of petition for rain.

In the *Kharf* season, the time of rain, the Daiyin, Mashājir, Bā 'Ubaid, and other near-by tribes come in succession on a visitation to Shaikh Sa'īd at Qaidūn on *Na'ā'im*, i.e. about 2 July, to ask for rain for their country. It is their custom, on arriving at the top of the Jōl overlooking Qaidūn on the Thursday evening, to shout, *Lā ilāha illa 'llāh* ! (*yuhallilūn*), saying, *'Umūm, 'umūm, yā Shaikh Sa'īd* ! By this phrase they mean that the saint should give them general floods and rain in all their valleys (*shi'āb*), and the chronicles mentioned above (pp. 68–72), do actually make an entry from time to time of ' general rain '. (One says in colloquial Arabic, *fi sahn 'umūm al-raḥmah* ' in expectation of general rain '.) Then they descend the *'aqabah*, chanting *zāmils* with verses describing their travel and the long distance they have come to perform the visitation to Shaikh Sa'īd because their land is suffering from drought, and ' it would be shameful of you were we to return without a miracle ' (*'ār 'alaik idh raja'nā bi-lā karāmah*). In composing these *zāmils* the poets play an important part in the ceremony.

On arriving at the foot of the mountain the Bedouin proceed to the tomb of Shaikh Sa'īd and circumambulate the wooden ark-like cover of the grave (*tābūt*). Some bring ghee which they pour on the *tābūt*, and others even jump up on top of the *tābūt* itself to pour the ghee on it. They also circumambulate the other *tābūts* at the tomb, all the time chanting *zāmils*. Their votive offerings of sheep, money, or grain, they deliver to the *khaṭīb*, i.e. the *qā'im* of the Bā Rāsain tribe who are (hereditary) *khaṭībs* of the *Jāmi'* mosque ; (these latter also send out messengers at appropriate times to collect tithes (*'ushūr/zakāt*) from the tribes). When they have made the visitation they enter the mosque chanting *zāmils*, and, ascending the minaret, they cry out, *'Umūm, 'umūm,*

[171] So in *Gloss. daṯ.*, with fuller explanation.

yā Shaikh Sa'īd! So they continue, without resting, all day and night of Friday, returning to their countries on the Saturday. The *khaṭīb* entertains them on the night of their arrival.

If several years pass without rain or they think Shaikh Sa'īd is reproaching them, they bring an *'aqīrah*, i.e. a cow or camel in procession with *zāmils* to the door that leads to the Shaikh's tomb, where they hock it and cut its throat, shouting, *Yā Shaikh Sa'īd, baḥr-ak naṭlub baḥr-ak*! By this they mean, *baḥr burhān-ak*. This might be rendered as, ' We want the ocean (-like extent) of your influence and miracles '. Then they drag the *'aqīrah* away to a house, and lock the door, those persons eager for meat falling on with their knives and, with clamour and menaces they cut it up, some coming away pleased with a large piece, but others disappointed, the largest share in any case going to the prominent people of the village. If there is a lot of meat they prepare it, dry it, and keep it for the coming days. The tribes may bring quite a number of *'aqā'ir*, and the author comments that on a certain occasion they brought *'aqā'ir* to Shaikh Sa'īd to petition him for victory over their foes the Ḥālikah.

It might be commented that this sacrifice of the *'aqīrah* is a method of putting a compulsion on a tribesman to take a certain action in aid of the man who offers it, and in this case the Bedouin are, as it were, trying to place Shaikh Sa'īd in a position where he must help them. Needless to say, Saiyid 'Alawī b. Ṭāhir strongly disapproves this practice, and indeed I should not be surprised to find that nowadays it is discontinued. Other vestiges of pagan belief with regard to petitioning for rain can be found in the sections on *istisqā'* in the Ḥaḍramī collections of *fatāwā* such as the *Bughyat al-mustarshidīn*.

IX

OBSERVATIONS ON IRRIGATION IN SOUTH WEST ARABIA

The vital labour of irrigation naturally affects the life of every Yemeni and is reflected in the idiom and proverbs of day-to-day life, in the ḥakamī/ḥikmī verse or in ḥumaynī poetry and the folk lore couplets of ᶜAlī.b.Zāyid, Saᶜd al-Suwaynī, Ḥumayd b. Manṣūr and many others. It is treated of in the law books, in the collections of legal opinions (fatāwā), etc., and is governed in practice by custom. In this paper I draw on my experience in Ḥaḍramawt, Dathīnah and the Yemen, as also on the four occasions when I worked as consultant to engineering firms on the customary law of irrigation. My Ḥaḍramī material is published (Serjeant 1964) but the Dathīnah data are still in my 1954 notebooks. In 1964 I came upon the Lahej Qānūn al-Majlis al-Zirāᶜī which is an attempt to codify the customs of water distribution of the Wādī - I looked briefly at this complicated system in the field but decided to pass over the problem of examining it in detail along with the lease of land there, to Dr ᶜAbdullāh Maqṭarī as a thesis topic and his study, printed by the Cambridge University Press, provides a useful basis for further study of wādī irrigation (Maktari 1971).

Here I shall try to outline types of systems known to me along with something of the law affecting them in the YAR and the former Aden Protectorates now known as the PDRY. I shall say something of the flood courses of the Yemen mountain chain where the spate waters flow with great and sudden impetuosity into the western and southern Tihāmah plains to water the alluvial lands, something of the Ḥaḍramī Masīlah, then the ghayls which correspond to the Omani falaj, the Iranian kārīz and the Libyan foggarah. I shall say a little about well irrigation and at this point I should mention the large cisterns of the Yemen which one comes across even at Shahārah about the very top of this lofty mountain, and the tanks of Aden, about which Professor H.T. Norris has published a study, although I do not intend to discuss them here. Nor do I intend to deal with the hot springs, used for bathing and also for irrigation, that one finds at various points in the Yemen, in the former Wāḥidī sultanate, at Tabālah near al-Shiḥr, etc.

The commonest method of conducting spate water from a wādī onto cultivable land is to construct a deflector thrusting out from the wādī bank at an angle so as to head off the spate in a channel to the fields - from the main channel minor channels will conduct the spate to other fields lower down. The owner of the top fields in a system builds a temporary barrage of mud, stone and/or brushwood across the channel so as to divert the spate to his land. When they have drunk, as they say, the barrage is broken and the water passes to the barrage across the channel leading to the next group of fields, and so on, to the lowest fields in the system. The allotment of shares in the water is made in various ways, but, if the spate is poor, it is obvious that the fields at the lower end of the system may receive little or no water at all. There may however be a main barrage (maᶜqam) across a wādī or channel - e.g. in Jīzān of the Saᶜūdī Tihāmah, where it is called ᶜAqm al-difāᶜ, which is of a permanent nature, and to which all holders of land must contribute cash or kind for maintenance in proportion to the extent of their holdings or on some other basis. Within any given field system there may be a number of holdings and the same principle operates - the upper field or fields drinking first and in sequence. In some wādīs there is a permanent or semi-permanent flow - this irrigation experts know as 'base-flow', and this is the case in Bayḥān, Wādī Rima^{c}, Wādī Mawr, to name only a few of the best known cases.

Of the wādīs that debouch into the Tihāmah should be mentioned Banā serving the fertile Abyan plain, Laḥj/Lahej, Zabīd, Rima^{c}, Surdud, Mawr, and over the northern border in Sa^{c}ūdī Arabia, Jīzān. When I first went to Dirgāg in Abyan, to pay our small garrison there, it was a mere waste with the ruins showing of its mediaeval wealth of agriculture, when it produced grain and cotton. The perpetual bickering over the nāzi^{c}ah water channel by the lower Yāfi^{c} and Faḍlī tribes had turned it into a desert. The Aden Government however obtained a grant from the Colonial Welfare and Development Fund, at the time a quite considerable capital sum, to lay out a new irrigation scheme. This turned out so successful that the Abyan Board was actually able to repay the grant, and, I was told in Khartoum, the quality of the cotton produced was so high that, had the output been larger, it could have rivalled Sudanese cotton. It seems however to have been badly damaged by heavy floods a few years ago. The Colonial Welfare and Development Fund also financed the construction of a weir above Lahej at Rās al-Wādī at the point where it divides into al-Wādī al-Kabīr and al-Wādī al-Ṣaghīr - this too, I have heard, has suffered bad flood damage - but whether by simple mishap or through bad management I do not know.

In the late mediaeval period of the Rasūlids it was the Zabīd province which returned the largest revenue of the agricultural districts in the Tihāmah - nearly a million dinars, plus 700,000 dinars of tax on crops, plus 11,800 dinars of tax on palms. Of the other Tihāmah valleys the richest was Wādī Mawr with a revenue of 575,000 dinars, plus 510,000 dinars of tax on crops. The Rasūlid sultans were very interested in the Zabīd district which has to this day a vast area of arable land and palm groves which the histories call nakhl al-mudabbī. The Zabīd water channel, still today called Sharīj al-Jurayb, was known by this name to the Fāṭimid historian ^{c}Umārah in the early twelfth century and historians who flourished about 1300 A.D. A flood sweeping down the Jurayb channel on April 11th 1393 destroyed part of Māti^{c} and Ṭurqūh villages, both of which are still marked on modern maps. The Rasūlids even kept a governor/administrator (^{c}āmil), and overseer (mushārif) at al-Jurayb, perhaps to look after the channel as well as collect the tax on its crops. So far I have come across no references to the other sixteen channels there. In 1971 a Hungarian firm, Tesco-Vitziterv, carried out a feasibility survey of Wādī Zabīd and its channels and the scheme they devised for weirs and diversion structures was implemented. Reference is made in their report on lands and water rights to the allocation of water to each channel made by a celebrated faqīh, Ismā^{c}īl b. Ibrāhīm al-Jabartī who died in 875/1470. His ruling seems to be in force to this day but I did not see the original text though they brought me a document upon which they rely, said to be in essence al-Jabartī's decisions - a ṣulḥ, i.e. a conciliation agreement made by the Zaydī Imām Ṣāḥib al-Mawāhib and dated 1704 A.D.[1]. This Imām's ruling lays down the watering periods (muddah) for each channel of the Upper, Middle and Lower sectors of Wādī Zabīd - al-Jurayb lies in the Middle sector. A channel has so many days of water alloted to it at fixed times of the year, the dating being by the Rūmī calendar in which January 1st = Kānūn Awwal 19th. This (Julian) calendar was in use in the Yemen before Islam and in it May is called by the Ḥimyarī name 'Mabkar'. But while villages in the Wādī Mawr district also use this Rūmī calendar, I came across two villages, Nāshiriyyah and Ḥumāsiyah, where, curiously enough, the 28 star calendar was in use.

The landed proprietors of Wādī Zabīd mostly live in the city, the land, according to the Tesco report, being cultivated by 'middlemen', but these turned out to be agents (wakīl), really farm managers, though the proprietors seem also to make contracts directly with the share-cropper farmers. It is significant that 55% of the land in this district is said to be waqf of the various categories known in the Yemen[2] and 10% is Mīrī land.

The Wādī Zabīd channels are beset by the problem of silting. Silt, called ^{c}abār in Wādī Rima^{c}, is regarded there as a valuable asset, but the flood in Wādī Zabīd deposits a heavy coarse gravel and a gritty sand of the kind used in parts of the Tihāmah for making concrete. This must be removed by lorries - not an economic proposition. The landowners would like to have the former system of earth and bush barrages brought back to replace the concrete and masonry spillways and carry off the silt. Yet this cannot have been entirely satisfactory either for formally they piled the surplus silt from the fields onto the banks of the channels and fields. When I asked them what they did when field levels rose too high to be irrigated in the existing system, they said they would extend the top of the deflector (muḥammal) in the wādī up-stream, thus taking off the flood at a higher level. This suggests a reason why, in neighbouring Wādī Rimaᶜ the channels (mashrab, as they are called there) are so long and curve round to the wādī in their upper reaches until they become almost parallel with it. It is as if the original channels there had been extended, at some quite distant past, to take off water from the wādī at higher levels than the much shorter original channels.

An extreme case of the difficulty of disposing of silt - or of the inability to discover a method of doing so - is to be seen outside the Ḥaḍramī city Shibām where fields have been reduced to relatively small areas which are surrounded by banks over, I suppose, ten metres high and very wide. Again, in 1940, while on camel patrol in the West Aden Protectorate, I came on a large area of abandoned fields at a point where a wādī debouches onto the Tihāmah plain. My enquiry as to why this had happened brought a tale that tribesfolk at the upper end of the area had feuded with those at the lower part till the land went out of cultivation. This would be typical but it also looked to me as if silting might have so raised the field levels that irrigation would be impossible.

Erosion is doubtless of course the commonest cause leading to land being abandoned. I cite the case of the well-known pre-Islamic town known as Naqab al-Hajar which I first visited in 1947. Naqaba means 'to hollow out, pierce' etc. and the Naqab of the name is in point of fact a deep and impressive cutting made in a rocky shoulder lying along the right bank of the wādī, so as to lead off the spate to a great area of arable land below the cutting - this area would have provisioned the town, al-Hajar, itself. When I looked at the naqab cutting in 1964 the bed of the wādī above the opening to the naqab had become so eroded that it was too high to receive any spate water. The Irrigation Department of the South Arabian Federation was at that time exploring the possibility of re-opening the Naqab so as to exploit the arable land below it that it once served, but after the withdrawal from Aden in 1967 I have no further information about it.

In the Tihāmah valleys - indeed practically wherever there is a flood water or perennial streams - there are long-standing disputes over the distribution of water. This brings to mind the experience of a friend, a young political officer, Peter Davey, sadly murdered by Ḏāliᶜ in 1947. Just before World War II he set out to resolve the age-old disputes in the Wādī 'l-Maᶜādin (Valley of the Springs) of the Ṣubayḥī area. I suppose they were over allocations of water from both base-flow and spate. He called in the holders of land irrigated from the Wādī and persuaded them to agree on a new allotment of water, based on the size of their holdings, and to cancel earlier decisions over which they had remained at loggerheads. This was put into effect. But no sooner did the first floods come down but all the shaykhs were back (in Lahej, I suppose, capital of the ᶜAbdalī sultan), demanding the return to the previous practice. It turned out that the wily landowners, suspecting that, as they were asked to give particulars of their land, a ruse had been devised to increase the taxation on the crops, had so under-stated the extent of their holdings that they received substantially less water than formally.

In Islamic law water is regarded as mubāḥ, i.e. not owned, until preferential rights in it are acquired by, say, the construction of a channel to a piece of property; but he who has acquired it is legally obliged to offer water surplus to his needs to others. Water in artificial channels is held in common by all those contributing to their construction in proportion to their contributions to the work. The priority in the sequence of water distribution given to the top field(s) 'drinking' from the channel, followed in turn by those lower down is termed in the Arabic maxim, al-aᶜlā fī/bi-'l-aᶜlā.

When spate water does arrive, especially if it be by night, as Dr Maqṭarī tells me, swift action must be taken to divert it on to the fields - so the niceties of the various entitlements to the flow may be infringed - this certainly leading to vociferous disputes.

In Wādī Mawr a field can be filled up with water up to the thigh (ḥadd fakhidh), but no higher; in Wādī Rimaᶜ the water fills the field up to the top of the bund (zabīr) which, in the land near the flood-bed is 2 to 3 metres high, but in fields further away the bund is only one to one and a half metres high. In Wādī Mawr the fields are given a second watering if sufficient water is available. Where water is passed from a field held by owner A to one held by owner B by cutting the field bank (zabīr) the owner of the lower field must cut (yifjar) the bank of the upper field at a specified place - should he cut it at another place not laid down by local custom he is responsible for any damage his action causes. Here the system was defined to me by the phrase, al-aᶜlā bi-'l-fajr wa-'l-asfal bi-'l-zabr, i.e. the top field breaks the barrage (ᶜaqm/maᶜqam) to obtain water, but the fields in sequence below it obtain their water by building a barrage across the channel - the barrage in each case being broken when the field to which it has diverted the water has received its due allotment. Alternatively, as I have stated, the owner of the lower field may obtain his water by cutting into the bank above him, but the owner of the upper field should be present. Yet even if he is not present for some reason the owner of the lower field, should he see a large flow of water coming down, may cut into the bank. It was however stated to me that a man cannot open an access for water to his land without permission of the water master. Upper fields in a system seem usually to have permanent masonry spillways (manfadh) for water to drain into lower fields. I was told also that it is the constructor of the barrage (ᶜaqm) who must cut it to release the water to flow onwards to the next diverter ᶜaqm.

In Mawr the water master is called wakīl al-ᶜaqm, in Rimaᶜ and Wādī Zabīd he is known as shaykh al-sharīj or shaykh of the channel, while on the north bank of Rimaᶜ he is called ma'mūr or mas'ūl al-mashrab, mashrab is a channel. In Lahej/Laḥj there are mashāyikh (pl. of shaykh) al-aᶜbār, ᶜubar being the local name for a channel there.

In Mawr the water master settles all disputes but if, for any reason he fails to settle a case, perhaps over an issue too large for him to deal with, he goes to the district officer (mudīr al-nāḥiyah) but the latter, I was told, would simply refer the matter back to him and he would state that the plaintiff is in error. It looks as if therefore, there is in effect no appeal there against the water master's decision but perhaps in technical matters such as this the district officer feels unable to judge.

The big floods are the water master's chief problem. His assistant, the qayyim, a sort of watchman, summons those responsible for sharing the work of maintaining the ᶜaqm (and channels ?) - they turn out with yokes of oxen and help to provide the brushwood used in construction and repair. People quickly respond to the summons when there is damage. If, in place of oxen a bulldozer is used to repair the ᶜaqm the costs are distributed in the same way and they pay the cost immediately. His other assistant is the khādim (of a social class peculiar to the Yemen, often employed elsewhere as sweepers) who cuts brushwood (ḥayjah) and brings it with oxen or camels to the ᶜaqm. In this way the water master returns the ᶜaqm to operation. He and his assistants also remove the debris the floodwaters (dufūᶜ) bring into the channels - bushes, vegetation, etc., but I have not checked whether this included the gritty sand which in Mawr is called jall.

Payment is made at the harvest (ṣarīb) to the water master and his assistants after deduction of the zakāt tax (normally 10% on irrigated land), the land owner and the share-cropper reckoning among themselves over the payment. The water master receives one measure (qadaḥ = 64 nafars) in twenty of any sort of grain crop (after deduction of the zakāt). A water master with whom I talked said that he received one in every ten measures of grain and Hayj, the noted shaykh in the north Yemeni Tihāmah, put the figure at one in twelve for his district. The watchman (qayyim) receives half a measure in every ten plus two sheaves of sorghum with stalk and ear (maḥzamayn dhurah, ^{c}ajūr ma^{c}a 'l-^{c}adhaq) and a basket of soghum heads (jabb/juzū ^{c}adhaqah). The khādim receives two sheaves with stalk and ear of sorghum, a basket of heads and the ^{c}arṣ, i.e. the debris of grains and the rubbish left on the threshing floor after the owner has beaten out his sorghum, amounting to about a qadah measure. He gets a good return because he does the larger part of the work. At Sūq al-Thalūth the qayyim was said to receive only one quarter measure in twenty. In Wadi Mawr as in Jīzān there are three croppings, called surū^{c}, of sorghum from the original sowing, the second being called khalfah and the third ^{c}aqbah.

The water master is responsible for the main ^{c}aqm but there is a water master for each branch (far^{c}) of the channel. He assesses the cost of re-making the ^{c}aqm and distributes the expenses among the property owners including the share-cropper proprietors - in proportion to the extent of each land-holding. He compels them to share in the construction of it and maintain the irrigation works, either by working it themselves with their beasts, or by payment, or by contributing in some other way. He distributes the water, so many turns (dawl, pl. adwāl) between the share-croppers (shurakā) acting in the capacity as a kind of overseer. If a man refuses to open an ^{c}aqm when he should the water master will open it despite him, or, in other cases, close an ^{c}aqm when a field has received its due allotment.

The construction of great concrete weirs to control the spate water of the large Tihāmah valleys from Jīzān southwards must result in far-reaching change in the agricultural economy of the country district and the lives of the farmers. Not all of these developments are of great benefit. The dam at Jīzān is a case in point. It created a great lake that exposed an extensive surface of water to evaporation, and the heat in coastal Sa^{c}ūdī Arabia is intense, so that this in itself made for a considerable loss of water. Furthermore it seemed that for various reasons less underground water might reach the sea - which could lead to seepage from the Red Sea and salinity in the wells. This might happen in the Yemeni Tihāmah. Thee were also other problems with the Jīzān dam where I did some consultancy work in 1971 but I do not know if its problems have since been solved.

The creation of permanent water distribution channels, with the control of the flow and elimination, at any rate in part, of the system of earth barrages to which earth had to be shifted by labourers with teams of oxen, and their replacement by bulldozers, means that men and animals will no longer be required. The main Tihāmah crop is sorghum (dhurah) the long stalk of which is fed to oxen, donkeys and camels. Obviously mechanical transport has already cut down the demand for dhurah stalk (qasab). In the Yemeni Tihāmah it will be some time, I suppose, before the full effect of the new controlled water systems is fully felt. They are of course not without benefits and in Rima^{c} piped water to villages along the Wādī banks is certainly a great boon. The effect of change in the human situation is also masked in a way by the volume of remittances coming in to the Yemen and the shortage of labour arising from the enormous (if mostly temporary) emigration to Sa^{c}ūdī Arabia.

In the Tihāmah there has been much investment in motor pump operated wells - this, I understand, has already had the effect of lowering the water table there. In Ḥaḍramawt by the time of my 1953-54 tour the introduction of motor pumps had led to a new type of contract between the share-cropper (sharīk) and the patron or capitalist, the ṭabīn, which was replacing the forms used in the 15th century forms of plea which I published in the Rivista (Serjeant 1955), traditional forms that would have been followed up till some four decades or so ago.

In 1964 when travelling with Stephen Day, the political officer in the Faḍlī sultanate at that time, just above the ^{c}Urqūb pass we came upon a well in the final stage of excavation, owned by a certain Ṣiddīq Aḥmad. It was to be operated by motor pump and my notes appear to indicate it was ten dhirāʿ deep. On the well three labourers were working, belonging to Bā Jābir, a group of Ḥigris (pl. Ḥugūr), a class differing ethnically from the Arabs of the country. Of this group it is said, Al-Ḥigrī bilād-uh al-khuḍrah, which seems to mean that the Ḥigri's home/land is wherever there is greenery(3). Their working hours, I noted, were from 6 a.m. to 6 p.m., though this would be expressed differently in terms of the Arabic day. They used the stick (maḍrab), clubs (harāwī, sing. hirwah), mattock (misḥāh/ḥijnah) and baskets (marbashah or qafīl). For each qāmah, height of a man, excavated they received in payment 25 riyāls, equivalent at that time to 125/- East African, along with breakfast (qurʿah), tea (shāhī) and lunch (ghadā). The well must also have been some ten dhirāʿ in diameter. It had at ground surface level an apparatus to which they referred as maqam with a pulley wheel (ʿajalah) for lifting the earth and there was a male camel in rut, muzzled, helping. To have the basket lifted and drawn to the surface they said, 'barraʿ !'.

At the time we arrived water had just been struck and we saw an animal being slaughtered about the place where it had been found. At each stage of the operation there is a sacrifice (fidyah), the first before the well is actually started, then when water is struck and finally when the masonry lining (ʿaṭārah(4)) is completed - one says, yaʿṭir al bīr, he gives the well a masonry lining. A fourth animal is sacrificed if a makīnah, i.e. mechanical motor pump is installed - making in all, four ghanam (sheep-or-goats).

The first animal is slaughtered fawq al-midmāk, i.e. the stone upon which the water drawer stands, directly on the edge of the well(5), the second fawq al-diʿamah, on the wooden post of the well(6), the third fawq al-maqām, which seems to mean the place of the bucket, on the side of the well (ṣubr al-bir)(7).

When dawshāns(8), and other classes reckoned inferior, slaughter a sheep-or-goat they received as a perquisite the skin (al-dīm), or, as another man put it, the head, skin of the neck and entrails (al-rās, al-dīm, al-raqabah, al-amāṣīr). The tribesmen (qabīlī) who slaughters gets none of these perquisites, only a share (qismah) in the meat.

It is curious how these ancient customs persist. The sacrifices at the various stages of construction are parallel to those at the various stages of building a house.

Ḥaḍramawt is very reliant upon well irrigation - Ḥaḍramīs tend to say, 'I have a well', when they mean the land which is irrigated from the well - others would say that they owned ṭīn, literally 'clay'. The water was lifted onto the fields by the labour of men, often including women, and animals, oxen, donkeys, camels. They would run down the inclined ramp from the well-head drawing with them the well ropes - when the bucket reached the top of the well it was jerked by a special rope so as to spill its contents into a cistern from which it flowed by shallow channels on to the fields. This was gruelling though not unhealthy work, but the well-worker (sānī) not infrequently received a rather unfair share of the crop he worked so hard to produce. The poet Saʿd al-Suwaynī (Saʿd the little well-worker) has, attributed to him, two couplets, one of which expresses the hatred of the sānī share-cropper for his ṭabīn, the other in which the sānī congratulates himself on having a good ṭabīn. The ṭabīn, a word known to the Sabaic inscriptions, is the patron-capitalist who provides the cultivator share-cropper with food and clothing plus seed etc., for which he is reimbursed of course from his share in the crop. There are many variant types of share-cropping contract depending on the input of the ṭabīn This sināwah as it is called had disappeared in the main Wādī Ḥaḍramawt when I was last there in 1964 and, sadly, one no longer hear the sānī singing to his beasts, 'Yā māl yā mālī', the squeak of the well-pulley and the splash of water as it tipped into the cistern. Now you hear only the comfortable chuff chuff of the motor pump that has relieved the sani of so much labour. In the hot nights of Ramaḍān which, in 1947, fell in July-August, you would hear the sānī working at the well through the night, for to work through the heat of day would have been insupportable. In Ṣanʿā, you can still see the long well ramps, called there mirna^{c} (Ḥaḍramī maqūd), but as in Ḥaḍramawt the sānī is heard no more.

Ḥaḍramīs tell you that in former times you could travel from Haynin at the upper end of the Wādī Masīlah down to the tomb of the prophet Hūd, shaded by palm groves the whole way. Today the lower Wādī, long before you reach Hūd is desert with only the rare shepherd to be seen and there are long gaps between villages with their fields in most of the rest of the Wādī but, travelling in the desert part one comes across, near Ḥuṣn al-^{c}Urr, a wide stretch of plain with many ruins of stone revetments of dams, channels and spillways - still more more striking is that in the lower area if you go up the side wādīs you find vestiges of spate breakers, dams or barrages in series. Each of these barrages pushed the spate-water into the fields behind or above it on each side of the wādī, allowing the surplus spate-water to continue on its way to the main Wādī, gradually losing its impetus and volume on its way[9]. So when the spate reached the main wādī bed, the Masīlah (which Professor Walter Müller showed us, some meetings back, is its name also in a pre-Islamic inscription it was already tamed to some extent. It seems fairly certain that further spate-breakers were constructed in the Masīlah itself though this would have to be demonstrated by a careful survey. A low barrage is today called maḍla^{c}ah, a word known to the inscriptions but not to the Arabic lexicons, and a smaller one is a marṣad. In the Wādī Dhahabān lies the mawza^{c} of the town of Shibām, so-called because it yuwazzi^{c} al-mā', distributes the water - cf. Mawza^{c}, the name of the Tihāmah town.

So it is far from impossible and, in my view even likely that the Wādī Ḥaḍramawt and its side entrant wādīs were all carefully provided with series of barrages with the double function of irrigating and spate breaking that extended from the upper reaches of the Masīlah down to the perennial river at Qabr Hūd. So the folk tradition, which may also have some Ms. support, of a continuous line of palm groves all the way, might have some foundation in fact. This however raises two questions. How was the system initiated and how was it maintained ?

Probably truces would be declared to allow of the repair of barrages or construction of new ones; there might be inviolate months or star seasons - truces were certainly declared by mansabs for various purposes until about the time of 'Ingrams' peace'. I have recorded (Serjeant 1964) that a manṣab, lord of a sacred enclave will proclaim a wakad, i.e. a corvée for the repair of a large dam or deflector. It then becomes compulsory for labourers to come and work at a much lower rate that they are normally paid. Apart from the wakad mansab, the two other types of demand made for this compulsory labour were the wakad dawlah, i.e. the sultan's demand, and wakad qabīlah or the tribe's imposition of compulsory labour for some project[10]. The theocratic element in the rule of the country is represented in the office of the mansab, the secular element by the sultan and tribe. The corresponding term in the Yemen for the corvée is sukhrī. This is likely to have been the procedure for initiating and maintaining irrigation works in the pre-Islamic era and it may be compared with the data provided by the Ma'rib Dam inscriptions.

In the lower reaches of the Wādī Ḥaḍramawt there are, to the best of my knowledge, no Arabic inscriptions as yet discovered except at Qabr Hūd and this might point to the commencement of the decay of the irrigation works as have taken place before Islam. But I would tentatively offer the hypothesis that the demand for manpower at the time of the Arab conquests was so great and the prospect of easier living so attractive, that migration brought, if not depopulation, at any rate insufficient labour to maintain the system - one might compare the shortage of manpower in Hadramawt after the World War II famine followed by the migration to the Sa^{c}ūdī oilfields. The departure for the wars in the 7th century A.D. is the single recorded event that might explain how the decline started. However decline continued during the Islamic period for the chronicler Shanbal covering the period 500-920/1106-1514 shows as well-populated, places that today and for several centuries earlier, have become waste. This in part due to the Nuqrah, a deep erosion of, say, 50 to 60 feet, in the clay bed of the Masīlah, running for some miles and commencing just below the village of ᶜInāt - it is a sort of miniature canyon which leaves the fields on each side of it high and dry. It was said to be still eating its way up-wādī. As a relief measure in the war-time famine the Aden Government made a great earth barrage at the lower end of the Nuqrah, employing dozens of teams of oxen to drag the soil to it in the traditional fashion - it was hoped that silt would gradually pile up beyond it. In 1947 we could swim across the lake that formed behind it - unfortunately it was swept away by floods a few years later. In the list of floods I compiled from the chroniclers it is possible that the devastating Sayl al-Hamīm that took place at the star al-Ḥūt (October 1-13) in 1712 might have been the cause of starting or enlarging the Nuqrah, but I know of no certain dating for it.

It is thought that the qanāt, known in the Yemen as ghayl, the underground canalization of springs with vertical shafts to the canals at intervals, spread from Iran (called there kārīz) to the Yemen, probably via Oman, and thence to the Ḥijāz[11]. At Ghayl Bā Wazīr I have described (Serjeant 1964) the curious natural craters in the rocky plateau from which underground channels of the qanāt type have been cut, bringing their waters to lower ground - the channels are called miᶜyān and the shafts naqab while, incidentally, ghayl here of the placename is used in its sense of a running stream. There may be other qanāts along the coastal strip of Ḥaḍramawt but I have not come across any in Ḥaḍramawt of the interior.

It is in the YAR that ghayls have been most developed and quite a number of towns there used to have their water supplied by them. In Ṣanᶜā' (Serjeant & Lewcock 1983) we noted that the first ghayl of which we have incontrovertible information is Ghayl al-Barmakī, constructed by the ᶜAbbāsid Barmecide governor of the city some time after 800 A.D. as a public benefaction but the literature suggests there was an earlier ghayl known as Ghayl ᶜAlib, probably near the present-day village ᶜAlab. Ṣanᶜā' had three ghayls rising at different places to the south of it, Ālāf, al-Aswad and al-Barmakī. We looked at such traces of the Ṣanᶜā' ghayls as Dr Costa had come upon (1975); they seem to have included collecting systems en route. With the massive extension of building much evidence of their courses is probably now destroyed, but in the road just outside the Mutawakkil Mosque I saw, a few years ago, part of the ghayl channel being dug up near the washing place for women, the Misbanah, during road repairs. Ghayls provided water for mosque ablution places, laundry and for irrigation. Dr Ḥusayn al-ᶜAmrī has published (Arabic and Islamic Studies, 1983) a deed of sale of al-Barmakī and al-Aswad to Imām al-Mahdī ᶜAbbās dated 1180/1767[12].

Popular saws indicate that the diversion of water from its source for use in another district was resented by those living by the source. Bahrain island which has also qanāts/ghayls has also curiously enough, a proverbial saying expressing the same sentiment - a modern poet quotes it to complain that foreigners, not Bahrainis benefit from the island's wealth. The Encyclopaedia of Islam article qanāt/kārīz by Professor Lambton provides ample data on its construction and maintenance in Iran but, as far as I know, no such data for the YAR is available. On what legal basis can a right be acquired to make a ghayl from a spring ? How are its waters and maintenance costs distributed ? Documents relating to the Ṣanᶜā' ghayls exist (Serjeant & Lewcock 1983:25) but they should be discussed with shaykhs and farmers. Iran qanāts had a mīr-āb chosen by its jount users - we can assume Yemeni ghayls had a water master. Who owned a ghayl ? It could be an imām as already seen. The study of the ghayl organisation in the Yemen should be undertaken as soon as possible, since most have already gone out of use.

* * * * * * *

NOTES

1. Copy in my possession. Cf. Tesco report.
2. Ṣanᶜā', 151b seq., glossary 597a.
3. Khudrah in classical Arabic is 'green plants' etc. Qafīl, panier pour porter (Glossaire daṯînois, 2518).
4. Word unknown to classical and colloquial dictionaries consulted.
5. See E. Bräunlich The well in ancient Arabia, Leipzig, 1925:48; Gloss. daṯ., mur, muraille, parapet, with reference to Tāj al-ᶜArūs; Serjeant 1964:43, al-ḥabl al-aᶜlā the top course of stones.
6. Braunlich 1925:32, the two wooden posts (diᶜāmatayn) holding the beam carrying the pulley-wheel. Gloss.daṯ.,margelle en bois du puits, with reference to Ibn Sīdah.
7. Kazimirski Dict., maqām, lieu...où on se lève. Gloss.daṯ., ṣubr, flanc; ᶜalā ṣubr al-wādī, on the side of the valley.
8. The drummer-minstrel, often used as a herald.
9. Cf. diagram, Serjeant 1964:50.
10. Cf. Ibn al-Athīr, al-Nihāyah, Cairo, 1211 H., IV, 227, wakada fulān-un amr-an...idhā qaṣada-hu wa-ṭalaba-hu.
11. Encyl.Islam (2), IV, 532.
12. See Bidwell & Smith 1983:29.

REFERENCES

Al-ᶜAmrī, Hussein Abdullah (Ḥusayn ᶜAbdullāh). 1983 'A document concerning the sale of Ghayl al-Barmakī and al-Ghayl al-Aswad by al-Mahdī ᶜAbbas, Imām of the Yemen, 1131-89/1718-85'. In Bidwell, Robin & Smith, G. Rex eds. Arabian and Islamic Studies; articles presented to R.B. Serjeant, 29-37.

Bidwell, Robin & Smith, G. Rex 1983 Arabic and Islamic Studies: articles presented to R.B. Serjeant. Cambridge.

Maktari, A.M.A. (ᶜAbdullāh Maqṭarī). 1971 Water rights and irrigation practices in Lahj. Cambridge.

Norris, H.T. & Penhey, F.W. 1955 An archaeological and historical survey of the Aden tanks. Aden.

* Serjeant, R.B. 1954 'Star-calendars and an almanac from South-West Arabia'. Anthropos 49:433-59.

-"- 1955 'Forms of plea, a Šāfiᶜī manual from al-Šiḥr'. RSO 30:1-15.

* -"- 1964 'Some irrigation systems in Ḥaḍramawt'. BSOS 27:33-76.

Ib. and al-ᶜAmrī, Ḥusayn. 1981 'A Yemenite agricultural poem'. In Wadād al-Qadī, ed. Studia Arabica et Islamica: Festschrift for Iḥsan ᶜAbbās, 407-27. Beirut.

Ib. and Lewcock, R. 1983 Ṣanᶜā'. Cambridge.

Tesco Vitziterv. 1971 AGL:SP/YEM, I. Technical report no.9 YAR Land and Water rights, Vituki, Budapest.

Reports on water law and custom for Jīzān, Mawr and Rimaᶜ. Copies of the Yemen reports were lodged with the Tihāmah Development Authority, Hodeidah/al-Ḥudaydah.

ADDENDA

p. 153: references for Serjeant 1954, see article III above; for Serjeant 1964, see article VIII above.

X

Customary irrigation law among the Baḥārnah of al-Baḥrayn

In 1963 I came upon the *Qānūn miyāh al-nakhīl*, printed in al-Baḥrain in 1379/1960, probably given me at Dā'irat al-Awqāf al-Ja'fariyyah when I was studying the *ḥaḍrah* fish-trap.[1] The preamble to this booklet runs, 'In view of the existence of laws (*qawānīn*) handed down by custom (*muta'āraf 'alay-hā*) in al-Baḥrain from ancient time concerning what is connected with palm trees[2] and their waters, it seemed appropriate, in order to facilitate their application, to collect these laws in a small booklet to which it is possible to refer as required. So it came to the preparation of this booklet, thanks being due to His Highness Shaikh Salmān b. Ḥamad Āl Khalīfah [1942–1961], who generously printed it at his own expense.'

In Abu Dhabi I was helped to understand some of the text by Sayyid 'Alī al-Tājir, who advised me to get in touch with Mr Ṣādiq al-Baḥārnah of International Agencies, Head of the Shī'ah Waqf. By good fortune his office was visited during one of our sessions by Shaikh 'Amr b. Sulṭān Āl Bin 'Alī, Nāẓir Amlāk Sumuww al-Amīr, who had participated in compiling this very booklet and was therefore able to clear up some difficulties for us. It is my pleasure to acknowledge my indebtedness to these gentlemen for sparing much of their valuable time to explain the text to me.

Though for purely practical purposes it was worthwhile over twenty-five years ago to print this collection of thirty articles, it now seems largely obsolete with the decline of palm cultivation upon which water is not now expended and with the decline of agriculture due to the attraction of other work in al-Baḥrain; one might add also that with the great reclamation of land many of the fish-traps have now disappeared from the coastal waters. The late James Belgrave[3] points to an even earlier decline. Formerly, he points out, al-Baḥrain was irrigated from natural springs and surface wells. In the twenties and thirties Major Frank Holmes drilled many artesian wells – which completely changed al-Baḥrain's system of irrigation. This increased the area that could be cultivated but later the water head went down because so many wells were drilled and some of the gardens in the south of the island dried up. The booklet, however, may be said to have an historical interest, even if today conditions have so much changed; and of course circumstances were altering when it was published, for the introduction of pumps is mentioned.

In past times agriculture was fundamentally based on the cultivation of the date palm, lucerne being the other main crop, and the season ended when the dates had been collected.[4] The palm was used for almost every purpose and the disposal of its various parts is carefully regulated in this document. The trunk was cut down the middle and could be used for building or burning; fences, houses, the *ḥaḍrahs* themselves were made of palm *jarīds* – even the fish-pots (*gargūr*, *garāgīr*) of various sizes were made of date stalks (*'idhq*) and were yellow in colour, but for some time now they have been made of wire net. Nowadays these parts of the palm tree seem to be used for hardly anything at all. Nevertheless where dates are concerned Shaikh Rāshid Ibrāhīm Āl Khalīfah told me in 1981 he was laying out land to palms in the expectation that dates would command a good price. The channels leading to agricultural land on the north west had been recently refurbished and improved. Shaikh Rāshid was planting the red *khinayzī* date, *khalāṣ* – a white date said by al-Nabhānī[5] to be the best variety, white *barḥay*, red *khaṣāyib 'aṣfūr*,[6] and *mawāchī*, which is an early fruiting red date that sells well. Another Baḥrainī date that al-Nabhānī qualifies as the best kind is the *marzbān*, and Lorimer picks out the *murzbān* and *khanaiza* varieties.[7] The dates of al-Ḥasā' listed include *khalāṣ*, *khunayzī* and *marzbān*.[8] Al-Nabhānī states that at one time there were 800 varieties of date in al-Baḥrain, whereas Lorimer more modestly estimates only 60.[9] Dates, he says, were exported to the island, some consumed locally and the rest re-exported. He mentions a *gallah* weight of 37½ lbs used as a date measure. The continuing reputation for date production from early times is evident from the proverb *Ka-nāqil al-tamr ilā Awāl*, cited by al-Nabhānī,[10] and the more ancient form of it, *Ka-nāqil al-tamr ilā Hajar*, cited by Vidal for al-Ḥasā', equivalent to the English, 'Carrying coals to Newcastle.'

In times past there was a seasonal migration to every palm grove (*fī kull nakhal maẓ'an*,[11]) from one district to another. 'During the date-picking season . . . temporary villages of palm branch huts sprung up among the date gardens when people from other parts of Bahrain and elsewhere in the Gulf came to pick the crop';[12] or, as I was told, a man, his wife and family went to work on the palms, setting up a sort of summer dwelling made of *sa'f*, palm frond. Al-Nabhānī also mentions seasonal migration to springs at the coast from mid-Rabī' to the first days of Kharīf, date-harvest time.[13]

THE RENT CONTRACT (ḌAMĀN)

In 1976 I was informed that the period covered by the tenancy or renting contract (*ḍamān*) commences from the close of the Ṣayf season (*al-ḍamān yabtadī min ākhir al-ṣayf*), which ends in al-Baḥrain with Suhayl (August 25 to September 7 according to the local almanac[14]), but the date quoted to me was September 23 – when the ripe dates (*ruṭab*) come to an end the *ḍamān* terminates. Yet more recently I was told that nowadays the *ḍamān* terminates on September 20 and the new *ḍamān* commences on October 10. The peasant would reckon this season following the star calendar. The poet 'Alī 'Abdullāh Khalīfah in his *'Aṭash al-nakhīl: mawāwīl* published in al-Baḥrain in 1970,[15] describes the *ḍumān* (sic) as '. . . a sum which the *fallāḥ* pays in return for the garden remaining in his charge for a specified period. He cultivates (*yizra'*) it and sells its produce (*maḥṣūl*) and takes upon himself the proceeds, the proprietor also taking what he requires (*kifāyata-hu*) of the crop as a present.' The *'akkār* (class. Arabic *akkār*[16]), called by the poet *fallāḥ*, cultivates the palm garden and is represented by him as saying, 'I pay a rent-charge perforce (*arfa' ḍumān ib-qahar*).'[17] This would be paid part in kind, part in cash. The peasant expresses his grievances in such verses as a piece quoted to me:

Asqī 'ala 'l-bambarah[18] *wa-'l-tīnah,*
Wa-'sqī 'alā man jaraḥ qalbī.[19]
I water the sebestens and the figs,
And I water for him who has wounded my heart.

Ṣādiq al-Baḥārnah informs me that no cereals were grown but vegetables, tomatoes and fruit trees were cultivated.[20] The fruit crop does not mature at the same time as the dates, but ripens after the contract between the peasant renter-tenant and the proprietor-owner (*mālik*, *ṣāḥib*) has come to an end – hence the arrangements for evaluating the fruit crop in articles 25 and 26 of the printed *Qānūn* below. I have not noted who is responsible for making the valuation.

THE BAḤĀRNAH OF AL-BAḤRAIN

Ibn al-Mujāwir (7th/13th century) states that on the island that he calls Uwāl there were 360 villages, all but one being Imāmī (Shī'ah).[21] This figure looks like an error, but could be true of the island and Shī'ah districts on the mainland. In a paper given at the Arabian Seminar in 1979, M. A. Tājir makes out a case for the descent of the Baḥārnah from the Arab Banū 'Abd al-Qays of Rabī'ah,[22] but while the literary evidence shows that they moved into the area it is not proof that they were the ancestors of the Baḥārnah, because the

Arabic writers may be referring only to tribal overlords or landowners, ignoring the Baḥārnah merchant or peasant population. It seems that there were groups of Baḥārnah villages known as *sayḥah*[23] and formerly each group had a headman known as *wazīr* – in the island as a whole there were four, sometimes five, of these officials. The obviously non-Arabic names of a number of these villages I submitted to my colleague, Dr I. Gershevitch, and he found they could be Middle Persian (Pahlawi) but that there is no evidence to show that they are not Islamic Persian.[24]

Most of the agricultural lands seem to be in the Baḥārnah villages, and the *Qānūn* translated below related especially to the villages on the north-west side of the island, notably Kirzakkān. It was here that Mr J. D. Homfray examined some of the *qanāt* type of water channels. James Belgrave says that the *qanāts* were first open channels that were then roofed with slabs of sea-rock called *farsh* (plural *furūsh*).[25] For al-Ḥasā', Vidal has recorded, 'The *qanāt* system once used on the edges of the oasis was apparently a late introduction that did not meet with popular favour and was subsequently abandoned, even though it could still be used with advantage at present.'[26]

THE BAḤĀRNAH AND TAXATION

A poll tax called *ṭarāz*, paid to the Shaikh of al-Baḥrain by the agricultural Baḥārnah is mentioned by Lorimer[27] and it would be interesting to know more about this, as also corvée (*sukhrah*), once found in most Arab countries, and only abolished in the Yemen after 1962. There is no tax on land and no tithes.

With the Baḥārnah here the law (*shar'*) is that *zakāt* is charged on the crops of dates, grain and barley (*ghallāt al-tamr wa-'l-ḥabb wa-'l-sha'īr*)[28] as well as upon gold that is not being used – but if a woman has necklaces stowed away in a box and wears them for a month she is not due to pay anything on them. With regard to the *khums*[29] I was told that this is paid in a cash sum (*mablagh 'aynī*) on *ruṭab wa-tamr*, ripe dates and dates, but not on the crop (*ghallah*). Theoretically the *khums*, fifth, should go to the Imām of the country, but in general in al-Baḥrain, after all expenses are paid and a sum deducted for trading next year, a fifth is paid to some good man to be employed for charitable purposes – it might include any kind of service that is appropriate. The Baḥārnah give 2½% of their dates to the Imām in the village to distribute as he thinks fit, probably to the poor.[30]

In Saudi Arabia, it was stated, the proprietor pays *zakāt* on the crop, but there is no fifth (*al-mālik fi 'l-Sa'ūdiyyah yusallim al-zakāt – mā fīh khums*). There the renter pays, or formerly paid to the proprietor, a sum of money in addition to a part of the crop (*ghallah*) – dates and ripe dates, and he also pays on chickens and ghee (*dihin*).

THE SPRINGS OF AL-BAḤRAIN

The springs from which water is conducted to irrigate fields or plantations on a lower level are listed in detail by Lorimer[31] along with the villages that depend on their water; some of the more important are named by Ṣalāḥ ʻAlī al-Madanī and Karīm ʻAlī al-ʻUrayyiḍ[32] who refer to a report made in 1953 on a survey of the natural springs (*al-ʻuyūn al-ṭabīʻiyyah*), adding that a preliminary survey reckoned that there no less than 155 of these. The famed sweet water springs in the sea are called *chawchab/kawkab*,[33] and a *dardūr*, a Baḥārnah word only current in the villages, means a continuous flow of water, a little stream in a channel, or a small quantity of water from a *chawchab*.

The best known of the springs is ʻAyn ʻAdhārī (Virgins' Spring), but ʻAyn Qaṣṣārī close to al-Manāmah and Umm al-Shuʻūm are also well known, the latter near al-Māḥūz village, south of al-Manāmah, and now so much built over as to be part of the city. In Sitrah are ʻAyn al-Raḥā and two springs known as al-Mahazzah,[34] al-Kabīr and al-Ṣaghīr. ʻAyn Wādiyān is in the village Wādiyān, ʻAyn al-Ḥakīm lies on the east coast opposite the island of al-Nabiyy Ṣāliḥ,[35] south of al-Manāmah. One of the oldest and most powerful is ʻAyn al-Khaḍrah in al-Khaḍrah district, a perennial stream that reaches the sea in Khawr Maqṭaʻ Tūblī. Others are ʻAyn al-Karāyim, ʻAyn al-Sayyid and ʻAyn al-Ḥamasah (Turtle Spring) in Tūblī village. James Belgrave[36] mentions a long channel running in land from Ṣadad to Kirzakkān from the spring called Umm Ijrayrī (Lorimer, Ijraʻī) a mile and a half to the north, and another from the spring of Mālikiyyah/Mālichiyyah to the village of the same name.

Fish are to be found in all these freshwater pools, some of those in ʻAyn Qaṣṣārī being occasionally a foot and a half long. Umm al-Shuʻūm indeed is named after the *shaʻim* (plural *shuʻūm*) fish in which it presumably abounds.

Sāb Abū Zaydān, near Bilād al-Qadīm, another powerful spring according to al-ʻUrayyiḍ,[37] was a place at which votive offerings (*nudhūr*) were made, and it used to be thronged with bathers and persons with votive offerings from most districts of al-Baḥrain, but it has now lost much of its standing due to the influence of education and the growth of housing near it.

An inscription has been photographed and reproduced by James Belgrave[38] that records a *waqf* of the spring known today as the Dhobi Pool (ʻAyn Qaṣṣārī); unfortunately the first part of it is missing. This I propose should be read, provisionally at least, as:

Ayḍ-an yuḍāf ʻalā niṣf Ḥamagān (حمكان) maʻa ṣurmat/ ṣurmati-hi Fūliyān li-'l-janūb ʻAyn al-Qaṣṣārīn al-Ṣughrā al-gharbī waqf-an sharʻiyy–an ila 'llāh taʻālā 10 Ṣafar 695 (?).

أيضا يضاف على نصف حكمان مع صرمة/صرمته فولسيان للجنوب عين المقصارين الصغرى الغربي وقفا شرعيا إلى الله تعالى ١٠ صفر سنة ٦٩٥ للهجرة .

Also there is adjoined to the half of Ḥamagān, along with the palm plot of/its palm plot, Fūliyān, to the south, the Lesser Spring of the Washermen, on the west, as a legal *waqf*-endowment to Allāh, 10th Ṣafar, 695/20 December 1295.[39]

If my reading of the date be correct the Dhobi Pool, this name being obviously a translation of ʻAyn al-Qaṣṣārīn, was constituted a religious property coinciding with the accession year of the Īlkhānid Ghāzān in which he became converted to Islam and took measures to promote the religion. Under the Muslim Īlkhāns *waqf* landownership expanded and Ghāzān was active in this respect.[40] The island was probably an Īlkhānid dependency at this time.

ʻAdhārī is popularly supposed to have been excavated in the Umayyad period, but there is no known evidence to support this belief.[41] A well-known proverb is current in al-Baḥrain, *ʻAyn al-ʻAdhārī tasqi 'l-baʻīd wa-tukhalli 'l-qarīb* (ʻAyn al-ʻAdhārī waters far (land) but neglects what is near at hand). If one says of a person merely, "*Adhārī*", it implies he neglects his relatives for strangers. This spring being in one of the higher parts of the island its water is led to lower ground at some distance away, neglecting, as ʻAlī ʻAbdullāh Khalīfah says,[42] what lies in its vicinity. He cites the proverb, with a political undertone, to complain that foreigners cull the fruits (*al-ajānib yaqṭifūn al-athmār*) of its production, not the local people. Similar proverbs are on people's tongues in the northern Yemen about the *ghayls* of Ṣanʻā' and other places, deploring that people living in the locality of the spring feeding the channel conducting water through a *qanāt* (*ghayl*) to the city derive no benefit from it.[43]

IRRIGATION

Lorimer distinguishes three types of irrigation,[44] date palms being classified accordingly into three categories – *nakhal al-sayḥ*,[45] *dūlāb*, and *nakhal al-gharrāfah*. (Of these the first kind is watered by gravitation from flowing channels [*sayḥ*], the second by lift of one or two skins raised by bullocks or donkeys walking down a slope, and the third by a *gharrāfah* or lever and skin with a counterpoise . . . the counterpoise is generally a basket of earth.' The *gharrāfah* is handled by one man. As a *dūlāb* is properly a waterwheel I suspect Lorimer is mistaken and should have written *dāliyah* of which in fact he gives an illustration.[46] This may be compared with el-Sāmarrāie's description[47] of the system used in 'Abbāsid Iraq, 'Cultivated lands were irrigated either directly (*yusqā sāiḥan*) or by machines (*yusqā bil sāniah aw ad-dāliya*).' Belgrave's description,[48] that must apply to the *gharrāfah*, runs, 'A man stands near the edge of the well and draws up the water in a sheepskin bag attached by ropes to a counterbalanced pole. In larger wells donkeys and oxen are used, walking endlessly up and down a ramp pulling the water to the surface from where it runs down into irrigation channels.'

A *MAWWĀL* ON 'AYN 'ADHĀRĪ FROM *'AṬASH AL-NAKHĪL*

To Mr Taqiyy al-Baḥārnah I am indebted for his help in elucidating and commenting upon this poem.

عذاري لي متى تسگين ذاك النخل لبعيد

عطاشة نتنحي بجي ونرفع صوتنا ونعيد [50]

نشوف الگيظ عيّد بالنخل واحنا بليا عيد [51]

عذاري لي متى لها الساب تجري الماي منه دوني ؟

نشف ريج العشب، ياعيب تركي زرعج الدوني [53]

عجب عكس الوفا تعطين! ياأهل الخير ودوني.. [54]

شمالي العين. كنت يمها وراحت تسگي اللي بعيد. [55]

1. 'Adhārī, for how long will you water that far palm grove?
2. We, close beside you, are droughty, raising our voice once and again,[56]
3. Seeing the summer heat gift the palm grove with feast (clothing) while we are without a gift.[57]
4. 'Adhārī, how long will she have a channel whose water passes me by?
5. Parched is the sap[58] of the bushes – o shame you leave your near-by crop unwatered!
6. Astounding that you should give contrary to (your) obligation! Take me, good folk,
7. North[59] of the spring next which I was – yet it went to water what is far away.[60]

THE CONTENT OF THE *QĀNŪN MIYĀH AL-NAKHĪL*

The undoubtedly ancient customary laws assembled together in this document aim at regulation of the distribution of water to various properties. Nothing is said about individual holdings and one would expect that landowners possess deeds accurately defining the extent of their land watered from a spring or springs and the number of watering periods to which each parcel of land is entitled. Perhaps these may be found already in al-Baḥrain Museum, at al-Muḥarraq at the time of my last visit to the island, and it would be surprising if there are no land and water cases in the Bahrain Court Records in the India Office Library.

٢٦ ـ نخيل تشتمل على اشجار من جميع الاصناف كالاترنج والرمان واللوز وغيره فاذا دخل الضامن على هذه النخيل بتمين الاشجار فله الحق عند خروجه في تثمين ثمرة الاشجار كما جرى في السابق واذا دخل الضامن بدون تثمين فليس له حق في ذلك بعد انتهاء الضمان •

٢٧ ـ ضمان النخيل العادى يبتدى في وقت ابتداء الضمان وينتهى في وقت انتهائه في المدة المعينة المتفق عليها فاذا تَبَقّى شيء من الثمار المتاخرة بعد انتهاء مدة الضمان فللضامن من القديم الحق في تثمين الثمار الباقية •

٢٨ ـ لا يجوز اخذ الكرايف أى الضَريب من منجيات النخيل أو السواقى الداخلة فيه أو الخارجة عنه الا برخصة من المالك أو المالكين على ان يكون ذلك تحت اشراف المالك أو من يقوم مقامه كالضامن أو غيره من المشتركين من المالكين أو الضامنين المشتركين في المنجيات والسواقى لئلا يقع ضرر على أهل الاملاك من خراب السواقى والمنجيات •

٢٩ ـ لا يجوز اخذ الرمال والمَدَاد من نخيل أو اراضى مملوكة أو جوانب سواقى أو منجيات الا برخصة من المالك أو المالكين للسواقى والمنجيات والنخيل والاراضى المذكورة •

٣٠ ـ اذا كانت ساقية أو منجى أو ساب داخل نخيل مملوكة أو خارجا عنها في ارض مملوكة فيجب على مالك الملك الذى فيه هذه الساقية أوالمنجى أو الممر الايُحدِث فيها أى ضرر كالدفن أو التضيق أو رمى بعض الاشياء كالكرب أو السعف أو الليف وغيرها فاذا حدث فيها شيء مما ذكر فهو ملزوم بازالة ما احدثه من ضرر لانه هو المسئول عنه •

حرر في ٩ شعبان سنة ١٣٧٩

الضامنين ان يدفعوا له ما يخص نخيلهم من الايجار وكذلك مالك النخيل التى ليس لها ضامن ان كانت النخيل منازيل أو مرافيع فيجب دفع ما عليهم من الايجار .

١٥ – دالية تسقى سابقا بالغَرْف كغرف الدلو باليد أو بالزاجرة وهو قديمى بالتأكيد فيبقى على حالته الاولى واذا صار اصلاح فى العين التى يغرف منها والنساب فعليه ما ينوب غيره بموجب الحاصل من تمر أو زرع يقدر بثمن معلوم ويدفع ما ينوبه بموجب التقرير أما اذا بَدّل الغرف بالمضخة فليس له حق الا بموافقة اهل الاملاك المشتركين معه .

١٦ – دالية غرف حَدَث فليس لمالكها حق الا بموافقة أهل النخيل التى تسقى نخيلهم من العين أو الساقية التى يريد يغرف منها وليس لصاحب الدالية ان يزيد مساحتها عما كانت عليه سابقا اذا حصل الموافقة من المذكورين لئلا تضر بأهل الاملاك التى تسقى نخيلهم منها .

١٧ – ساقية أو عدد من السواقى أو محل مُجَمَّع لفوهات النخيل فيجب على كل ضامن أو مالك نخيل متوليها بنفسه اذا اراد اخذ الماء لنخيله أو اراد ان يسكّر على الباقين فعليه ان يسُدّ الفوهات سَدا مُحْكَما بالكرايف الموجودة في المحل أو

بالقرب منها ولا يجوز فوهة نخيل أو عدد من الفوهات من الرمال أو الارض الفوقية المجاورة للفوهة المقصود سدها أو الفوهات حيث ان ذلك يوقع الضرر على مجارى فوهات النخيل الاخرى .

١٨ – عمار النخيل أو الزراعة بموجب المقاولة بين المالك والمقاول اذا كان العمار بالمَغَرّس عن -/١ روبية أو أكثر واذا تعهد المقاول على ان يكون العمار ضَرْبة واحدة أو أكثر عن ايجار معلوم وحدث خَلَل في الشروط التى التزم بها المقاول على نفسه أما في عدد الضربات أو فَرَاعة النخيل أو كثافتها فبعد التحقيق يجرى للمقاول أو المقاولين نصف القرار الذى تقرر سابقا جزاء عما حدث من الخلل .

١٩ – شروط الضمان وما يتعلق من اصلاح النخيل أو الزراعة اذا ضمن أى شخص من احد ملاك النخيل أو زراعة والتزم على ان يكسّح النخيل أو الزراعة في كل سنة مرتين ويضرب السواقى الخارجية والداخلية والمنجيات التى تتعلق بها كل سنة مرتين كما ذكر أو أكثر بموجب الشروط المقررة في ورقة الضمان وكذلك الحُظْران التى تحفظ النخيل أو الزراعة فاذا اخلّ بشيء من ذلك فعليه أكمال أو تثمين ما تركه

وبموجب تثمين أهل المعرفة فهو ملزوم به .

٢٠ – شروط التقليج وهو قطع الكَرَب اليابس على الضامن قطعه في احد وقتين الاول بعد جمار النخيل بعد النبات ويجب على الضامن ان يترك دور واحد يابس عن ضرر النخيل وفي الوقت الثانى في اول بِشْرة الرطب على ان يترك دورين أو ثلاثة واذا ترك الضامن التقليج في الوقتين المذكورين فللمالك الحق ان يؤجّر على تقليج الكرب اليابس من النخيل ويحسب المَصْرَف على الضامن لانه ملزوم به كشرط اساسى الا اذا سمح عنه المالك فليس على الضامن حق فيما ذكر .

٢١ – شروط قطع السعف اليابس يكون في اول طلوع الثريا أى وقت ابتداء موسم البارح واذا سمح المالك بقطعه قبل ذلك فلا بأس على ان يقطع السعف اليابس فقط ويترك الاخضر الى وقت ما يكون يابس والوقت الثانى في نفاط اثمار النخيل فللضامن الحق في قطع السعف اليابس فقط واذا تعدى وقطع سعف اخضر يعاقب الضامن على قدر جُرمه أما بحبس أو غرامة حسب ما تراه المحكمة .

٢٢ – النخلة الساقطة والميته ليس للضامن فيها حق التصرف الا بعد كَشْف المالك أو من يقوم مقامه واذا عمل

بخلاف ذلك فهو مسئول للمالك عن ثمن النخلة فيما لو كانت حية تثمر وأما اذا تعمد المالك بعدم الحضور على الكشف فللضامن الحق في رفع القضية الى المحكمة .

٢٣ – عين طبيعية قد جفّ ماؤها وهى مشتركة بين عدة نخيل لها اوضاح فاذا اتفق اهل الملك على وضع مضخة على العين بينهم بالشراكة على ان يعطى كل واحد منهم ما يخصه من الماء فلا بأس واذا لم يتفقوا على ذلك فلكل واحد منهم الحق في ان يضع مضخة لاستعمالها لأوضاحه الخاصة .

٢٤ – كذلك اذا دعت الحاجة لضرابة العين الطبيعية فهى على الجميع فاذا احد المشتركين ابى أن يدفع ما ينوبه من المصرف فللمشتركين الحق في اخذ الماء الذى يخص شريكهم حتى يدفع ما عليه من المصاريف .

٢٥ – يوجد قاعدة معمول بها على ان مالك الزراعة لا يضمّن الا بتثمين الزرع عند دخول الضامن الجديد يبقى عليه كدين للمالك فاذا انتهت مدة الضمان فللضامن الحق في تثمين الزرع فاذا زاد ثمنه عن الاول فله الزيادة وان نقص فهو مسئول عن النقص يدفعه حالا قبل خروجه واما اذا اتفق المالك والضامن بخلاف ذلك فشأنهم لانهم مُخَوّلون فيما يريدون .

قانون مياه النخيل

البحرين

١٣٧٩هـ - ١٩٦٠م

«مواد القانون»

١ - العين تشتمل على اربعة عشر وَضَح اذا كانت لها ساقية واحدة أو اكثر ، والوضح من طلوع الشمس الى غروبها ان كان صيفا أو شتاء،ونصف الوضح من طلوع الشمس الى وقت الزوال ليكون نصف اليوم اذا كان نهارا ، وكذلك من وقت الزوال الى غروب الشمس،وليلا من غروب الشمس الى نصف الليل الاول أو من نصف الليل الاخير الى طلوع الشمس وثلث الوضح ثلث النهار أو ثلث الليل ، وكذلك ربع الوضح ربع النهار أو ربع الليل •

٢ - النخيل التى تسقى بالاقدام مثلا اذا كان من فيّ سبعة أقدام صباحًا أو أقل أو أكثر أو من في سبعة أقدام بعد الظهر أو اثنا عشر قدم العصر أو أقل أو أكثر فهذه الاوضاح تبقى كما كانت سابقا بالاقدام حسب الترتيب القديم •

٣ - المرافيع والنفوخات صفتها واحدة لان سواقيها تكون مرتفعة تسمى نفوخات ومرافيع تسقى اذا ارتفع الماء لغيرها فيكون ليس عليها ربط •

- ١ -

٤ - الذوب على حسب تفاصيله الآتية وهي كما يلى :-

اذا كان نخيل يسقى من الذوب خاصا له فليس لاحد ان يتعدى عليه الا اذا كان له شراكة فيه وكيفيته اذا كان الماء في وقت معين الى نخل معين في أى يوم من الاسبوع فبعد انتهاء سقى النخيل وصار الماء الى غيره فيكون فاضل الماء يسمى ذوب أو هدة الساقية ، كذلك الشواخل التى من وراء سكار النخيل والمياه التى تأتى من المَنجَيّات كل ذلك يسمى ذوب ، ويمكن ان يكون النخل واحد أو أكثر بالشراكة فيها بين النخيل •

٥ - العين التى يسقى منها نخلان أو ثلاثة واوضاحها محصورة للنخيل المذكورة فيكون مُصَرَف اصلاحها بالوضح وأما العين العمومية التى تتوزع اوضاحها على نخيل كثيرة فيكون مصرف اصلاحها على الحاصل من أثمار النخيل التى تسقى منها •

٦ - اصلاح السواقى اذا كانت العين لها ساقية واحدة فمصرف اصلاحها على النخيل التى تسقى منها واما اذا تفرعت منها سواقى متعددة فكل ساقية يكون مصرف اصلاحها على ملاك النخيل التى تسقى نخيلهم من هذه الساقية •

- ٢ -

٧ - ممر الماء الذى يمر في نخل أو نخيل متعددة الى سقى نخل أوجملة نخيل ليس للمالك حق ان يبدل الممر الذى هو ساقية بموجب الاصطلاح السابق ولا يغيرها عن موضعها الا بموافقة اهل الاملاك التى تسقى نخيلهم منها فاذا حصلت الموافقة من المذكورين فليس له حق ان ينزّل الساقية أو يرفّعها عن مستواها السابق •

٨ - ساقية تسمى صُوَّدَق وهى تمر على ساقية أخرى اذا اختل بناؤها فلصاحبها ان يبنيها على قاعدتها السابقة بدون تنزيل أو ترفيع •

٩ - المنجيات القديمة والحديثة ليس لاحد ان يسدها بل تبقى كما كانت ما عدا صاحب الجادى الذى يسقى نخيلة من المنجيات فله حق في ذلك بدون ضرر على النخيل التى تنجّى •

١٠ - لو صار زيادة في الماء عند مُلّاك النخيل وليس بجواره أرض خالية لاجل تخريج الماء الزائد وبجواره نخيل أو أرض مملوكة فيجب على صاحب الملك المجاور أن يسمح له بحفر منجى في اطراف ملكه من احدى الجهات التى يختارها مالك النخيل أو الارض والمصاريف على طالب المنجى •

١١ - الجَصة هى ارضية الساقية وجوانبها من جميع

- ٣ -

الجهات ان كانت مشتركة أو خاصة بنخيل واحدة فلو حدث فيها خلل من الاقامة في بنائها السابق فيجب اصلاحها كما كانت سابقا ان كانت مشتركة فعلى جميع المشتركين وان كانت خاصة بنخل فهى على مالك النخيل •

١٢ - لو كان عددٌ من النخيل تتساقى مع بعضها البعض في آن واحد فليس لاحد من مُلّاك النخيل المشتركة ان ينزّل ساقية نخيلة عما كانت عليه سابقا بل تبقى كما كانت لئلا يقع ضرر على النخيل التى تتساقى معه واذا حدث الضرر على بعضها من جراء ارتفاع الساقية وعدم الرَى فيجب قسمة الماء بموجب مساحة النخيل •

١٣ - يوجد بعض النخيل نازلة تتساقى مع نخيل مرتفعة وعلى النخيل النازلة سابقا يوضع في ساقيتها عديل أو اصدار حسب العادة السابقة بدون ضرر على النخيل النازلة أو المرتفعة واذا حدث الضرر فيجب قسمة الماء على حسب مساحة أرض النخيل التى تتساقى جميعا •

١٤ - لو كانت سَيحة نخيل تسقى من عين بالتسكير أى بالربط والهَد كمثل عَيْن عَذَارى أو غيرها ولها شخص معين من قِبَل مالكى النخيل يسكّر الاوضاح بالايجار فعلى جميع

- ٤ -

SUMMARY

Clauses 1, 2, 3 deal with the definition of watering periods and how they are estimated.

Clause 4 affirms the right of palm gardens irrigated from water surplus to the requirements of higher properties or from seepage water, to that water.

Clause 10 affirms the right of drainage from an upper plot through an adjacent plot.

Clauses 7, 8, 9, 12 affirm that no alterations may be made to existing channels except by general consent.

Clauses 5, 6, 8, 11, 24, 30 require costs of maintenance and repair of channels and springs be shared among proprietors benefitting from them.

Clause 13 stipulates that an upper palm plot is to be irrigated before one below it.[61]

Clause 14 arranges for payment of a man to supervise the management of irrigation according to each plot's entitlement.[62]

Clause 15 shows the changes already taking place in the island with the replacement by pumps of human or animal water lift. This perhaps gives emphasis to clause 16 concerning bringing new land into cultivation.

Clause 23 makes provisions for the introduction and sharing of water among pumps.

Clause 17 lays down rules as to how stoppings should be made to channels.

Clause 18 lays down regulations for contracts of hire of labour for tilling.

Clauses 19, 20, 21, 22 stipulate the responsibilities of the renter of a palm plot.

Clauses 25, 26, 27 define how an incoming tenant/renter shall take over existing crops from an outgoing one.

Clauses 28, 29 prohibit the removal of surface spoil or debris from palm plots without the proprietor's permission.

Note: In style the construction of the *Qānūn* is rather loose and the wording imprecise, partly because of the weak syntax tending to the colloquial and lack of punctuation. It is not drafted with the care of a professional composing a government document – nevertheless the sense comes through fairly well. Grammatical slips appear on p. 1, line 1, and p. 2, line 8; on p. 3, line 11, and p. 4, line 7 نحيلة should be read نخيلة; and on p. 6, line 9, *kathāfah*, thickness, instead of *kishāfah*.

THE ARTICLES OF THE LAW

1. The spring comprises fourteen *waḍaḥ*-periods, whether it has one or more channels, the *waḍaḥ*-period lasting from sunrise to sunset, be it summer or winter. The half *waḍaḥ*-period lasts from sunrise up to the time of the meridian, so as to be half a day if it be daytime, and likewise from the time of the meridian up to sunset; and, at nighttime, from sunset up to the first half of the night, or from the latter half of the night up to sunrise. The third of a *waḍaḥ*-period is a third of a day or a third of a night. Likewise the quarter of a *waḍaḥ*-period is a quarter of a day or a quarter of a night.

2. Palms are irrigated by feet. For example, if there are seven feet of shade [cast] in the morning, more or less, or seven feet of shade [cast] after midday, or twelve feet, more or less, at *'aṣr*-time. These *waḍaḥ*-periods continue [to be reckoned] by feet just as they formerly were, following the old arrangement.

3. *Warāfi'* and *unfūkhāt* are [treated in] the same way – because the channels irrigating them are high up they are called *nufūkhāt* and *marāfi'* (raised), which are irrigated when water rises to others and there is no stopper on (the entrance to the channels leading to) them.

4. *Dhawb* (surplus water) is dealt with in accordance with the particulars about it, as follows: If palms are irrigated exclusively from surplus water (*dhawb*) no one has any right to infringe on it unless he has partnership in it. The way it works is that if water [flows] at a certain time to a certain palm plantation on any day of the week, when the irrigating of the palms is finished and the water goes elsewhere, the surplus water is called *dhawb* or *haddat al-sāqiyah* (the release/opening of the irrigation channel).

Likewise the seepage water (*al-shawākhil*) which comes from behind the stopping (*sikār/iskār*) [of the irrigation channels] of the palms and the waters which come from the drainage channels – all this is called *dhawb*. The palm plantation may be a single (plantation) or more [than one] by sharing in the [waters?] among the palms.

5. [In the case of] the spring from which two or three palm plantations are irrigated, the *waḍaḥ*-periods [of irrigating] being restricted to the aforesaid palms, the cost of maintaining it in good repair is [fixed] according to [the number of] *waḍah*-periods [belonging to the individual plantations]. As for the public spring of which the *waḍah*-periods are distributed among many palms the cost of maintaining them [the springs] in good repair is [fixed as a charge] on the date harvest of the palms irrigated from it.

6. The maintenance in good repair of the irrigation channels when the spring has [only] a single channel [leading from it] – the cost of repairing it is (charged) on the palms irrigated by it: but, if a number of channels branch out from it [the spring], the cost of maintaining [each channel] is a charge on the proprietors of the palms whose trees are irrigated from this [particular] channel.

7. The channel for the passage of water passing through a palm plantation or number of palms to irrigate a palm plantation or group of palms – the proprietor has no right to alter the channel for the passage of water [passing through his property], which, in accordance with the preceding technical usage, constitutes an irrigation channel; nor may he change it from its siting, except by agreement on the part of the property owners whose palms are irrigated from it. If agreement on the part of those aforesaid does take place he has [nevertheless] no right to lower or raise the irrigation channel from its previous level.

8. An irrigation channel called *ṣūdaq* when it crosses over another irrigation channel – if its structure develops faults – its owner has the right to rebuild it on its existing foundation without raising or lowering [the level of it].

9. Old and new drainage channels – nobody may block them up. They shall, on the contrary, remain as they were, except the *ṣāḥib al-jādī* who irrigates his palms from the drainage channels, for he has a right to this, but without [causing] detriment to the palms they serve to drain.

10. Should an excess of water develop in [the holding of any of] the palm proprietors with no vacant land adjacent to it for the discharge of the surplus water, there being palms or land under ownership adjacent to it, the owner of the adjacent land is obliged to permit him to dig a drainage channel along the outskirts of his property from one of the places to be selected by the owner of the palms or land, the costs to be borne by the party requesting [permission] for the drainage channel.

11. The *jaṣṣah* is the floor and sides of the irrigation channel in all directions/areas whether held in shares or [allotted] exclusively to one [group of] palms. Should some fault due to the passage of time appear in it in its original structure it must be restored as it was previously. If it be held in shares this is the responsibility of all those sharing in it and if [allotted] exclusively to a [single] palm plantation the responsibility lies with the proprietor of the palms.

12. Should a number of palms be irrigated along with each other at one time no one among the palm proprietors sharing [in an irrigation channel] has the right to lower the irrigation channel to his palms below its existing [level]. On the contrary it shall stay as it was so that no detriment be caused to the palms irrigated along with it. In the case where detriment to some of them does occur on account of the rising of the [level of the] irrigation channel the water must be shared out in accordance with the area of the palms [to be irrigated].

13. There are some palms on low (ground) irrigated along with palms on high (ground). A regulator or deflector will be placed first on the palms on low (ground) upon the irrigation channel to them in accordance with the existing custom, without [causing] detriment to the palms on the low or high (ground). If detriment does enter the water must be shared out in accordance with the extent of the ground area of the palms which are irrigated together.

14. If a district [containing] palms should be irrigated from a spring by *taskīr*,[63] i.e., by stopping up and opening [the irrigation channels], as, for example, [in the case of] 'Ayn al-'Adhārī or other springs, and have a person appointed by the palm proprietors engaged for a wage to close the *waḍaḥ*-periods [of watering], all the renters must pay him that portion of [his] wage applicable in respect of their palms. Likewise the palm proprietor with no renter, be the palms on low-lying or high (ground), is obliged to pay the same [proportion] of his wage as that for which they are responsible.

15. A small plot of land (*dāliyah*) irrigated formerly by scooping, such as scooping with the leather bucket, by hand or by the *zājirah* [bucket and pulley with animal traction], this being olden of course but remaining in its original state – if repair is made to the spring from which he scoops [water][64] and the *sāb*-channel he is obliged [to contribute] the same share as others in proportion to [his] harvest from dates or crops evaluated at a fixed price, paying his share in accordance with what is established; but if he replaces scooping by the pump he is only entitled to do so by agreement of the proprietors sharing with him [in the water from the spring].

16. A small plot of land (*dāliyah*) scoop–[irrigated], newly [brought into cultivation] – its proprietor is only entitled [to irrigate it] by agreement of the palm-owners whose palms are irrigated from the spring or irrigation channel from which he wishes to scoop [water]. The owner of the *dāliyah*-plot may not extend its area beyond that existing at the time when the agreement by the aforesaid [palm-owners] took place, lest it cause detriment to the palm-proprietors whose palms are irrigated from it [the spring or irrigation channel].

17. An irrigation channel, number of channels, or a gathering tank with [several] exits [for irrigation channels] to palms – each renter or proprietor of palms of which he is in personal control, if he wishes to take water for his palms or to close [the exits] to the rest, is responsible for stopping up the exits soundly with the bush to be found in the locality or in the near vicinity of them. It is not permissible [to stop up an][65] exit, or a number of exits [of a channel to] palms with sand or the surface soil next to the exit or exits which it is intended to stop up, since this causes detriment to the water-courses of the exits [leading to] the other palms.

18. Cultivating palms or tilling [land for growing lucerne, vegetables, etc.] is (carried out) in accordance with the contract between the proprietor and contracting party. If turning over the soil [is priced] at rupees 1 or more per *maghras*-plot, and the contracting party undertakes that the turning over of the soil shall be one turning over or more for a fixed hiring, and a shortcoming appears in meeting the stipulations responsibility for which the contracting party took upon himself, either in respect of the number of times the soil should be turned over or in [planting] the palms too far apart or too closely together, [proven] after investigation, the contracting party or parties shall have imposed (on him or them) [deduction of] half of the [wage] fixed originally, by way of penalty for the shortcoming which took place.

19. The stipulations regarding rents and what relates to them in the matter of management of the palms or tilling [land for growing lucerne, vegetables, etc.] – if any person engages to rent from one of the proprietors of palms or [land for] tilling [lucerne, vegetables, etc.], taking responsibility for clearing [the channels to] the palms or tilling [land for growing lucerne, vegetables, etc., of grass and the like] twice per annum, and for clearing the external and internal irrigation channels and drainage channels connected with them [of debris], twice per annum or more, as stated, in accordance with the stipulations defined in the lease, as also to [maintain] the fence protecting the palms or [land for] tilling [lucerne, vegetables, etc.] – should he fail to implement any of this he will be obliged to carry it out in full, or to have valued what he has neglected to do, he being held liable in accordance with whatever valuation the experts make.

20. The stipulations for *taqlīj*, i.e., cutting the withered stubs [of the palm branches off the trunk] – the renter must cut them off on one of two occasions, the first being after the ripening of the palms following fertilisation, the renter being required to leave one round of withered [palm stubs] so as not to damage the palm trees, and on the second occasion at the first sign of the ripe dates, with the proviso that he leave two or three rounds [of stubs]. If the renter neglects to cut [them] off, the proprietor has the right to hire [a man] to cut the withered stubs off the palms and to charge the cost to the renter's account since the stipulation basically is that it is he who is responsible for this – except if the proprietor relieves him [of it] – in which [last] eventuality there is no obligation on the part of the renter for what is mentioned (above).

21. The stipulations for cutting down the withered palm frond – this takes place at the commencement of the Ascension of al-Thurayyā (the Pleiades), i.e., the beginning of the season of the Bāriḥ (North) wind. If the proprietor permits it to be cut down before that there is no objection to cutting down the withered palm frond only, leaving the green (frond) until such time as it withers. The second occasion is at the close of the fruiting of the palms. The renter has only the right to cut down the withered palm frond; if he goes further and cuts down green palm frond the renter will be punished in proportion to [the gravity of] his offence, either by imprisonment or a fine, as the court sees fit.

22. A fallen or dead palm – the renter has no right to dispose of it except after examination [of it] by the owner or someone standing in for him. If he acts contrary to this he is liable to the proprietor for the value of the palm tree in the way of the fruit it would produce were it living, but if the proprietor absent himself deliberately from the examination [of it] the renter has the right to take the case to court.

23. A natural spring the water of which has gone dry, it being shared among a number of palms holding [rights to] *waḍaḥ*-periods in it – if those who own property [in it] agree on mounting a pump on the spring [shared] by partnership between them on the basis that each one of them shall be given the water to which he is entitled, this is in order. If (however) they do not agree on this then each one of them has the right to mount a pump [on it] to be used during his own *aḍaḥ*-periods.

24. So also if necessity dictates the cleaning out of a natural spring this is the responsibility of the whole [body of holders of rights in it]. If one partner refuses to pay the charge falling to his share the [rest of the] partners have the right to distrain upon the water to which their [defaulting] partner has an entitlement until he pays the charges for which he is responsible.

25. A custom/principle exists, followed in practice, whereby the proprietor of [land for] tilling [vegetables, lucerne, etc.], lets [to the incoming renter], with valuation only, of the crop, at the time of entry of the new renter, this remaining outstanding as a debt owed by the latter to the proprietor. If the term of the letting comes to an end the renter has the right to have the [existing] crop valued. If its value exceeds that of the previous [crop] he receives the balance. If it be less he is responsible for the deficiency to pay it outright before he quits. However if the proprietor and renter agree on something else it is their affair because they have the right to do as they wish.

26. [The term] *nakhīl* (palms) covers trees of all sorts such as the citron, pomegranate, almond, etc. So if the renter takes over these '*nakhīl*' with valuation of [the crop on] the trees, he has the right, at [the time of] his quittance to have the fruit on the trees valued as took place previously. If the renter takes over without valuation he has no right to it when the letting terminates.

27. The customary renting/letting of the palms commences at the time of the beginning of the renting/letting and terminates at the time it terminates, at the specified time agreed upon. If any fruit is left over after the termination of the period of the renting/letting the former renter has the right to have the fruit left over valued.

28. It is not permissible to take the *garāyif*, i.e., *ḍarīb* (silt and weed) from the drainage channels of the palms or the irrigation channels entering or coming out of it [the plantation?] without permission of the proprietor or proprietors, with the proviso that this [be carried out] under the supervision of the proprietor or someone standing in for him, such as the renter or other partners of the proprietors or renters sharing in the drainage and irrigation channels so that the owners of properties may not suffer detriment through dilapidation of the irrigation or drainage channels.

29. It is not permissible to take the sand and *madād* (mud, earth, silt and dung) from the palms or lands under ownership or from the sides of irrigation or drainage channels without permission from the proprietor or proprietors of the irrigation or drainage channels and palms and lands aforesaid.

30. If an irrigation channel, drainage channel or *sāb*-channel be [either] within [a group of] palms held in ownership or outside it, in land held in ownership, the proprietor of the property in which this irrigation or drainage channel or channel for the passage of water (*mamarr*) lies must not introduce into it any (thing which causes) detriment [to it], such as filling it in, making it narrower, or throwing such things as palm stubs, frond or fibre. If anything of the kind aforesaid happens to them he is obliged to remove any [cause of] detriment he has introduced because he is responsible for it.

Written on Sha'bān 9, 1379/February 8, 1960

COMMENTARY

Article 1

The *waḍaḥ* is a time unit of twelve hours (*sā'ah*, plural *suwa'*), varying a little in time between summer and winter, there being fourteen *waḍaḥs* in a week. It is also used at sea (cf. my 'Fisherfolk and fish-traps in al-Baḥrain,' *BSOAS*, London, 1968, XXXI, iii, 513, quoting a local almanac). I have incorrectly called it a 24-hour period in al-Mukallā in 'Star calendars and an almanac from south Arabia,' *Anthropos*, (Freiburg, 1954), 438. The seasons in Baḥrain are *Ṣayf* (24 June), *Kharīf* (25 September), *Shitā'* (23 December), *Rabī'* (22 March). The length of time during which a plot may be irrigated is calculated in *waḍaḥs* or fractions thereof. Presumably the *waḍaḥ* is also equivalent to a certain number of *qadams* (for which see article 2). The important *waḍaḥ*-period is that extending from dawn to dusk (*shurūq al-shams ilā ghurūb-hā*). A cultivator will receive water in accordance with the document in his hands (*ḥasb al-wathīqah (al-mulk) fī yadi-hi*), and he may have any of the periods of watering mentioned.

Article 2
A stick is set up to serve as a sundial, the shade it casts being measured by feet (*qadam*). For the sundial observed at Rās al-Khaymah see Appendix.

Article 3
Al-marāfī' (seemingly with no singular) would be palms planted on high places which only 'drink' when the water rises to their level. The *nufūkhāt/unfūkhāt* would seem to coincide with the definition in *Tāj al-'arūs, al-nafkhā' min al-arḍ ... qīla hiya arḍ layyinah fī-ha 'rtifā'*, soft land, raised. If there is little water the *nufūkhāt* receive what there is available; when the spring (*'ayn*) is low in water they would receive what remains behind the *iskār* (stopping of a channel) or what flow there is. *Rabṭ*, stopper, is when they put down a bag or cloth (*yaḥuṭṭūn kīs aw qumāsh*) – not cloth usually but *khayshah*, jute. See article 14, *al-rabṭ wa-'l-hadd*, stopping and releasing.

Article 4
Dhawb, from *dhāb, yadhūb*, to flow; *haddat sāqiyah*, presumably from classical Arabic, *hadda*, to demolish, *haddayna 'l-mā'*, we have released the water. *Shawākhīl* (singular *mishkhāl?*) is waters seeping from a *sidd* a (temporary) dam or stopping usually made of gunny-bags. One says, *yishkhal minnah* = *yaẓhar minnah*, it issues from it. Cf. *Tāj al-'arūs, ṣaffā-hu*, to filter, and Jalāl al-Ḥanafī, al-Alfāẓ *al-Kuwaytiyyah*, Baghdād, 1383/1964, 196, *shakhkhālāt, ḥujar khāṣṣah li-ḥiyāḍ al-mā' al-mu'allaqah fī buyūti-him*. A *sikār* is a stopping to close the flow of water to any place, usually of mud of stone (also *iskār*) = *rabṭ*. *Munajjā(h)*, plural *munajjayāt*, a drainage channel (*majrā*) which takes surplus water to the sea. F. S. Vidal, *The Oasis of al-Hasa*, Aramco (Dhahran?), 1955, glossary, a canal from one garden to another.

The surplus water after the palms are irrigated goes to waste as drainage; it is not owned. This water would seep through the stopping at the end of the drainage channel or probably be released. When palms are 'quenched' (*rawiyān*), i.e., have received sufficient water, further watering would leave them spoiled (*talfānah*).

Article 5
It was commented that, *maṣārif al-'ayn tu'tabar 'alā ḥāṣil al-nakhīl*, the costs of (maintaining) the spring are estimated in accordance with the date-crop. *Maṣraf*, expense.

Article 7
A *mamarr* is a channel passing through one's land to another place.

Article 8
A *ṣūdaq* is a channel raised above the level of the land, coming from a natural spring (*'ayn ṭabī'ī*), from which water is released onto the ground below it. It crosses over the *sāqiyah* or *sāb* channels like a bridge. For *sāb* see article 15.

Article 9
Ṣāḥib al-jādī is defined as *al-shakhṣ alladhī lah ḥaqq yishrab min al-munajjayāt*, the person entitled to irrigate from the drainage channels. This entitlement must be recorded in a *wathīqah* document. Sometimes the water also seeps up through the ground to his land. So far I have not found *jādī* in the lexicons, but Fāliḥ Ḥanẓal, *op. cit.*, 134–5, suggests various senses for this root.

Article 11
The *jaṣṣah* is made of gypsum on top of stone (*al-juṣṣ fawq al-ḥajar*), a sort of cement made of stones coming from Qaṭar which they burn, then it is beaten (*aḥjār tātī min Qaṭar yuḥarriqū-hā thumma tudaqq*). Al-Nabhānī, *al-Tuḥfat al-Nabhāniyyah*, Bahrain, 1342 A.H., 38, alludes to a *ma'dan al-juṣṣ* in al-Bahrain and, 27, to workers in *juṣṣ* (*jaṣṣāṣūn*) on al-Qaṭṭārah island. Lorimer, *op. cit.*, II A, 235, refers to a gypsum mine at Umm Na'asān. The depth of the channel was stated to be usually about four inches (*hinsh*).

Article 13
'Ādil and *iṣdār/ṣidār* both mean deflector. Cf. classical Arabic *'adala*, to deviate, deflect, *aṣdara*, to cause to return. When a channel gets below its proper level to irrigate the land a sort of bar is made in it, described as a sort of dam (*sidd*), in order to raise it to its former level – nowadays it is usually heavy cement. It was also described as *naw' min al-sikār*.

Article 14
For 'Ayn al-'Adhārī, see p. 476. *Manāzīl* is said to have no singular.

Article 15
The *dāliyah* is a bucket and pulley, but comes to mean the land irrigated by means of it. It is a widely-used term. For *dāliyah* in Iraq see Ḥusām Qawām el-Sāmarrāie, *Agriculture in Iraq during the 3rd century A.H.*, Beirut, 1972, 25 *passim*. The *zājirah* according to my notes from Ṣādiq al-Baḥārnah is, or has, a bucket of *jild al-ḥayawān*, animal hide, and is basically the apparatus for scooping/lifting water, but according to 'Alī Tājir it means the whole system of watering by animal power. Oxen used to be used but both 'Alī Tājir and Belgrave say donkeys were also employed. Cf. C. Reinhardt, *Ein arabischer*

Dialekt gesprochen in Omān und Zanzibar, Stuttgart-Berlin, 1894, 44, *zayjrah*, place for drawing water, and verb, p. 147. See Lorimer, *op. cit.*, IIA, illustration of well apparatus, opposite p. 230. A *sāb* (plural *sībān*) is a small narrow water channel, = *majrā* or *jadwal* below ground level, coming straight from the spring (*min al-'ayn ra's-an*), both *sāb* and *sūdaq* being *sāqiyahs*. *Sāb al-may*, the water flows (*jarā*). Cf. class. Arabic *sīb*, water-channel. One speaks of Sāb 'Adhārī, as in *'Aṭash al-nakhīl*, *op. cit.*, 8, 122. By *zar'*, rendered 'crops', any field crop is meant, even tomatoes, but *zirā'ah* means vegetables, etc.

Article 17
The *maḥall mujammi'*, rendered 'gathering tank', is shaped and is something like a distribution tank. *Garīfah* (plural *garāyif*), *kāf* presumably written for *gāf*, was stated to be grass/bush (*ḥashīsh*) growing nearby which is cut from underneath and a solid mass of it used to stop up the channel.

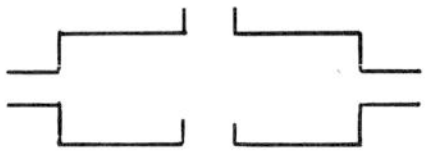

Article 18
'Amār, rendered here as 'cultivating', was said to mean turning over the earth with the hoe (*taqlīb al-arḍ bi'l-sakhkhīn* (Vidal, *op. cit.*, glossary, *ṣakhkhīn*)). Vidal, 167, gives *'amar* as fertilizing, but all these senses clearly derive from classical Arabic *'amara*, to cultivate. For the sense of *zirā'ah* see note to article 15. The *maghras* (*maghāris*) was stated to be 18 × 18 feet, i.e., 324 square feet. Vidal, 174, gives the size of the *maghras* in al-Ḥasā as 15 *dhirā* square.

A young palm taken from its nursery to be planted out elsewhere is called *najīlah* (*najāyil*) according to *'Aṭash*, *op. cit.*, 112, 22 (= *naqīlah*?). *Farā'ah*, spacing out of palms. In each village there are special families with animals and ploughs whose occupation is to plough the ground. They are hired by the man who farms land. Cf. the Baqqārah of Ḥaḍramawt, specialists in ploughing. Vidal, 148, says there are no ploughs in al-Ḥasā.

Article 19
Clearing (of palm channels) is done with the hoe, as is clearing of the irrigation channels (*al-kisāḥah bi-'l-sikhkhīn waka-dhālika ḍarb al-sawāqī*). *Yaḍrib al-sawāqī* = *yuzīḥ min-hā mā saqaṭa fī-hā min al-ashyā' al-mutarassibah*, to remove the deposit (*rawāsib*) of any kind, mud, vegetation, etc., which impede the flow of the water. Nowadays this clause is obsolete and no longer followed. *Ḥaẓrān*, fence.

Article 20
Jummār is the ripening of the palms (*nuḍūj al-nakhl*); in classical

Arabic it is the heart of the palm. *Al-nabāt* is the fecundation by the male of the female palm. Vidal, glossary, gives *siff* (*sufūf*) as the spikelets of the male palm. The *dawr* is a round of dry palm stubs (*karab*), the broad base of the palm branch). *Bishrah*, the first appearance – one says *yubashshir qabl*, it appears early.

Not mentioned in the *Qānūn* is a phrase I saw in a document in the Dā'irat al-Awqāf, *ṣarm al-muḍar* (perhaps to be read *muḍirr*?) *wa-'adam al-ḍarar*. *Muḍar* was said to be *al-ḥiml/ḥamal al-zāyid*, excess of fruit while the dates are still green, so that some must be cut off in order not to weaken the palm and to ensure a good crop in the following year. Vidal, *op. cit.*, 166 has *ṣaram*, date harvest, and *ṣarm* in Kuwait means *al-fasīl min al-nakhl*. The phrase would thus mean, 'cutting off excess fruit and no detriment (in consequence) to the crop.' One says, *Ashtakī ḍarar*, I make a plea of damages (to my property etc.). See also 'Abd al-Qādir Bāsh A'yān, *op. cit.*, in note 6, 101.

Article 21

According to the Baḥrain almanac in my 'Fisher-folk and fish-traps in al-Baḥrain,' *op. cit.*, 513, *al-Thurayyā* is June 8–20. H. R. P. Dickson, *The Arab of the Desert*, London, 1949, 249, gives *Bāriḥ al-Thurayyā* as April 15 to May 27, and *Bāriḥ al-Jawzā'* or *al-Kabīr* as July 9 to 16, but Jalāl al-Ḥanafī, *op. cit.*, 29, says it falls in June. It begins with a strong *shamāl*, north-west wind – this 'Alī Tājir called *mūsim al-Bāriḥ*. Cf. Lorimer, *op. cit.*, 236.

Nafāṭ (= *nafādh*) when the dates are fully ripe, becoming *tamr*. The *maḥashsh* is the curved knife used for cutting palm fronds, lucerne (*jatt* = classical *qatt*) and grass. *Kharaf al-nakhlah*, is to cut the fruit of the palm (*'Aṭash*, *op. cit.*, 5, 121). The *minjal barrānī* is said to be used for trees. Cf. Vidal, *op. cit.*, glossary. Jāḥiẓ, *Rasā'il*, Cairo, 1384/1964, 195, mentions a *minjal Baḥrānī* used to behead a man; the editor understands this as 'of Baḥrain.'

Article 23

'Ayn ṭabī'iyyah is a flowing spring. *Jaff* means the water has receded, gone deep down in a well, whereas *naduba* would mean that it has *no* water.

Article 25

Qā'idah, rule, principle, etc., was stated by Ṣādiq to mean *'urf*, custom.

The existing crop which could be lucerne (*qatt*), vegetables (*khuḍār al-ḥashā'ish*), ladies' fingers (*bāmiyah*), is assessed. The incoming renter (*ḍāmin*), may take over the crop as assessed, or the landowner may do so – in which case he pays the outgoing renter (*ḍāmin*) and takes it. He sells to the incoming renter. Or the outgoing renter may deal directly with the incoming renter, any difference in value being later settled between them.

Mukhawwalūn, rendered 'they have the right,' was said by 'Alī Tājir to mean 'they are authorised, have the authority.'

Article 27
For the beginning of the letting (*ḍamān*), see supra p. 473.

Article 28
Ḍarīb = *al-khurūj*, what is brought out. Cf. article 19, *ḍarab al-sawāqī*, to clean the channels, and *ḍarbah* (plural- *āt*) a turning over of soil (article 18); *ḍirābat al-'ayn*, clearing the spring and deepening it if necessary.

Article 29
Madād is fertilising sand and muck. If the top level of a field is carried off, *madād* is brought to replace it ('Alī Tājir). Lorimer, 242, says fish manure is used to fertilize the date groves.

SUMMARY

Obviously this paper is little more than exploratory; it is circumscribed by the fact that circumstances have not permitted me to carry out the fieldwork that should go hand-in-hand with studying this type of document.[66] Little has been said about palm and date cultivation.[67] A wide lore expressed in rhyming saws on agriculture, the seasons, the star calendar must surely exist, as, for example:

> *Idhā ṭaḥa 'l-kinār, tsawa 'l-laylah wa-'l-nahār.*
> When the *nabaq* (fruit of *Zizyphus spina christi*) falls night and day are equal.[68]

Peasant ditties and shanties could also tell much of life in the villages. The sundial, the *dāliyah* and the *zājirah* with their no doubt extensive technical vocabulary, need further enquiry. I have already referred to the documents relating to lawsuits and the information they are likely to contain. If, however, such studies are to be undertaken at all they must be carried out while the memory of traditional agriculture still remains with the villagers of al-Baḥrain. In the October number of *al-'Arabī* one can see how difficult this will soon be, with the threat to agriculture owing to the vastly increased consumption of water and the migration of so many villagers to the urban areas.

NOTE

As this paper, when written, was based on the author's work in al-Baḥrain carried out prior to 1983, no account has been taken of the literature that has appeared since the latter date. The Bahrain Court Records have, I understand, been calendared since that time and could be consulted. The papers of the agricultural history conference held at Kuwait immediately following the 1983 conference at al-Baḥrain are not of direct concern for this paper and it is mostly only by chance that one comes upon local Arabic publications in the Gulf and elsewhere that are. Of special interest, however, if only for comparison, may be cited A. A. Brockett's *The Spoken Arabic of Khābūrā* (1985) for its rich agricultural vocabulary collected actually in Oman.

1. See 'Fisherfolk and fish-traps in al-Baḥrain,' *Bulletin of the School of Oriental and African Studies*, London, 1968, XXXI, iii, pp. 486–514.

2. *Nakhlah*, plural *nakhīl*, palm tree; *nakhal* is a palm garden in al-Baḥrain.

3. *Welcome to Bahrain*, London, 52 seq., and 3rd ed., p. 64.

4. Belgrave, *op. cit.*, p. 52; J. G. Lorimer, *Gazetteer of the Persian Gulf, 'Oman and central Arabia*, Calcutta, 1908 and 1915, II A, p. 241. Ṣādiq al-Baḥārnah said no cereals were grown.

5. *Al-Tuḥfat al-Nabhāniyyah*, 2nd ed., al-Baḥrain, 1342 A.H., p. 29.

6. Fāliḥ Ḥanẓal, *Mu'jam al-alfāẓ al-'āmmiyyah fī Dawlat al-Imārāt al-'Arabiyyah al-Muttaḥidah*, Abu Dhabi, 1978 (?), p. 194, notes *khaṣṣābah* for the palm and its date, and two varieties, black and white. Jalāl al-Ḥanafī, *Mu'jam al-alfāẓ al-Kuwaytiyyah*, Baghdad, 1383/1964, p. 115, *al-khṣāb* dates at al-Jahrah. Cf. for Iraq, 'Abd al-Qādir Bāsh A'yān al-'Abbāsī, *al-Nakhlah sayyidat al-shajar*, 1383/1964, pp. 67–8, list of 86 names of dates at Baṣrah.

7. *Op. cit.*, I, ii, pp. 2296–7.

8. F. S. Vidal, *The Oasis of al-Hasa*, Aramco (Dhahran?), 1955, p. 163.

9. *Loc. cit.* Lorimer's *qallah* is probably the same as classical *jullah* (plural *jilāl*) cited by al-'Abbāsī, *op. cit.*, p. 47, quoting the lexicons and adding that it is still used in Basrah. It is made of *khūṣ*, palm-frond, and is apparently synonymous with *qawṣarah*. In the UAE it is cited as *jillah*.

10. With a variant *bambarā*, sebestens.

11. *Ma'ẓan* in class. Arabic, a journey to seek herbage.

12. Belgrave, *op. cit.*, p. 53.

13. *Op. cit.*, p. 25.

14. 'Fisher-folk and fish-traps . . .', pp. 512–3.

15. P. 121. 'Abdullāh b. 'Abd al-'Azīz al-Duwaysh, *Dīwān al-Zuhayrī*, Kuwait, 1971, p. 16, lists the types of *mawwāl* to which he adds another two. *Mawwāls* in al-Baḥrain were said mainly to relate to love, but also agriculture and pearling.

16. For the *akkār*, palm cultivator in Iraq, cf. Jāḥiẓ, *al-Bukhalā'*, Cairo, 1948, p. 121, etc.

17. *Op. cit.*, notes to p. 12.

18. Persian *sipistān*, plum-like fruit of the tree *Cordia*.

19. Cf. the sentiments of grievance expressed by the peasant woman in *Dīwān al-Karkhī* ('Abbūd), 2nd ed., Baghdad, 1956. Perhaps this is an established theme in colloquial verse. The metre seems to be a type of *mujtathth*.

||–|––˘–|–˘––||––|––˘–|–˘––

20. But see infra, p. 474 and n.28.

21. *Descriptio Arabiae meridionalis* ... *Ta'rīkh al-mustabṣir*, ed. O. Löfgren, Leiden, 1951–4, p. 301.

22. See also his *The Baḥārnah dialect of Arabic*, London, 1982. Cf. Lorimer, *op. cit.*, II A, 208; and 'Fisher-folk and fish-traps ...', pp. 488–9.

23. Cf. Vidal, *op. cit.*, glossary, *sayḥ*, flowing water. Ḥuṣām Qawām el-Sāmarrāie, *Agriculture in Iraq during the 3rd century AH*, Beirut, 1972, p. 25, 'Cultivated lands were irrigated either directly (*yusqā saiḥan*) or by machine.' See below, p. 476.

24. In the earliest Islamic documents the inhabitants of al-Baḥrain, meaning the whole area of the island and coast that went under that name, are called Majūs.

25. *Welcome to Bahrain*, Manāmah–London, 1973, p. 94.

26. *Op. cit.*, p. 148.

27. *Op. cit.*, II A, p. 252. He also refers to a tax, *nōb* on the gardens of private individuals, but this does not seem to have applied to any particular group (p. 251). For *sukhrah*, *op. cit.*, p. 248.

28. This appears inconsistent with Ṣādiq's statement (p. 473 of this article, and n.20).

29. See 'Fisher-folk and fish-traps ...,' p. 489.

30. Muḥammad Ghānim al-Rumayhī, *Qaḍāyā al-taghyīr al-siyāsī fi 'l-Baḥrayn 1920–1970*, is reviewed in *al-Khalīj al-'Arabī*, Basra, 1977, VII, p. 105 seq.; it appears to have data on cultivation and Shī'ah–Sunnī differences.

31. *Op. cit.*, II A, 230 seq., p. 236; cf. Muḥammad Aḥmad Ḥasan 'Abdullāh, 'Masādir al-miyāh fi 'l-Baḥrayn', *Dirāsāt al-Khalīj wa-'l-Jazīrat al-'Arabiyyah*, III, April 1st, 1979.

32. *Min Turāth al-Baḥrayn al-sha'bī*, Beirut, n.d., p. 68.

33. *Loc. cit.*

34. Shaikh Rāshid b. Ibrāhīm says it is not called 'Ayn al-Mahazzah, simply al-Mahazzah, as in *al-Tuḥfat al-Nabhāniyyah*, p. 75.

35. Al-'Urayyiḍ writes *al-Nabīh Ṣāliḥ*, but Lorimer, *Nabiyy*, presumably the prophet who is buried in so many places.

36. *Welcome to Bahrain*, 3rd ed., 1973, p. 94. Some of the springs appear on the map. Shaikh Rāshid remarked that it is curious that the tortoise (*ghaylam*) that had been absent for ten years recently reappeared in the sweet waters of the island.

37. *Op. cit.*, p. 71.

38. *Welcome to Bahrain*, ed. 1970, p. 70; illustrated p. 77.

39. The vocalisation of Ḥamagān and Fūliyān is conjectural. *Ṣurmah* was described to me in 1963 as a palm plot. Belgrave read *sarmah*, which he calls a line of trees along a water channel. A term *khīsah* was cited to me as a place where palms have grown naturally.

40. See *Cambridge History of Iran*, V, 1968, pp. 516–7.

41. *Min Turāth al-Baḥrayn*, p. 70; cf. *'Aṭash al-nakhīl*, infra, spelling *qarīb* as *jarīb*.

42. *'Aṭash* ..., p. 20, p. 122.

43. See Serjeant & Lewcock, *Ṣan'ā'; an Arabian Islamic city*, London, 1983, p. 26.

44. *Op. cit.*, IIA, p. 242. For irrigation in general cf. Vidal, *op. cit.*

45. See note 23.

46. *Op. cit.*, II A, p. 242. Vidal, *op. cit.*, glossary, says the *gharrāfah* is also called *'iddat ghurf* = *shadūf*, of which he gives an illustration.

47. *Op. cit.*, p. 25, from Abū Yūsuf, *Kitāb al-Kharāj*.

48. *Welcome to Bahrain*, 3rd ed., p. 64.

49. Metre – – – ⏑, repeated four times, but the last foot of verse 1 as printed seems out of pattern. Note, in verse 7, *kint* is one long syllable. Spellings reflect local pronunciation, *gāf* = class. *qāf*, *chīn* = *kāf*, *yā'* = *jīm*.

50. *Intakhā*, to ask for help, said to derive from *nakhwah*; *yamm-ich* = *janb-ik*.

51. *Al-gayẓ* = *qayẓ/qayḍ*.

52. *Sāb*, see comment to article 15 (pp. 489–90). *Min dūnī* = *qurbī*, near me.

53. *Rīj* = *rīq*, *al-dūnī* = *al-dānī*, the near at hand, *tark-ich* = *tark-ik*, *zar'-ich* = *zar'-ik*.

54. *Al-wifā* = *al-wāfī*, explained as *alladhī yafī bi-'l-wa'd wa-yu'ṭi 'l-qarīb thumma 'l-ba'īd*.

55. *Yam-hā* = *janba–hā*.

56. As if to make a deaf 'Adhārī hear.

57. *Al-gayẓ, giyẓ* in Fāliḥ Ḥanẓal, *Mu'jam* . . ., p. 487, is the hot summer period which, he says, is from the ascension of al-Thurayyā up to the ascension of Suhayl. This is also the same in Kuwait. The summer heat has made the palm groves ripen with dates, as if the fruit were new clothes that one receives at the two feasts (*'īd*) – *'īd* can also mean a feast gift.

58. *Rīq*, lit., saliva. *Rīq* in class. Arabic means the saliva of the beloved. This is a reference to *'ishq*, love – 'Adhārī is not loving.

59. 'Adhārī flows northwards, to water northern groves or gardens. If there be any political allusion at all in this verse, perhaps one should think of Kuwait and the special treatment it accords native Kuwaitis.

60. Taqiyy al-Baḥārnah referred me to similar themes in the *Dīwān* of Ibn al-Muqarrab. Cf. al-'Uwaysī's ed. Damascus, n.d. 303; Cairo ed., 1383/1963, p. 255, verses 28–29 in which the poet complains that it is better to leave al-Qaṭīf where all he receives of pearls is the oystershell (*maḥār*) and of bunches of dates, only the stalk!

61. In all irrigation systems known to me the higher plots must be watered first. This is also *sharī'ah* law.

62. In south-western Arabia the 'water-master' called in Wādī Rima' for example, *shaykh al-sharīj* appears to have an importance not accorded him on the island.

63. *Taskīr* is stopping the flow until the level of the water is raised.

64. Perhaps the passive should be read, 'the water is scooped.'

65. '*Sadd*' should be inserted before *fūhāt al-nakhīl*.

66. At Abu Dhabi I was advised to consult Ḥājjī Muḥammad Āl Ḍayf of Damistān village for *ḥaqq al-saqy*.

67. Lorimer, I, ii, p. 2294, does refer to earlier studies on the date palm and of course there is a body of modern literature upon it.

68. Al-Nabhānī, *op. cit.*, p. 29. The *nabaq* appears *fī burjay al-i'tidālayn al-Ḥamal wa-'l-Mīzān*.

APPENDIX

For comparison the division of the 24-hour day into watering periods, as given to me by Mr Muḥammad A. Muḥammad of the Agricultural Trial Station at Rās al-Khaymah in 1964, is quoted below. Here each garden has a long established customary right to a known number of *qiyāsāt* (singular *qiyās*), measures, the supervision of which is entrusted to a *saqqā'ī*. The latter receives payment of a maund or two of dates depending on the area of the garden concerned. The day is reckoned from sunset, and there are 48 *qiyās* in 24 hours, but they are not of equal length.

Al-Sāriḥ, after *maghrib*, the sunset prayer till midnight, divided into two periods, 1 *maghrib al-shams* to *Raqīb al-Mīzān*, 3 *qiyāsāt*, 2 *Raqīb al-Mīzān* to *maṭla' Suhayl*, 3 *qiyāsāt*.
Rub' Ṭāli' al-Shams, divided into two periods, 1 *maṭla' Suhayl* to *maṭla' al-shams*, 6 *qiyāsāt*, 2 *maṭla' al-shams* to next following, 6 *qiyāsāt*.

Rub' Khalaf (*ṣubḥ*), 9–12 a.m., 6 *qiyāsāt*.
Al-Rāyiḥ, from midday (*ẓuhr*) till before the *'aṣr*, late afternoon, 6 *qiyāsāt*.
Al-Rawāḥ, before and till after the *'aṣr*, 6 *qiyāsāt*.
'Ashwānī, after the *'aṣr* till about half an hour before *maghrib al-shams*, 6 *qiyāsāt*.
Al-Ghabshānī/Ghibshānī) (*ghabash*, darkness at the beginning of night (classical Arabic)), 5 minutes at the time for the call to the sunset prayer.
Nuṣṣ Laylah, up to *Raqīb al-Mīzān*, 6 *qiyāsāt*.

Al-Ghabshānī is normally not claimed but goes to the earlier or later period alternately.

During the day the sundial (*mizwalah* or, locally *'badeh*,' perhaps classical Arabic *biddah*, share, portion?) is used, and the following would be the pattern – *Rub' Ṭāli' al-Shams*, till the sun's shade is 14 feet, *Rub' Khalaf*, till the shade is 1 foot before noon, *al-Rāyiḥ*, till the shade is 7 feet during the afternoon, *al-Rawāḥ*, till the shade is 30 feet, *'Ashwānī*, till the shade is approximately 50 feet. During the night the rising of a series of stars each marks the beginning of a *qiyās*, eleven in all. These are *al-Quṣṣah*, *al-Thurayyā*, *Najm al-Thurayyā*, *Sābiq al-Mīzān*, *al-Mīzān*, *Najm taḥt al-Mīzān*, *al-Sābiq*, *Arba'at Suhayl*, *Najm warā al-Najmayn*, *Arba'at Insūr*, *al-Nusūr*.

XI

A YEMENI AGRICULTURAL POEM

R. B. SERJEANT and ḤUSAYN ᶜABDULLĀH AL-ᶜAMRĪ

IN COMPANY WITH Mr. Tom Barrett, R.B. Serjeant, acting as consultant to the British Montane Plains and Wādī Rimāᶜ Project in the Yemen Arab Republic, visited Šayḫ Muḥammad Zayd of Rubāṭ ᶜAmrān on September 14th, 1974. The Šayḫ showed them a Ms. of mixed contents relating to agriculture and calendars. He kindly lent them the Ms. for zerox copying and the copy is now at Cambridge.

The Ms., of 228 pages, contains comparative tables between the Hiğra and Rūmī calendars, an almanacal poem of Ğābir al-Ġaffārī, a piece by Qāḍī Zayd b. ᶜAbd Allāh al-Akwāᶜ,[1] a list of Jewish feasts, Coptic months, medical data, eclipses, a mediaeval almanac, etc. There are diagrams of the stars of the Mansions. There are tables by al-Ḥusayn b. Amīr al-Muᵓminīn al-Manṣūr (the latter early 11th/17th century) for 1040/1629-30. Much of the data are repetitive. The collection is specially useful for studying agriculture in the Yemen and the traditional lore connected with it, a good deal of which is also to be found in the *Buġyat al-fallāḥīn* of the Rasūlid monarch al-Malik al-Afḍal.[2]

The *urğūza* reproduced here from this Ms. was composed by a certain Aḥmad b. Abī Bakr al-Zumaylī[3] who is unknown to the printed Yemeni biographical works. On the title-page he is however called "the faqīh the very learned scholar". In verses 180-81 he states he has drawn his information partly from "trustworthy authorities", partly through experience (*tağriba*). His citations mostly appear to derive from popular literature, perhaps oral sources, proverbs and saws, but we have not attempted to trace more than a few of these. Al-Zumaylī is obviously a Šāfiᶜī, most likely from the Zabīd region. His style in fact is what is known as *šiᶜr al-fuqahāᵓ* in the Yemen — explained as *naẓm laysa fīhi šāᶜiriyya*, and this "faqīh" style is also to some extent apparent in the prose of *Buġyat al-fallāḥīn*. The *faqīh* of the Lower Yemen villages has certain religious functions and teaches children, but would rarely, if ever, be a scholar of the type of the great Šāfiᶜī or Zaydī *ᶜulamaᵓ*. The *ᶜulamaᵓ* often compose both in Ḥakamī and Ḥumaynī verse employing in the latter colloquial forms and vocabulary, but al-Zumaylī's versifying is of an emphatically different category from the Ḥumaynī of, say, al-Ānisī; he was probably not a man of great education despite his references to such authorities as al-Ġazālī and al-Māwardī.

1 Another member of al-Akwaᶜ house, the faqīh ᶜAlī b. al-Ḥasan b. Muḥammad, is author of a *ğadwal* of the Arab and Rūmī months for the years 1201-1380 (British Museum Ms. Sup., 771).

2 The English translation of part is used here, R. B. Serjeant, "The Cultivation of Cereals in Mediaeval Yemen," *Arabian Studies*, I (London-Cambridge 1974), 25-74. This relates to the 8th/14th century. *

3 Sayyid Aḥmad al-Šāmī prefers the reading al-Zumaylī to al-Zamīlī, citing *Tāğ al-ᶜarūs*, *Zumayla ka-ğuhayna, baṭn min Tuğīb minhum al-Zumaylī al-muḥaddiṯ*, but does not know if this family has members in present day Yemen.

The Arabic of the *urǧūza* resembles that of the pilot Ibn Māǧid — they share in numerous conventions and clichés that seem to form the stock-in-trade of this class of versifier. This mnemonic semi-colloquial verse style has a long enough ancestry side by side with classical Arabic verse.

Al-Zumaylī follows colloquial Arabic in his treatment of *hamza*, using or ignoring it as suits the metre. He nearly invariably writes the *hamza* where the metre requires it. Thus we have *intihā* in one place, *intihāʾ* in another, and the *hamza* of *al* may be retained where classical Arabic demands *waṣla*. *Hāḏihi* can be – ⌣ –, and the suffix *hu* has generally the quantity of *hū*. He often disregards the vowels of classical grammar, as for instance in *muštahar* and *muᶜtabar* where *muštahir* and *muᶜtabir* are correct. He spells *al-Ǧawzāʾ* in various ways to suit rhyme or metre. Clichés also in some cases found in Ibn Māǧid are: *yā ṣāḥ*, *fa-ftin*, *fa-fham mā ḏukir*, *fī ṭūlihā wa-l-ᶜarḍ*, *ṭūl al-dahr*, often used to pad out a line. The verses tend to be rather trite and often descend into the banal.

The orthography of the Ms. is retained, but we have supplied nearly all the vowels, mainly on the basis of classical Arabic.

Whatever attitude be taken over the quality of the Arabic, this didactic compendium of agricultural wisdom is of real value both in respect of the information it contains and of what it reveals of the mentality of a farming society. Furthermore it contributes new vocabulary, including sometimes ancient words that escaped the classialc lexicographers. The month Mabkar for instance is "Ḥimyaritic", and *ᶜād*, as used in v. 162, is found in the *Ǧāmiᶜ* of Ibn Wahb[4] in a verse attributed to the Prophet himself. Some verses (e.g., 128, 147, 154) are not agricultural at all, but recommend appropriate seasons for certain actions — identical entries are found in ancient and contemporary almanacs.

The Rūmī (Byzantine) or Syrian Christian months may be assumed, with near certainty, to have been introduced into the Yemen from Byzantium before Islam, but the Ḥimyaritic month Ḏū/al-Mabkar is retained as an alternative to Ayyār. This need occasion no surprise since Greek and/or Roman influence in the pre-Islamic Yemen is so evident in the coinage, bronzes, sculptures and even such names as Decianus and Romanus. Ḥaydara remarks that the Rūmī agricultural year commences with Āḏār (commencing March 14th).

The following sources have been used:

Abū ᶜAlī al-Marzūqī, *Kitāb al-Azmina wa-l-amkina*, (no place of publication 1389/1969).

Al-Bīrūnī, *Kitāb al-Tafhīm li-awāʾil ṣināᶜat al-tanǧīm*; *The Book of Introduction in the Elements of the Art of Astrology*, transl. E. Ramsay Wright (London 1934).

A. M. A. Maktari, *Water Rights and Irrigation Practices in Laḥj* (Cambridge 1971).

Muḥammad Aḥmad Ḥaydara, *Ṭawāliᶜ al-Yaman al-zirāᶜī* (Taᶜizz, for 1391/1971).

* R. B. Serjeant, "Star Calendars and an Almanac[5] from South-West Arabia," *Anthropos*, 44 (1954), 133-59 (to which some small corrections now need to be made).

[4] Ed. J. David-Weill (Cairo 1939-48), II, 32.

[5] A translation of Ḥaydara's almanac for 1365/1945-46.

CONTENTS

1-6 Praise of Allah and the Prophet

7-19 Preface with list of chapters

20-52 *Chapter I*
(The virtues of agriculture as practised by the pre-Islamic Prophets, Abū Hurayra and the Companions. Nawawī, Māwardī, and *ḥadīṯs* are cited)

53-75 *Chapter II*, on lands and tilling them
(Irrigation, land law, quoting Ġazālī, advice to clear and bund land but not over-water it)

76-87 (A) Tilling
(Appropriate seasons for ploughing)

88-100 (B) Manure
(Manuring is necessary. Saᶜd b. Abī Waqqāṣ's practice)

101-107 (C) General

108-116 (D) Allah's bounty

117-134 *Chapter III*, on sowing
(Seasons of sowing in certain districts according to the Rūmī calendar)

135-145 (A) Directions for sowing according to the star calendar.

146-159 (B) Further directions for sowing according to the Rūmī calendar and certain stars in *ṣayf* and *ḫarīf*

160-170 (C) Sowing is to be done promptly in *ṣayf*

171-179 (D) Care must be given to preparing the land

180-184 (E) Further directions on sowing and certain types of millet

185-189 (F) Sowing certain types of millet

190-201 (G) Sowing in Zabīd and Qurtub

202-208 (H) Further on seasons for sowing

209-211 (I) Sowing in Kānūn I

212-216 (J) Sowing in Kānūn II

217-226 (K) Sowing in Šubāṭ and Āḏār

227-231 (L) Concluding injunctions

232-246 *Chapter IV*, on the management of what is dispensed
(Generous action at harvest-time, and self-restraint)

247-253 (A) Moral considerations about land

254-262 (B) Virtue of toil, crafts practised by David and Solomon, earning is a *sunna*

263-276 (C) Certain moral injunctions

277-278 (D) Commends the Rūmī calendar

279-285 *Chapter V*, on the four seasons
(Introductory)

286-291 (A) Method of determining the day in the Rūmī month

292-300 (B) Rūmī months and Zodiacal signs, their correspondence

301-318 (C) Appearance of Zodiacal signs at certain dates in the Rūmī months

319-325 (D) Method of determining dates of the passage of the sun through the constellations of the Zodiac

326-331 (E) Method of determining in which constellation of the Zodiac the moon is

332-336 (F) How to determine the Mansions, commencing with their dawn ascension

337-364 *Ḫātima/Conclusion*
(General on heat, cold, winds etc.)

365-369 Praise of Allah and blessings on the Prophet

[٨٣] **كتاب غاية النَّفْع في نشر فضائل الزَّرْع**

تأليف

الشّيخ الفقيه العالم العلاّمة

شهاب الدّين أحمد ابن أبي بكر الزّميلي

كان الله له ونفع به

ولا حول ولا قوّة إلاّ بالله العليّ العظيم

وصلّى الله على سيّدنا محمد وعلى آله الطيّبين الطاهرين

[٨٤] بسم الله الرحمن الرحيم
وبه نستعين

الحمد لله اللطيف الهادي … ربّ الأنام رازق العبادِ
حمداً يوافي ما له من النِعَمْ … ومن يد المواهبات[١] والكَرَمْ
سبحانه من باسطِ للرزقِ … ومن لطيف بجميع الخلقِ
جعلْ معايشْ خلقه بالماءِ … فصبّه لُطفاً من السماءِ
أحيا به الأشيا جميعاً فحَيَتْ … وكلّ خيرٍ منه في الأرض نَبَتَتْ
ثمّ الصلاة بعد حمد المنعِمِ … على النبيّ المصطفى من هاشمِ

وبعد فافهم أيّها المطالِعُ … وكلّ من غَرَضُهُ ينتفعُ
أنّي نظرْتُ في أصول الكسبِ … بعين راسي ثمّ عين قلبي
والله قد علّمنا الإكسابا … بفضله وسبّب الأسبابا
وقسّم الأسباب ـ هذا زارعُ … وتاجرٌ هذا وهذا صانعُ
هذهْ أصول جملة المكاسبِ … ما غيرها أسباب للتسبّبِ[٢]
فاخترتُ منها مكسباً محمودا … مهّد⟨ت⟩ به لأهله تمهيدا
رسمْتُه في خمسة أبوابِ … منها جمعتُ جملة الكتابِ
أوّلها في نشر فضلِ الزرعِ … وذكره وما به من نفعِ

١ لعل المؤلف يقصد « المَوْهِبات » وزاد الألف لإكمال الوزن .

٢ في الأصل : المتسبب .

وباب في فضل الأراضي فاعلَمِ وكلّ ما تحتاجه من خِدَمِ
وثالث الأبواب في التِّلامِ جميعه إلى تمام العامِ
ورابع الأبواب في التدبيرِ للنفقات دائم الدهورِ
وباب خامس في الفصول الأربعَهْ إلى انتهاء كمالها متابعَهْ
وفي حلول الشمس ثمّ القمرِ في أيّ برج في جميع الأشهرِ

الباب الأوّل

ولْنبتدي بأوّل الأبوابِ إعلم هداك الله للصوابِ
بأنّ خير الكسب كسب الزَّرْعِ لأنّه كسب عميم النفعِ
وحرفةٌ للأنبيا[٣] والرُّسْلِ وكلّ أصحاب التُّقَى والفضلِ
فاستمسكوا بحرفة الزراعَهْ فإنّه من أكبر النفاعَهْ
أثنى عليها ذو الجلال المنعمِ بنصّ آيات الكتاب المُحكَمِ
وكرّر المعنى بها تكريرا متابعاً في وَحيه مرارا
وكم أتى عن صفوة الخيارِ محمّدٍ في الزرع من أخبارِ
وكم ذكرْ فيها من الفضايلِ والأجْرِ والخير الكثير الطايلِ
[٨٥] لأنّها حرفةْ أبينا آدَمِ[٤] صلّى عليه ربّنا ذو الكرَمِ
علّمَه جبريل مقتضاها وكلّ ما نحتاج من أشياها
ومن جنان الخلد خير منزلِ أخرجْ له بذر الطعام والحلي[٥]
وكان يحرثْ باليد الكريمَهْ بثور احمر صحّ في الروايَهْ
وقد زرعْ أيّوب وابْراهيمُ والكلّ منهم مرسلٌ كريمُ
وقد زرعْ أيضاً أبو هُريرَهْ وكان يختار التِلام بكرَهْ
وعايشهْ من آمامة الطواهرِ وحفصة زوج النبيّ الطاهرِ
اخترنَ زرع الأرض يوم القِسْمَهْ بخيبرٍ من جملة الغنيمَهْ
وكنّ يزرعن الزروع الفاخرهْ وهنّ من عُمّار دار الآخرهْ
وهذه دلالةٌ تُتَّبَعُ تدلّ أنّ الأنبيا قد زرعوا
وبعض أصحاب النبي محمّدِ وهم أجلّ قدوةٍ للمقتدي
فبان أنّ حرفة الزراعهْ مختارةٌ بها قوامُ الأمّهْ
وصار أفضلْ جملة المكاسبِ وأشهر الأشيا بأصل المذاهب[٦]
وقد بذلك صرّح الماوَرْدي وغير ذلك بعد رزق الآدمي
وقال أيضاً هكذا النّوَاوي[٧] مفتي العلوم معدن الفتاوي
وكم روايات أتت في الكتبِ مشهورةٌ مرويّةٌ عن النّبي
تشهدْ بفضل الزرع طول الدهرِ وأهلها[٨] وما لهم من أجرِ

٣ في الأصل : الأنبياء .
٤ « آدمِ » بالكسرة يستقيم مع القافية .
٥ لعل المقصود « الخلّ » .
٦ الظاهر أن الوزن يقتضي « المذهب » .
٧ يعني « النووي » .
٨ في الأصل : أهلهم .

ذُكرْ بأنّ اسم الإله الباري ... مكتوب في الأوراق والثماري
وقال ثلث الزرع رزق الآدمي ... ومــا بقي للطينِ والبهايم
وقال تُكتبْ حسناتٌ عَشْرا ... للآدمي حين يبــدا بذرا
بكلّ حبّــهْ ثمّ مهما نبتَتْ ... كأنّ نفساً مؤمنهْ بها حيَتْ
وحينما يكيلــهُ ويعقــدُ ... تمحى الذنوب مثل يوم يولدُ
وحينما يفــرّح النينــا ... تُكتَبْ له عبــادة أرْبعينا
[٨٦] من السنين ثمّ حين يُشْبِــعْ ... منــه الفقير أو يواسي الجائــعْ
بِأمن عذاب القبر طول الدهرِ ... في هــذه الدنيا وبعد الحَشْرِ

الباب الثاني في الأراضي وخدمتها

فبــان بعد ما ذُكرْ من حالِ ... أنّ العقــار أفضل الأموالِ
لأنّــه محلّ أصل الزرعِ ... بــه قوام أصل كلّ نفــعِ
وخِيرة الساقي على الغيولِ ... ومــا قربْ من موضع السيولِ
من الضواحي في بقاعِ الأرضِ ... جميعها في طولهــا والعرضِ
لأن كثر السيل يأتي بالزّبَدْ ... ويولــد النّيْس الكثير والزّبَدْ
وذاك يُغنيهــا بكلّ فصلِ ... طول الزمان عن حصول الذِبْلِ
وقــد نها إمامنا الغزالي ... من بيَــع ذي الأموال للأموال
وقــال بيع الأرضِ نقص القوّهْ ... نقصٌ عظيمٌ يهتك المروّهْ
ردّوا بهــا شهادة الإنسانِ ... حكّــام شرع الله في البلدانِ
حكــاه في كتابه الوَسيطِ ... المنتقى من دفتر البَسيطِ
وحرّموا في الشرع مــعْ الأرضِ [٨أ] ... أرض اليتيم نظر⟨اً⟩ للحفظ
ولم يقيمــوا غير من فضّهْ ... ولا ذهبْ غــالي نفيس القيمهْ
مقــامها لِما بهــا من نفــعِ ... وليس رأيٌ فوق رأي الشرعِ
وقــال من أحيا من الخرابِ ... أرضاً ثقهْ بالمنعم الوهّابِ
بارِكْ له فيهــا وفيما قد فَعَلْ ... وكان عونَهْ ربّنا عزّ وجَــلْ
وقــال رزق الناس في الخبايا ... وهي وِهاد الأرض والزوايا
يعني إثارةْ كلّ أرضٍ صالبَهْ ... وحرثهــا حتّى تصير طيّبَهْ
صلّى عليه ربّنا من ناصــحِ ... ونــاظرٍ للخلــق بالمصالــحِ
فبــادروا للأرض بــالعمارهْ ... وكلّ شغلٍ نافــعٍ وخِدْمَهْ
وبطشهــا بكــثرة التطييبِ ... وحرثهــا والذِبْلِ والتقشيبِ
وجَدْرهــا في موضع التجديرِ ... وحَرٍّ مــا يحتــاج للحَرورِ
[٨٧] وحين مــا تِشقــي فخففوها ... لا تحرثوا حتّى تجنّبوهــا
حَذاركم من الخُلُبْ طول الأبَدْ ... فمــا هلاك الأرض إلاّ بالرَبَدْ

[٨أ] كذا في الأصل ، ولعل الصواب : بيع الأرض .

(ا) فصل في الحرث

فالحرث أحسنْ وقته في الدهرِ   وخَيَرْهُ عنـــد طلوع النَّسْرِ
وذاك قبـــل موقف الشمسِ   بارْبعْ ليـــالٍ قِيل أو بخمسِ
أوّلهـــا فإن حصلْ توانيِ   أو غفلـــةٌ فـــلا يفوت الثاني
ففيـــه حفظ الزرع والرطوبَهْ   وقَصْر زرعٍ ثمّ كثر سُنْبُلَهْ
سمعتُه من بعض أهل المَخْبَرَهْ   فقلتـــه على سبيل التذكِرَهْ
والشمس تبلغ حدّهـــا في اليمنْ   خامس عشرْ كانون أوّلْ فافطَنْ
وذاك وقت للوقوف مشتهرْ   عنـــد العوامّ قول عُرْفٍ معتبَرْ
ولا وقوف عند أهل المعرفَـــهْ   بـــل تنتهي ولا تصير واقفَـــهْ
وذاك غايهْ قَصْر أيّام السنَهْ   كم قد بذاته أوّلَتْهُ الألسنَهْ
قالوا وحرث مدّت[٩] العَجوزِ   ذ⟨ا⟩ك السَّعَدْ[١٠] وذاك خير فوزِ
وقد سمعتُ من كلام ناصحِ   بأنّ شقّ الأرض غير صالحِ
في وقتهـــا وتلك سبْعٌ تعتبَرْ   ثلاث من شباط وأرْبعْ من أذارْ

(ب) فصل في الذِبْل

والذِبْل خير خِدْمةٍ مُجَرَّبَهْ   بذاك قـــال أهل التجربَهْ
بـــه صلاح جملةً الأطيانِ   تنبتْ بـــه كالطبّ للأبدانِ
لأنّهـــا تَضْعُفْ بكلّ عامِ   من الزروع دايم الدّوامِ
تمتصّ مـــا فيها من الرطوبَهْ   وكلّ شيءٍ نافـــعٍ وقوّهْ
ولا تُقَوّى قوّة الأطيـــانِ   إلاّ بـــه في جملة الأحيـــانِ
وقد حرصْ عليـــه ربّ زارعِ   لِما رأوا فيـــه من المنافـــعِ
وكان سعد ابن أبي وقّاصِ   يحملـــه إلى مكـــان قاصي
من أرضه وذاك بعض العشرَهْ   العـــارفين المتّقـــين البَرَرَهْ
[٨٨] وكان يشري ذاك بالطعـــامِ   كيلاً بكيلٍ عـــام بعد عامِ
وقد سمعْت من ذوي الإيقانِ   بأنّـــه مذموم في نيسانِ
قـــالوا يذِلّ قوّة الترابِ   والأرض في الصحرا وفي الشِعابِ
والدبل خير ضَبْطه بالدالِ   من تحت لا من المكان العـــالي
أتا بذاك في صحيح الجَوْهَرِي   وذاك خـــير ضابطٍ معتبرِ

(ت) فصل

ثمّ آعلموا أعـــانكم باريكمْ   سبحانـــه في كلّ ما يعنيكمْ
بأنّـــه ليس بمخلوقٍ ظَهَرْ   رزقٌ من الأرزاق إلاّ بالمَطَرْ
والله قـــد أجرى لنا بحكمتِهْ   عوايـــداً بفضلـــه ورحمتِـــهْ
إنزاله لنـــا بوقت الصيفِ   لنفعنـــا ومـــدّة الخريفِ

٩ كذا في الأصل ، والصحيح : مدة .

١٠ « السعد » مشكّل هكذا ليوافق الوزن .

وذاك وقت للزروع معتَبَرْ … جميعها لكونه زمان حَرْ
تنشو[١١] بها الزروع ثمّ لا تقِفْ … وغيرها من الفصول تختلِفْ
طباعها بالبرد ثمّ اليبسِ … يصير نشْو[١١] زرعها بالعكسِ

(ث) فصل

فعظّموه وٱشكروه وٱحمدوا … كما جرت من برّه العوايدُ
ووحّدوه دائماً لا تُشركوا … به سواه فاعلاً فتهلكوا
وٱدْعوه ثمّ لا تقولوا وٱحذَروا … هذا من النجم الفلاني تكفروا
وإن رأيتم أنّه بالعادَهْ … لِوقته فذاك بالإرادَهْ
من ربّكم فلو يشاء يمْنعُ … كنتم ترون النجم ماذا يصنعُ
لكنّه سبحانه من كافي … مُغْني العباد واسع الألطاف
أتى به من غير سؤْلٍ وطَلَبْ … تفضُّلاً من كفّه اليُمْنى وَهَبْ
في وقت تحتاجه الخلائقُ … تبارك الخالق نِعْمَ الرازقُ
فليس وقتاً لانتفاع[١٢] الخلقِ … كالصيف يُرجى لاكتساب الرزقِ

الباب الثالث في التِلام

[٨٩] وقد رأيت الناس كلّ العالِم … زارعْ ذوَا حرصٍ على المتَالِمِ
لكونها من المهمّ الأكبرِ … وكل شغلٍ دون ذلكْ يَصْغَرِ
فقلت سوف أودع الأناما … بقدر ما أُلْهِمتُهُ كلامَا
إعلم هداك الله للصوابْ … بأنّ ز⟨ا⟩رع العَلَيّ من أصابْ
نيسان وقت المتلم المشهورِ … به صلاح جملة البذورِ
أكثرها في وقته بَذْروها … لزرعها والبرد يُقْدِموها
وليس لي علمٌ بذاك فادرِي … بمقتضى ترتيب كل بذرِ
وإنّما ذكرتها لمعنىَ … فهمتُ أنّ⟨ه⟩ عنه لا يُستغنىَ
وذاك ترتيب شهور العامِ … وذكر ما فيها من التِلامِ
حتّى يتمّ دورها وينْتهي … لينْتفعْ بذكره من يشتهي
وسلخ نيسان ⟨ف⟩قلت يتْلموا … أهل الجبال عندنا ويُسْهموا
وذاك شهر الفصد والحِجامَهْ … والأدويَهْ والنزْع للعمامَهْ
ثمّ التِلام في وُصاب السافلِي … جميعه في مدّةْ الأوابلِ
أوانه مدّةْ شهور المبْكرِ … أخصّ أوقات التلام والذَّرِي
وفي لسان الروم هو أيّارُ … ومنه أيّامٌ لنا نختارُ
فخير وقت للتِلام في الجبلْ … على الثريّا مدّة العشر الأوَلْ

١١ يعني بـ « تنشو » تنشأ ، وبـ « نشو » نشوء .

١٢ في الأصل : وقت الانتفاع .

كحدْحد وجبـل القحارِ[١٣]   وساير المعـاقل الكبـارِ
وقورٌ وسايــر ٱلمَعَاشِرْ   من كلّ أرضٍ ثابتٍ وحِمْيَرْ

(ا)   فصل

وعشره الوسطى هيَ المختارَهْ   لِكَونِهـا شديدة الحراره
تصلحْ لِزرع الغَربةْ والدُّخْنِ   يشْخُنُ فيها الزرع كُلُ ثُخْنِ
وأوّل الوسطى شروق المعتبَرْ   تبدو الثريّا بكرة الرابعْ عَشَرْ
وحين يغْربْ يطلـع الإكليلُ   فهو على طلوعهـا دليلُ
لأنّــه رقيبهـا دَوامـا   طلوع ذا غروب دامـا داما
[٩٠] وثمّ للأمّي دليلٌ مشتهِرْ   عند ٱحْمرارات السيول في الصَبِرْ
إذا أتا وقت المطرْ مَوْسِماَ   ولا ٱعتبـار إن أتـا متْقدّمَاَ
فهـذه عـلامةٌ حقيقَـــهْ   لمن خفى عليه عَدّ[١٤] الرومِيَهْ
وذاك خير وقتهـا في النزْح[١٥]   يفتحْ بـه الرحمن خير فتْح
أيّــام وقت المتلم المختـارِ   مُوافقاً مـدّة نعش الثَّورِ
وخيره سمعتهُ مِن من ذَكَرْ   من الثقات كلّها التاسعْ عَشَرْ

(ب)   فصل

وعشْرة الآخرْ تمام المَبْكَرِ   يصلح بحد غَرْدَقَهْ فيها الذري
وذاك سهم اللبس للكتّان   تبرّداً والشُرْب للألبـان
وفي حزيران تــلام سَخْمَلَ   معتــادةٌ من الزمان الأوّلَ
ولو تُلمْ سخمل بوقت غردقَــهْ   لا باس بهْ قد قال أهل المعرفَهْ
وخير متْلم سخملٍ في الجوْزِ   من نال ذلكْ فاز خير فَوْزِ
زمـان وقت طول الأيّامِ   عند ٱنتهاء الشمس حدّ الشامِ
وذاك وقت النصف من حزيرانْ   ووقت ما تحلّ بُرج السرطانْ
أخصّ وقت متْلم الجوزاءِ   عند ٱنتصاف الشهر بالوفـاءِ
وذاك شهر أكل لحم البَقَرِ   جميعهـا والأكل للصَّيد الطَّرِي
وذاك وقت ضرر المجـامَعه   والهَجْر للنسوان والمخادَعَهْ
مـا دام ذاك الشهر بالمَطالِ   حتّى يتمّ الشهر بالكمالِ
والبـاجِسي بعد تمام الجَوزِيَ   على دواخلْ عَشْر من تَمّوزَ
هذهْ متـالم كلّ وقت الصيف   والبعض من متـالم الخريف
هذا[١٦] عليهـا جُملة المُعوَّلَ   طول المـدا لآخرٍ وأوّلَ

١٣ هذه الأسماء للأماكن مجهولة، ونشك في صحة الأصل، و يجوز أن تقرأ «كجدجد» بالجيمين.
١٤ في الأصل : عدد.
١٥ في الأصل : النزح.
١٦ كذا في الأصل.

(ت) فصل

والقَدْمُ صالحٌ فاحْذروا وبادِروا وقت التِلام وٱحْذروا توٰاخروا
فالصيف ضَيفٌ والفتى من يُكْرِمُه عند الوصول بالقِرا ويغْنَمُهْ
[٩١] وقيل صَيدٌ والفتى من صادَهْ وٱهل الكَسَلْ كلٌّ يقول عادَهْ
وٱنطوا ولو قلّ الرَّوِي في الصيف ولا ترجّوا مطر الخريف
فزَجْرة الصيف القديم عِلّهْ قديمةٌ إذا ⟨أ⟩تمّها اللهْ
مُباركهْ وزرعها يأتي قوي وليس أمطار السنين تستوي
فغالِب السنين وقت جَحْرِي في غالب الأوقات طول الدهرِ
وتلك من أوّل حزيران فاعتبِرْ إلى وفا تمّوز فافْهَمْ ما ذُكِرْ
حَذاركم تُفرّطوا فتنْدَموا بل وقِّتوا وقت التِلام وٱتْلِموا
والزارع الرحمن ذو الهبات ومُنْبِت الزرع بالنبات
فقد ذكرْ أيضاً بذا مُبِينَا حقّاً وقال نَحْنُ ٱلزّارِعُونَا

(ث) فصل

وحين ينبتْ زرعكم ويصلَحُ ويستعِدّ الفَقْعُ هُمّوا واقْفَحُوا
فالكثْر مَحْقٌ ما به نَفاعَهْ والعَدْلُ أحسنْ حالة الزراعَهْ
إلاّ إذا كان كثير النَّيْسِ بالأودِيَهْ فما به من باسِ
لأنّ مجرا[١٧] النَّيْس والغُبارِ في الأرض يُخْمِدُ كثرة البذورِ
وأكرِموا زراعة الأودانِ بالرَّقْش والتنضيف للأطيانِ
وبالنّقا عن جملة الأشجارِ جميعها والحرث بالأنْوارِ
وكلما زدتم لها من خدْمَهْ وجدتموه زائداً في الغَلّهْ
هذهْ خِدَمْ كلّ الزروع الصَّيفِيَهْ وجملة المتالم الخريفِيَهْ
فأكرموها غاية الإكرامِ فذالك الإكرام للطعامِ

(ج) فصل

وهذه متالم الأوقات جميعها نَقْلاً عن الثقات
وبعضه عرفْتُه بالتجربَهْ لكلّ أوقات السنين الذاهبَهْ
مُرَتّبَهْ جميعها على الولا كما ذكرْت أوّلاً فأوّلا
حتّى يتِمّ الدور بالزمانِ جميعه إلى ٱبتدا نَيْسانِ
وقد ختمت الماضيَهْ بالباجسِ ونبتدى بالثالثي والخامس

[٩٢] (ح) فصل

فالثالثي من ٱنسلاخِ الشهرِ أعني به تمّوز طول الدهرِ
يذرا حُجَيْنا فيه والبَدَنْجا طول الزمان كلّ منْ إحتاجا

١٧ يعني « مجرى » .

ومن دخول الخمس والعشرينــا   في آب يتلم خامسي يقينَا
يذرا البذنجا فيه والغَريرا   قــد جرّبوها دائماً مِرارا
تطلع بــه بنات نعش الشامِيَهْ   ثمّ سهيــل أشهر اليمانِيَهْ

(خ) فصل

والسابعي من آبْتدا أيلولِ   إلى دواخــل عشر بالكمالِ
بحـد علْو الوادي من زَبيدِ   وفي جميــع هذه الحدودِ
وأوسط الوادي بحـد القُرْتُبي   منها إلى الثالثْ عَشَرْ يا صاحبي
وأسفل الوادي إلى الثالثْ عَشَرْ   وذاك وقت الأوّلين المعتبَرْ
بــه البَذَنْجا أوفق البذورِ   بذرٌ عليــه أكثر المَذَارِي
وتفْسَخ الزرّاع جُلْجُــلانا   من أوّل إلى آنتهــا عشرينــا
وعشْرةٌ أخــيرة الأيــامِ   للجلجــلاّن صالح التِــلامِ
وذاك أيضاً وقت إعْتــدالِ   ليومنــا وليلنــا بالطولِ
وشهر تشرين آبْتداء الأشهرِ   به صلاح كلّ أنواع الذَرِي
قــال بذاك تصلح الغَريرا   ويصلح المكّيّ والحُمَيْرا
والحَرْجيّ والجلْجِلان يصلحْ   يُذرا بــه قالوا ويأتي صالحْ
وذاك وقت القطــع للأخشابِ   منقول عن أكــابر الصحابِ

(د) فصل

ومَتْلَم الغَرْبي بتشرين العَقَبْ   وعشْرة الوسطى إليها قد نَدَبْ
يختــار كالوسطى بشهر المبْكَرِ   أخصّ وقتٍ مستحبٌّ للذَرِي
غرب الثريّا يطْلع الإكليــلُ   كما ذكرنا ذا لذا دليــلُ
وذاك يطلعْ بكْرة الثالثْ عَشَر   من شهر كانون الأخير المعتبَرْ [٩٣]
يــذرا الغريرا فيــه والأقيرعُ   والجلجلان والحُجَيْنَا تَتْبَــعُ
وذاك أحسنْ وقت غِشْيان النسا   في كلّ ليله والتدفُّؤْ بــالكسا
بــه دواءٌ للحُسوم يعتبَرْ   لا سيّما ذات الشباب في السَحَرْ

(ذ) فصل

وليس في كانون للتــلامِ   وقتٌ سِوَى لسنةٍ أيّــامِ
أوّله قالو⟨ا⟩ بــه للتَحَرّي   مــع البَذَنْجا مَتْلَمٌ لا أدْرِي
وذاك شهر طبعُهُ كما مضَى   موافقهْ فيــه الجماع يرتضَى

(ر) فصل

وشهر كانون الأخير إن مضَى   من ذاك نصف بالحساب وآنقضَى
يأتي دخول مَتْلَم البَطّيخي   ومتلم القِثّــاء والبطّيــخِ
قالو⟨ا⟩ وأوّلْ متلم الوكيري   ثلاث أخيَرْ نصفه الأخيري

وهو الّذي سُمِّي تلام الجَحْري   وقتاً به الحَمْرا أصحّ بَذْري
وذاك وقت الغرس للأشجارِ   وسقيها بالليل والنهارِ

(ز)   فصل

وحينما يدخلْ شباط فافهمْ   إلى دواخلْ عشْر خير مَتْلَمْ
حمرا ويدخل متلم الغُذَيْقي   ثالثْ عَشَرْ شباط بالتحقيقِ
وفي شباط الجلجلان يَسْفَحْ   قالوا ويأتي عِلّةً ويصْلَحْ
وقال أيضاً بعض أهل الهِمَمِ   كلّ شباط صالحٌ للتلَمِ
حتّى العجوز وَهْيَ عِنْدي صادقٌ   مُجَرَّبٌ وناصحٌ وحاذقٌ
وذاك شهر قال فيه بالعَطَبْ   من الشجرْ جميعه قَطْع الخشَبْ
وحين يدخلْ مبتدا أذارِ   يأتي دخول متلم البتيري
وفي أذار من دخول الثاني   قالوا دخول متلم اليَقْطينِ
وقيل في نيسان للبتيري   متلم وتمّتْ جملة الشهورِ
وذاك وقت ⟨ا⟩لإعتدال وآستوى   نهارنا وليلنا يأتي سَوَا [٩٤]

(س)   فصل

هذا متالمْ كلّ أوقات السنَهْ   جمعتُها أوضحتها مبيَّنَهْ
جعلتها منظومةً للحفظ   والله يفعلْ ما يشا ويقضي
فلْيقصد المخلوق باب الخالقِ   توكّلاً سبحانه من رازقِ
والزرع خير آلة التسبّبِ   لكلّ ذي توكّلٍ مُكتسبِ
فاعْنوا به وآدْعوا سريع اللطفِ   يَمُنّ بالغيث فهُو يُكفي[18]

الباب الرابع في تدبير النفقات

وحينما يأتي الثمار فاشكروا   ربّ العباد وآحْذروا لا تبْطُروا
وبعد يَبْس الحبّ وآشتداده   يكون ذاك مقتضى حصاده
وعجّلوا خَبيط أنواع الذرَهْ   والدُّخْن لا يضرّكم تأخرَهْ
وحينما يفرَغْ طعام الغلّهْ   كيلوه بالمكيال وآسْئلوا آلله
بأنْ يبارِكَ لكم فيما وهَبْ   سبحانه من فضله من كلّ حَبْ
وأنفقوه عِبْرةً واقتصدوا   فالنَصّ ما عال آمرءٌ ما اقْتَصَدْ
وعاملوا أنفسكم بالشَفَقَهْ   فانّ في التدبير نصف النَفَقَهْ
وآجتهدوا بالدُّقّ للطّحينِ   وأنعموا إجادة العجينِ
وآجتمعوا بصحبكم والليلِ   على الطعام عند وقت الأكلِ
وقنّعوا أنفسكم لا تتْبَعوا   مُرادها فإنّها لا تُقْنَعُ

١٨ في اللغة العامية يقال « يكفّي » بتشديد الفاء وذلك يوافق الوزن .

وحيثُ تطلبْ شهوةٌ لا تقْدِروا بنَحْبِها فصبِّروها وٱصْبُروا
لتستريحوا من هموم الدَّيْنِ فالمستريح مَن قَنَعْ بالدُّونِ
ولا تهينوا الخبز والطعاما حذاركم فتصْبحوا ⟨ا⟩لنَّدامى
بل قيِّدو⟨ا⟩ نعمتكم بالشكرِ لله ربِّ العرش طول الدهرِ
إذا تِشْتُوا[١٩] نعمتكم لا تنْفَدُ فهكذا علَّمنا محمّدُ

[٩٥]

(ا) فصل

ثمّ انظروا منافع العقارِ والأرض أغْنتْكم عن الأسفارِ
بحمد ربٍّ قادرٍ وهَّابِ إذ السفرْ نَوْعٌ من العذابِ
يستغرق القلب العزيز القدرِ بالهمّ والأفكار طول الدهرِ
والفكر في الأرباح[٢٠] والخسرانِ وزايد الأثمان والنقصانِ
فعاملوا الأرض بما تحتاج فإنّ في خدمتها رواج
ولا تُحبّوها ترِق[٢١] أرزاقكمْ فيها فإنّ الرزق من خلاقكمْ
بل إنّها من آلة التسبُّبِ للزارع المتوكّل المتسبِّبِ

(ب) فصل

والأكل من كدّ اليمين مُسْتَحبْ إليه خير المرسلين قد نَدَبْ
وكان يرعى غنماً ويحلبْ بكفّه الطاهرِ ثمّ شربْ
وكان د⟨ا⟩ود النبي قَنوعَا في الملك وَهْوَ يصنع الدروعَا
حتّى يكون القوت من كدّ اليدِ اللهُ أكبرْ يا لهُ من مَقْصَدِ
ثم ٱبنه ملكْ بقاع الأرضِ جميعها بطولها والعرضِ
وقد أطاعته الرياح الخافقَهْ والجنّ كلاً والطيور الصافقَهْ
وكان يصنعْ كلّ يومٍ قُفَّهْ تواضعاً لربّه وعفَّهْ
يبيعها لقوته وما حَصَلْ شرا به له طعاماً وأكَلْ
فبان أنّ الإكْتساب سُنَّهْ وقايداً إلى نعيم الجَنَّهْ

(ت) فصل

بشرط أن تهتمّ بالصلاةِ ولا تؤخِّرْها عن الأوقاتِ
ولا تغلّوا واجب الزكاةِ ولا تؤخِّرْها عن الميقاتِ
فما هلكْ مالٌ وضاع مالُ إلاّ إذا خالطها الإخلالُ
وأن تكون الأرض من حلالِ لا شبهةٌ فيها بكلّ حالِ
وأن تكُن[٢٢] تعرف ما يُعتقدُ وما يُذَمُّ فعلُه ويُحْمَدوا[٢٣]

[١٩] هكذا في الأصل، إلا أنه لا يوافق الوزن . ولعل القرءاة الصحيحة : « إذ تشكروا » .
[٢٠] في الأصل : والأرباح .
[٢١] يعني « ترقّ » .
[٢٢] بدل « تكون » لضرورة الوزن .
[٢٣] يعني « ويحمد » .

[٩٦] وأن تقيم حــاجة الإخوانِ   وساير الأهلــين والجيرانِ
وأن تجانبْ صحبة الأشرارِ   وتعتمدْ ذكر الإلـٰه البــاري
ولا تـَنـِمّ دايم الـدوامِ   كـَلاًّ ولا تغتاب[٢٤] ذا الإسلامِ
ولا تـَحيف عندمـا تـُوصّي   والثلث كالثلث لا تستـَقـْصِي
وقــد نـُهينـا مدّة الحيَوَاة[٢٥]   عن الضررْ ومدّة المماتِ
وعدّهُ من جملة الكبايرِ   نبيّنـا في شرعه المطهّرِ
فاسْعَوْا بما في مدّة الدنياءِ[٢٦]   بهـا خلاص مدّة الأخـْراءِ[٢٦]
فإنّهـا دار النعيم الباقي   جوار ربّ العزّة الخلاّقِ
والله يهدينـا وإيّاكم لِما   به الأمان من ضلالات العَمـَا[٢٧]

(ث) فصل

ثمّ اعلموا بأنّ علم الرُّومِي   مبـارَكٌ من أفضل العلومِ
علم يقين عنه لا يُسْتَغْنـَى   لأنّــه أصلٌ عليــه يُبْنَى

الباب الخامس في الفصول الأربعة

حساب أوقات الشتـا والصيفِ   ومــدّة الربيــع والخريفِ
مع انتقال الشمس في البروجِ   ومكثهـا ومــدّة الخــروجِ
وأيّ برج فيه قد حلّ القـَمـَرْ   وطالع الفجر الـّذي هُو يعتبَرْ
من المنازلْ كلـّها طول المدَا[٢٨]   وغير ذا وكم به من اقتــدَى
فاعـْنوا بضرْب أصله لتفهموا   في أيّ يومٍ أنتموا وتعـْلموا
وذاك معروف وقــد أوْدعـْتـُهُ   في بعض أبياتٍ بهـا نظـّمتـُهُ
وهذه الأبيــات فافـْهموهـا   لتعرفوا الأشيـا وتـُتـْقِنوهـا

(ا) فصل

وإن تـَرُمْ تعرفْ بأيّ يومِ   من أيّ شهر يا أخي من رُومي
[٩٧] دع عنك مكسورًا مع المُبينِ   وخذ جبور القرن من سنين
إحدا عشرْ فاضربْ وأنـْظرْ ما انتهى   زدْ فوقه يومـَيْـن قال ذو النُّهـا[٢٩]
واطـْرحـْـه روميـاً ومهما يبقـَى   تكمـّله رومي إليــه ترقـَـا[٣٠]
أضفْ إليه ما مضى من عـَرَبِ   واحـْفظـْه يا صاح وقلْ لي حـَسْبي
مابـْدَأَ من تشرين مبـْدا الرُّومِي   متى تساوى فـَهـْوَ هــذا اليومِ

٢٤ كذا لضرورة الوزن وحسب استعمال اللغة العامية .
٢٥ يعني « الحياة » ، ولا معنى لعبارة « ومدة الممات » .
٢٦ كذا في الأصل .
٢٧ يعني « العمى » .
٢٨ يعني « المدى » هنا وفيما يجيء بعد .
٢٩ يعني « النهى » .
٣٠ يعني « ترقى » .

(ب) فصل

فاجتهدوا وَاعنوا بهـذا الأصلِ — لتعرفوا دخول كـلّ فصلِ
وها أنا أذكرْ حلول الشمس كَمْ — تحلّ في بروجهـا حتّى تُتَمْ
إعلمْ هداك الله ياخي ذو العُلَى — بأنّهـا تحلّ شهرًا كاملا
في كـلّ برجٍ فحلول الحَمَلِ — خامسْ عشرْ أذار فافْهمْ وَاعْقِلِ
وذاك أوّلْ يوم بالحساب — للصيف فافْهمْ مقتضى خَطَابي
لا شكّ فيـه حِسْبَةٌ محقّقَهْ — تلقّبوه النـاس بالمُسبَقَهْ
والثّور ينزلْ برجه رابـع عَشَرْ — نيسان مـا ذكرْتُ من خبَرْ
وتنزل الجوزاء من أيّارِ — بالخمس والعشر بألأبْـآرِ
وحين تخرجْ ثمّ فصل الصيفِ — وتلك أوّلْ مـدّة الخريفِ

(ت) فصل

أوّله تحلّ برج السَّرَطـانْ — بخمسةَ عشَرْ يوم من حزيرانْ
قالوا وفي تمّوز في السادسْ عَشَرْ — تحلّ في برج الأسَدْ ألمشتهَرْ
وهكذا قـال الثقات النَّقَلَهْ — سادسْ عشرْ آب تحلّ السُنْبُلَهْ
وحين تخرج ينقضي[٣١] زمانهْ — فافهمْ ويأتي للشتـا أوانهْ
أوّل شتـا تحلّ بالميزان — سادسْ عشر أيلول بالإتقان
وهكذا تحـلّ برج العقربْ — سادس عشر تشرين فافهمْ تُصبْ[٣٢]
أوّل برج القوس تنزل فيـهْ[٣٣] — خامس عشر تشرينهم ثـانيَهْ
[٩٨] وينقضي وقت الشتا ويأتي — وقت الربيع أعـدل الأوقات
والجدْي تنزل فيه أيضاً فاعلَمْ — خامسْ عشرْ كانونها المتقـدّمْ
والدلْو تنزلْ فيه في الرابع عشَرْ — من شهر كانون الأخير تعتبَرْ
ولا خلاف في شباط يـأتي — ثالثْ عشرْ تحلّ برج الحوت
وحين تخرجْ منه ترجـع الأولْ — ودار ميقـات الزمـان وَانتقَلْ
هذهْ بروجٌ للفصول الأربعَهْ — إثْنَي عشرْ جميعهـا متابعَـهْ
ثلاثـةٌ تخصّ كلّ فصلِ — كما أتتْ من ضرْب فصل الأصلِ
سهّلتهـا لجملة الطـلّاب — ليهتـدوا منـاهج الحساب
وَ⟨أ⟩حْسبوها حسْبَةً حقيقيَهْ — لتعرفوا دخول وقت الروميَهْ
لأنّهـا هي الّتي لا تُنْكَرُ — أوقاتهـا أصلاً ولا تُغيَّرُ
طول المـدا والدهر يا فلانُ — لا زايـدًا فيـه ولا نقصانُ

(ث) فصل

وإن تشا تعرفْ حلول الشمسِ — في أيّ برجٍ يا صحيح الحِسّ

---

٣١ في الأصل : يقتصى .
٣٢ لعل المؤلف نطق كلمة « تصب » بفتح الصاد .
٣٣ نقترح تشكيل كلمة « فيه » على هذا المنوال لضرورة الوزن .

فانظرْ إلى ماضي الشهور الروميَهْ	زدْ فوقها خمسةْ عشرْ متْواليَهْ
وأسْقِط الجملةْ مثالاً ليبينْ	لكلّ برجٍ يا فتى ثلاثينْ
فحيث يُغْنِي عدد الأيّامِ	فالشمس في ذَكٍ[34] فاعتبرْ كلامي
وما نقصْ عن الثلاثين فاحْسبَهْ	وقدِّرهْ منها حساباً وٱنسبَهْ
وٱبْدأْ من الميزان في الحسابِ	وذاك وجْه الصدق والصوابِ
لأنّه آوّلْ للشهور الرومي	نقلْتُ ذاك من ذوي العلومِ

(ج) فصل

وإنْ تشا تعرفْ حلول القمرِ	في أيّ برج يا صحيح النظرِ
فانظرْ إلى ماضيك من شهر العربْ	أضْعفه وٱنظرْ ما ٱنتهى من الحسبْ
وزدْ عليه خمسةً أيّاما	تكملةً محسوبةً تماما
وأعْطِ كلّ برج خمْسهْ يا فتَى	وٱبْدأْ ببرج الشمس وٱنْظرْ ما أتَى
[٩٩] من دون خمسةْ فالقمرْ فيه قطعْ	مِن نِسْبت الخمسهْ بقدرما انقطعْ
وإن أتى خمسهْ تمام فاعلَم[35]	بأن ذاك البرج قطعه نَمَى

(ح) فصل

وإن تشا تعرفْ بأيّ منزلَهْ	فاحْسبْ حساباً لا يردّ قائلَهْ
وٱنظرْ إلى الماضي بشهر العرَبِ	زدْ فوقه ثلاثةً يا صاحبي
وٱبدأ من الطالعْ وقت الفجرِ	للمنزلهْ يوماً فخذْ بأمْري
فحيث ما ٱنتهى حسابكْ وكمَلْ	فافهمْ فتلك المنزلهْ فيها نزَلْ
هذا الّذي شرّطْته في الخُطبةِ	وقلته في أوّل الأرجوزةِ

خاتمة

وهذه خاتمةٌ حميدَهْ	مباركهْ ألفاظها مُفيدَهْ
في وقت ذكر البرد ثمّ الحرِّ	وطول أيّام السنهْ والقصْرِ
والإستْوا وكلّ ذي ٱعتدالِ	في جملة الأيّام والليالي
إعلمْ جُعلْتَ قدْوة للمقتدي	يا صاح أنّ البرد قالوا يبْتدي
تاسعْ عشرْ من شهر كانون العُقُبْ	وجدْته في كلّ تاريخ الكُتُبْ
كذا ٱشتداد البرد منْ عشر مَضَتْ	من شهر كانون الأخير وٱنْقضَتْ
قالوا ويَفْتُرْ في شباط إن مضَى	إثنا عَشَرْ نهار منه وٱنقضَى
وجاء مبْدا ٱلحرّ في أيّارِ	يا صاح من تاسِعْ عشرْ نهارِ
وحين يبقا[36] خمْس من تمّوزِ	يشتدّ أكثرْ من زمان الجوزِ

[34] يعني « ذاك » .
[35] هكذا ينطق « فاعلم » لضرورة القافية .
[36] يعني « يبقى » .

قالوا ويفْتُرُ ذاك بالحساب   لعشْرِ أيّامٍ مضتْ من آبِ
قــد ذكرنا أطول الأيّامِ   عند آنتهاء الشمْس حدّ الشامِ
وذاك في رابعْ عَشَرْ حزيرانْ   في وقت ما تحلّ برج السرطانْ
وأقْصر الأيّــام في الزمانِ   عنــد آنتهاء حدّها اليَمانِ
وذاك يوم النصف بالإتقانِ   من شهر كانون الأخير الثاني
وتـأخذ الأيّام في الزيادَهْ   إلى آنتها آلجَوزا على الحقيقَهْ
[١٠٠] كذا آعتدال الليـل والنهـارِ   في شهر أيلول وفي أذارِ
عند حلول الشمس في الميزانِ   برج الحَمَلْ يا صاح والميزانِ
وعندمـا تحلّ برج الدّلْو   طول المـدا يشْتدّ برد الجو
كذا آشتداد الحرّ في تمّوز قَدْ   قــالوه معْ حلولها برج الأسَدْ
وفي شباط الماء يخرجْ في الشجَرْ   في الخمس والعشرين منه يعتبَرْ
ورابع العشرين من أيّــارِ   يكون نقص الما من الأشجارِ
ويسقط الأوراق من كلّ الشجَرْ   في الشهر من كانون للثالثْ عشَرْ
وتكْمل الثمار من آب علَى   إثنين والعشرين منـه كمّلا[37]
وينتهي ريح الشمال إن حَسَبْ   نا في ثلاث سلخ تشرين العقَبْ
والبحر يعلى فيـه كلّ عامِ   طول المدامـعْ تاسع الأيّامِ
كذا آبتدا عواصف الرياحِ   من بعد عشرين بهِ يـا صاحِ
وتختفي الهوام يوم الثـامنْ   من شهر كانون الأخير فافْطَنْ
وفي شباط الغرْد قالوا تظهَرْ   إذا بقتْ ستّ بهِ تعتبَرْ

تمّتْ بحمد الله ذي الجلال   والكبْريـا والعزّ والكـمال
مقــاصد الأرجوزة الميْمونَهْ   من ربّنــا أتتْ بــه المعونَــهْ
فالحمـد[38] لله الـّذي أعانا   سبحانــه من مالك سبحانــا
ثمّ الصلاة بعـدحمد المـاجد   على النبيّ المصطفى محمّــد
لا تنقطـعْ ما دامت الأيّامُ   وآلــه الأطهـار والسلامُ

تمت الأرجوزة بحمد الله وتوفيقه والحمد لله رب العالمين وصلى الله على محمد وآله .

٣٧ بدل « كمل » .

٣٨ في الأصل « بالحمد » .

COMMENTS ON INDIVIDUAL VERSES

16 تِلام, ploughing of land for sowing. Cf. تِلم, أتلام, a furrow. Also chapter III, *passim.*

25 Cf. Qurʾān, Sūra 32: 27; 39: 21, etc.

30 For the difficult reading الخلي, one would have expected النخل.

34 ᶜĀʾiša is "the exemplar (*imāma*) of pure women".

35 Ibn Hišām records the distribution of dates to ᶜĀʾiša and wheat to the Prophet's wives in general, while Wāqidī says all his wives received dates and barley.

41 These statements have not been found in al-Māwardī's works consulted, but in *Adab al-wazīr* (Cairo 1348/1939), 12, he refers to *ṭabaqa li-l-zirāᶜa wa-l-ᶜimāra ağruhum ᶜalā l-inṣāf.* V. 10 also seems to derive from Māwardī.

42 The well known Šāfiᶜī jurisprudent Yaḥyā b. Šaraf, etc.

47 Aḥmad b. Ḥanbal, *Musnad* (Cairo 1311), II, 14, 4, *bi-kull ḥasana ᶜašr ḥasanāt.*

49 I.e. ties up the grain-sack الغرارة.

50 النين, the eye.

55 ساقي is irrigated land; ضاحي, rain-land.

57 رَبَد, stagnant water, الماء الراكد. Also v. 75. نَيْس, sand, gravelly soil, *passim.*

58 ذِبْل, manure of dung or ash. Also chapter II, especially v. 99.

62 Brockelmann, *GAL*, I, 424, Sup. I, 752 seq., *al-Basīṭ fī l-furūᶜ* and *al-Wasīṭ al-muḥīṭ bi-aqṭār al-Basīṭ,* neither of which seems to have been printed.

69 *Buġyat al-fallāḥīn,* 30a, notes that land is ploughed until it is cleared (*tuṭayyab*) of scrub and grass, so this must be the sense of *ṭayyiba* here. Cf. v. 72. *infra.*

71 خِدْمة means "tilling". Cf. Ḥaḍramī *ḫaddām al-naḫl,* the peasant who works on palms.

72 See v. 69. تقشيب seems synonymous with *taṭyīb,* to clean the ground.

73 جَدْر, (field) walls of mud, stone or *zābūr* (blocks of sundried mud), تجدير to make walls. Landberg, *Glossaire daṯînois,* 389 has much detail on حَرَّ, to build up a bund using the scraper-board drawn by oxen, the verbal noun being حُرُور. This term also figures in *Buġyat al-fallāḥīn.*

74 شَقَّي, to work, labour. Possibly *tuḥannibū-hā* is to be read at the end of the line, meaning "to make (the ground) sticky, clinging". This would be supported by v. 75, but *Gloss. daṯ.,* has only *ḥanib,* not *ḥannab.*

75 خُلُب, mud, also the mud used for making sun-dried brick.

76 Al-Marzūqī, *al-Azmina,* I, 355, gives the ascension of al-Naṣr al-Wāqiᶜ on 28 Tišrīn I.

79 Around Ṣanᶜāʾ most grain has a short stalk. A short stalk and full ears are desired.

82 الوقوف, the winter solstice.

85 Al-Bīrūnī, *al-Tafhīm,* 311-14, "The days of the old woman (ᶜ*ağūz*) are seven days beginning on the 26th of Šubāṭ. They are not free from ice, cold and winds nor of extreme changes in the weather, and consequently are called the cold days of the old woman". Ḥaydara's 1971 almanac gives Šubāṭ 25th for the commencement of this period, 10th March.

94 This tale of Saᶜd b. Abī Waqqāṣ does not seem to figure in the usual biographies.

95 Ibn al-Aṯīr, *Usd al-ġāba* (Cairo 1284-85), II, 290, numbers Saᶜd as *aḥad al-ᶜašara sādāt al-ṣaḥāba.*

96 "Measure for measure" means barter.

99 The allusion is to the convention common in Yemeni Mss. of placing a dot under the *dāl* to distinguish it from *ḏāl*.

100 *Al-Ṣiḥāḥ* (Būlāq 1282), II, 183, *dihl al-arḍ iṣlāḥuhā bi-l-sirğīn wa-naḥwihā.*

110 Cf. Ibn Ḥağar al-Hayṯamī, *al-Zawāğir ʿan iqtirāf al-kabāʾir* (Cairo 1951), I, 147, quotes a Tradition that the Prophet forbade people to say, "*Muṭirnā bi-nawʿ kaḏā wa-kaḏā*", and substituted "*Muṭirnā bi-faḍl Allāh wa-raḥmatihi*". Al-Bīrūnī, *op. cit.*, 87, says "The term *anwāʾ* is associated with the rains because the times of their occurrence are related to the setting of the mansions in the morning in the west". Cf. al-Marzūqī, *op. cit.*, II, 249: "Star Calendars," 439, etc.

117 المَتَالِم, sing. مَتْلَم, *passim,* time or season for sowing. Cf. "Cultivation of Cereals...," 53, 71.

120 Probably because the higher fields are irrigated first. Cf. al-Māwardī, *al-Aḥkām al-sulṭāniyya* (Cairo 1380/1960), 180.

127 أتلم, to sow.

129 Wuṣāb al-Sāfil, and above it Wuṣāb al-ʿĀlī, are in Wādī Rimaʿ, N.E. of Zabīd.

130 مَبْكَر is another name for Māyis/Ayyār. Cf. "Star Calendars," 441.

131 Mabkar is ancient South Arabian and Ayyār the corresponding name "in the Greek tongue"; see A.F.L. Beeston, "New Light on the Himyaritic Calendar," *Arabian Studies,* I (London-Cambridge 1974), 2, *ḏ-mbkrn* = May.

133 The first hemistich is apparently corrupt. Al-Hamdānī, *Ṣifa,* ed. Muḥ. al-Akwaʿ (al-Riyāḍ 1394/1974), 269, mentions a wādī Ḥubul of the Aṣbaḥīs of Ḥimyar; cf. D.H. Müller ed. (Leiden 1884-1891), 75 for Ḥubul near Šarʿab. Sayyid Aḥmad al-Šāmī suggests possibly, Ğadğad, hard level ground, and thinks al-Qiḥār might mean al-Quḥra, also called al-Quḥrī.

134 Al-Hamdānī, *op. cit.*, ed. Müller, 128, alludes to Quwwir bi-l-Maʿāfir — perhaps the latter should be read for al-Maʿāšir, though the latter may be correct. The *Ṣifa,* ed. al-Akwaʿ, 197, alludes to Banū Ṯābit in Radmān and Ḥimyar.

136 الغَرَبة or الغَرْب, a kind of good quality *dura,* bigger than bulrush millet (*duḫn*) but coming about the same season. It is mentioned by *Buġyat al-fallāḥīn.*

139 داما = دائما.

140 Ğabal Ṣabir of Taʿizz.

142 عَدّ الرومية, the reckoning of the Greek months.

144 By Naʿš al-Ṯawr is intended the four of the seven stars of the Great Bear (Ursa Major) which rise during al-Ṯawr, the Zodiacal sign for September.

146 غَرْدَقة is defined by *Tāğ al-ʿarūs* as إلباس الغُبار الناس. The dust-storms called *ġubār* are discussed in "Cultivation of Cereals," 31, which blow from about this time.

148 سَخْمَل, an unknown evidently Lower Yemen word, probably a sort of *dura,* but other cereals are also planted in Mabkar.

150 الجوز = الجوزاء as in v. 153.

155 المخادعة, being closeted within a *maḫdaʿ*, chamber.

157 الباجسي, what is sown at the star al-Bāğis. Ḥaydara's 1391/1971 almanac enters Nağm al-Bāğis, 13 days, as commencing on lst Tammūz/14th July, whereas our author places it during the first ten days of Tammūz. Ḥaydara notes that lentils and fenugreek (*ḥilba*) are sown at this star. See Ibn Māğid, pub. G. Ferrand, *Instructions nautiques* (Paris 1921-23), I, 11v, 36v, (where it is identified as a Yemeni name for al-Šiʿrā al-ʿAbūr), 95r, 113r (*al-mağīb al-bāğisī*), 144v. As a name for a crop sown at this star it is also known to Yemeni mediaeval almanacs. Cf. v. 184.

160 *Al-qadm* here seems to mean sowing early.

162 عَادَهْ, still, not yet.

163 أنطا, to give (class. and colloquial). We have no authority for the sense but it appears to mean "set to sowing".

164 زَجْرَة, driving rain. Sayyid Aḥmad al-Šāmī suggests to us that ʿ*illa* should read *ġalla* which seems acceptable. *Ṣayf* here seems to mean the *d̲ura* crop sown at this season.

166 جَحْرِي, i.e. *ğaḥr*, synonymous with *qayẓ*, a hot period — about May, June, July, with dust-storms (*ġubār*). It gives its name to a millet crop; cf. v. 215, sowing of *al-ğaḥrī*. See "Cultivation of Cereals," 31, 50, 70.

168 "Nay, observe the time of sowing and sow."

170 Qurʾān, Sūra 56: 64.

171 فَقَعْ, become yellow, ripening (class.). يِفَقْحَ فَقَحْ, in Yemeni Arabic means to remove the surplus sprouts of *d̲ura* for animal fodder ("Cultivation of cereals," 70). This sense seems appropriate here. It might however have the sense of *manaʿa* (class.), for as Ḥaydara advises in a note on the 25th of Ḥazīrān, "Do not irrigate the crop lest it become yellow."

174 غُبار, dust-storm. Cf. note to v. 166.

175 أوْدان, sing. وَدَن, fields. The word is known to al-Hamdānī, *Ṣifa*, ed. Müller, 199.
رَقْش to scrape clean (of stones, rubbish etc.) imperfect verb, *yuraqqiš*; it is similar to *tanḍīf / tanẓīf* which follows. *Aṭyān* means fields.

176 نَقَا, clearing (the ground of scrub etc.). Ḥaydara's 1971 almanac, notes on the 21st of Ḥazīrān, *tunaqqa l-zurūʿ min al-ḥašāʾiš*, and *tanqiya* is commonly used in the *Buġyat al-fallāḥīn.*

184 For al-Bāğisī see note to v. 157. Al-Ṯaliṯī and al-Ḫāmisī are named after the third and fifth stars of Ursa Major (Banāt Naʿš); see note to v. 144.

186 حُجَيْنا, a red millet known as Ḥamrāʾ the ears of which come up crooked, bent over, thus giving it this name. Cf. "Cultivation of Cereals," 50, and v. 206 *infra*. بَذَنْجا, a type of millet, and vs. 188, 194 *infra*. The readings Budayğā/Buğayda of *Buġyat al-fallāḥīn* are to be corrected to this in "Cultivation of Cereals," 48. بيضاء, white millet.

187 غريره / غَرَيرا, a type of millet, and vs. 199, 206.

190 السَّابعي, named after the seventh star of Banāt Naʿš. Cf. note to v. 184.

191 This must be al-Šarğ al-ʿUlyā of Wādī Zabīd as in "Cultivation of Cereals," 51.

192 Qurtub near Zabīd is called by al-Šarğī, *Ṭabaqāt al-ḫawāṣṣ* (Cairo 1321), 167, *min aqdam qura l-Wādī Zabīd.*

193 *Al-awwalīn* are the "ancients".

194 مَذاري, sing. مِذْرا, time of sowing. Cf. "Cultivation of Cereals," 31 (*mad̲rāt al-d̲ira*), 63, note 33.

195 يِفَسْخ فَسَخ, to sow by hand-scattering. This term is used in *Buġyat al-fallāḥīn.*

199 مَكّي, probably a type of *d̲ura*, but the word is not known to us. حُمَيْرا, a kind of *d̲ura.*

200 حَرْجي, a kind of millet described in "Cultivation of Cereals," 51.

202 غَرْبِي, presumably the same as *ġarba/ġariba*, a kind of millet ("Cultivation of Cereals," 46). *Tišrīn al-ʿAqab* (which in v. 341/1 must be read al-ʿUqub for the rhyme) is Tišrīn II. *Nadab*, to recommend, commend.

206 الأقيرع. This term is unknown to us, but of course suggests some sort of *qarʿ*, gourd pumpkin, but cf. *qurāʿ*, breakfast.

210 تحرّي (ه), to do the correct thing (class.).

214 الوكيري, a term unknown, but it is evidently a name for red Saḥrī millet. Cf. "Cultivation of Cereals," 51.

218 الغَذْ يَقي, a sort of millet eaten green and soft, parched as "*ğahīš*" (Cf. "Cultivation of Cereals," 47).

219 The sense appears to be that sesame (oil) overflows. If at first it does not do well (*ya*ʾ*tī* ᶜ*illa*), later it will turn out well and produce a good harvest.

222 عَطَب, dying, decay (class.).

223 البتيري, probably a kind of *d̲ura*, this being much sown at this time. A mediaeval Yemeni almanac records under 16 Kānūn I/29 December, *Fīhi btidā*ʾ *ṭulū*ᶜ *al-Nasr al-Wāqi*ᶜ *wa-fīhi matlam al-batīrī/butayrī wa-summiya l-malīḥ mulayḥ*. Al-Butayrāʾ is a name for the sun when it is high or spread out over the earth, but it may have no connection with this name which may be derived from an unknown Yemeni star-name. Cf. v. 225.

234 خَبيط, beating of the *d̲ura* cobs with flails. One says: *uḫbuṭu l-d̲ira.*

237 Aḥmad b. Ḥanbal, *Musnad*, I, 446,.

243 Cf. the Yemeni proverb, *al-dayn hamm bi-l-layl wa-mad̲alla bi-l-nahār*, quoted in chapter XII of *Ṣan*ᶜ*ā*ʾ*: an Arabian Islamic City*, forthcoming.

247 Class. ᶜ*aqār*, land or house property.

252-53 Do not be over-fond of land or your provision (from Allah) will diminish (*tariqq* = *taḫiff*), for land is only a means for gaining a livelihood.

270 Class. *namma*, to make mischief by calumny, reflects the Tradition in Muslim's *Ṣaḥīḥ*, Īmān, 168: *Lā yadḫul al-ğanna nammām. Iġtāba*, to speak evil of a person in his absence; cf. Qurʾān, Sūra 40: 12, *Wa-lā yaġtab ba*ᶜ*ḍukum ba*ᶜ*ḍan. D̲ā l-Islām* means a Muslim.

271 Do not be unjust in the way you bequeath the third of your property of which you are free to dispose.

277 By *al-Rūmī* the Greek months are intended.

290 By *min al-*ᶜ*Arab* the poet means the Hiğra calendar.

297 المُسَبَّقة appears to be a purely local term.

309 *Al-mutaqaddam* (to be so read) means Kānūn I.

341 See note to v. 202.

364 الغِرْد, bird-song (class.). Ḥaydara's 1971 almanac notes for the 19th of Šubāṭ the appearance of *suwayd*, small birds, and presumably therefore bird-song!

ADDENDA

p. 407, n. 2: see article VII above.
p. 408, l. 37: see article III above.

INDEX

The definite article al- and the usual abbreviation for ibn, b., are ignored in this index - though not the word ibn itself. Where spellings of the same word differ because of the different transliteration systems used in the articles, the main heading follows the spelling of the word as it is found first and cross references to this main heading are employed.

al-Abbas b. Ali: *see* al-Afdal al-Abbas b. Ali
Abbud: VIII 33
Abd-Aziz b. Mahdi al-Banna: I 486, 505, 506, 507, 509, 511
Abd al-Aziz b. Muhammad b. Abdullah Al Khalifah: I 486, 505, 507, 509, 511
Abd al-Kuri: IV 94; V 118
Abd al-Latif b. Mahmud: I 501–2
Abd al-Rahman, Saiyid: VIII 46
Abd al-Rahman b. Abdullah b. Ali Salim: I 506
Abdawh: I 488
Abdullah b. Amr: VIII 57
Abdullah b. As'ad al-Yafi'i: VII 31
Abdullah al-Haddad: VIII 71–2
Abdullah b. Ibrahim: I 501–2
Abdullah b. Ja'far b. Alawi: VIII 71
Abdullah b, Khamis al-Shuruqi: I 486–8, 491, 495, 497, 501–3, 506, 510
Abdullah Rahaiyam: III 436, 439; VIII 55, 66, 72
Abdullah b. Yilaiyil: I 501–2
Abu Ali: VII 45
Abu Ali al-Marzuqi: XI 408
Abu Arish: VII 29
Abu Bakr, the first caliph: III 445
Abu Dhabi: I 504; X 471
Abu 'l-Hasan Muhammad Taj al-Arifin al-Bakri al-Siddiqi: VIII 35
Abu Hurayra: XI 409
Abu 'l-Khair: VII 26
Abu Makhramah: *see* Ba Makhramah
Abu Muhammad: VII 45
Abu Ubaidah: VIII 57
Abyan: VII 48; IX 146
Abyan (Cotton) Board: IX 146
Abyssinia: VII 45, 52
Aden: I 512; III 440, 444–6, 449–51, 454–5, 457–8; IV 94; VII 27; VIII 57; IX 145–7
Aden Protectorate(s): II 193; III 440; IV 94; VI 307; VII 27; VIII 33, 54; IX 145, 147
Adnan: VI 322
al-Afdal al-Abbas b. Ali, al-Malik, Rasulid sultan: VII 25, 26, 39; VIII 46; XI 407
Africa: IV 98; V 121
Agaryshev, Anatoli: VII 29
Ahmad, Imam: VII 25, 28, 35
Ahmad b. Abi Bakr al-Zumayli: XI 407–8
Ahmad Faraj, Shaikh: VII 32
Ahmad b. Hanbal: XI 424
Ahmad b. Hasim Taitun: I 506
Ahmad b. Majid: *see* Ibn Majid, Ahmad
Ahmad b. Mansur al-Sa'di, Shaikh: VII 29
Ahmad b. Muhammad Ba Ras: VI 311
Aibun: VIII 49
Ain: VIII 69
'Aiša: XI 424
Ajam, Persians: I 487
Ajawid: I 486
Ajwadi: I 486
Al Abd al-Shaykh: VI 311–2
Al Abudan: VIII 54–5
Al al-Amudi: VI 311
Al al-Attas: VIII 49
Al Aydarus: VI 312
Al Ba Adail: VIII 54
Al Ba Awdan: VIII 54
Al Ba Huraish: VIII 54
Al Hamtush: VI 311–2
Al (al-)Kaf: VI 312; VIII 34, 35
Al Kathir: VI 311
Al Khalifah: I 502
Al Mirsaf: VI 311–2
Al Ninu: VIII 54
Al al-Shaikh Bu Bakr b. Salim: VIII 45

Al Tamim: *see* Tamim
Alawi b. Tahir, Saiyid: VIII 76
Ali b. Abd al-Malik al-Muttaki: III 439
Ali Abdullah Khalifah, poet: X 473, 476
Ali Abdullah al-Salman: I 510
Ali b. Abdullah: I 499
Ali b. Abi Talib: III 444–5
Ali b. Ali al-Yamani: III 443
Ali al-Dukhi: I 504
Ali b. Salim al-Attas, Saiyid: VIII 46
Ali al-Tajir, Sayyid: X 471, 489, 491
Ali b. Tawq: I 499
Ali b. Zayid: IX 145
Alwawh: I 488
Amakin, Wadi: III 433, 436, 438
Ambrosiana: *see* Biblioteca Ambrosiana
Amd, Wadi: VIII 69, 72, 74
Amir, sultan of the Yemen: VIII 69
Amr b. Sultan Al Bin Ali, Shaikh: X 471
al-Amri, Husayn: IX 152; XI 407
Annah: VII 46
Arabia: I 486–8, 495, 512–3; VII 32; VIII 65, 72
Arabic: I 487, 493; III 438; IV 94–5, 98; V 109, 111, 125; VII 28; VIII 47, 75; IX 148, 151; X 474, 488–91; XI 408
Arabs: I 488, 501–2; III 434, 439; IV 98; V 126
Aramaeans: I 488
Ard Bin Makhashin: VIII 72
Arhamah b. Muhammad b. Fadalah: I 499
Asba'un: III 438
al-Ashraf Umar b. Yusuf, al-Malik, Rasulid sultan: VII 26, 36
al-Asma'i: VIII 57
Atash al-Nakhil: X 477
Attas, Saiyids: VIII 43
Awad b. Sa'id: VIII 72
Ayal Yazid, and Jabal: VII 30, 32
Aydid: VI 312
Ayl: V 118, 122
Ayn (al-)Adhari: X 475–7, 489
Ayn al-Hakim: X 475
Ayn al-Hamasah: X 475
Ayn al-Karayim: X 475
Ayn al-Khadrah: X 475
Ayn Qassari/al-Qassarin: X 475–6
Ayn al-Raha: X 475
Ayn al-Sayyid: X 475
Ayn Wadiyan: X 475

Ba Faqih al-Shihri: VIII 67, 70
Ba Harun: VIII 47
Ba Itir: VI 311
Ba Hashwan, A.M.: VIII 58, 62
Ba Jabir: IX 150
Ba Jaray: VI 312
Ba Makhramah, Abdullah b. Ahmad: VII 29, 33; VIII 35, 55–6, 60
Ba Matraf: *see* Muhammad Abd al-Qadir Ba Matraf
Ba Rasain: VIII 75
Ba Sabrain, Ali b. Muhammad: VIII 56
Ba Sanjalah: VIII 69–70
Ba Sawdan: *see* Muhammad b. Abdullah Ba Sawdan
Ba Sumbul: VIII 54
Ba Tayi': *see* Said Ba Tayi'
Bab al-Bahrain: I 507
Badr, raid: V 122
Badr b. Ahmad al-Kasadi: V 122
al-Baghawi, Abu Muhammad al-Husain: VII 45
Baharnah: I 486, 488, 489, 506; X 471, 473–5
(al-)Bahrain: I 486–8, 490, 491, 496, 501–2, 504, 506–12; IX 152; X 471–3, 475, 478, 487, 491, 493
Bahraq: VI 308; VIII 61
al-Bahrayn: *see* (al-)Bahrain
al-Bahriyah, fish-trap: I 502
Baihan: VII 34, 35, 45; IX 146
Baisar: I 486, 487
Bait Hanami: VII 32–3
Bait Maslamah: VIII 69
Bait Muslimah: VIII 71
Bal-Haf, Wadi: VIII 71
Bana, Wadi: IX 146
Banadir: V 119, 123
Bani Arraf: I 503
Bani Hushaish: VII 33
Bani Murrah: I 503
Bani Tawq: I 502
Banu Abd al-Qays: X 473
Banu Sahl: VI 312
Banu Shihab: VI 312; VII 43, 53
Banu Zannah/Dannah: VI 307
Barawah: V 119, 123
Barr al-Khazayin: V 118, 122
Barrett, Tom: XI 407
Basra(h): I 486, 495; V 109, 123
Batavia: VIII 35
al-Batin: I 500
Battah: *see* Pate
Bedouin: I 487; VI 310; VIII 75
Beeston, A.F.L.: VII 31
Belgrave, Sir Charles: I 486, 488–9, 506
Belgrave, James: I 504; X 471, 474–5, 477, 489

Biblioteca Ambrosiana: VII 26; VIII 34
al-Bida, fish-trap: I 507
Bilad al-Qadim: X 475
Bin Ajwad: I 486
Bin Madhkur: I 498
Bint Kamal, fish-trap: I 510–1
al-Biruni: III 434, 438–9; XI 408
Bohras: III 444
Bombay: V 123
Bor: VIII 33, 69
Bornstein, Annika: VII 28
Bowen, R. LeBaron: VIII 45
Boxhall, Peter: IV 94
British: VII 28
British Museum: VIII 34
Brockett, A.A.: X 493
Bu Bakr b. Shaykh Al Kaf, Sir: VI 307
Buqlan, Wadi: VII 53
Bura': VII 53
Bustan Mansuriyah: VII 36
Bustan al-Rahah: VII 36
Bustan al-Shajarah: VII 35
Bustan al-Sharqi: VII 36
al-Buwaiti: VII 45
Byzantium: XI 408

Cairo: VII 26, 39, 51
Cambridge: XI 407
Christian(s): I 488–9
Colonial Welfare and Development Fund: IX 146
Costa, Paulo: IX 152

Dahr, Wadi: VII 32
Daiyin: see Dayyin
Dali': IX 147
Dammun: VIII 54
Dar es Salaam: I 490
Dathinah: VIII 33, 41, 43, 55; IX 145
Davey, Peter: IX 147
David: XI 409
Daw'an: III 438–9; VI 307, 309–11; VIII 56, 59, 61, 67–70, 72
Day, Stephen: IX 149
Dayjur: VI 311
Dayyin: VI 307; VIII 75
De Goeje, M.J.: III 439
Decianus: XI 408
Dhahaban, Wadi: VIII 71; IX 151
Dhobi Pool: *see* Ayn Qassari
Dhofar: IV 98
Dickson, H.R.P.: I 491, 512; X 491
Dirgag: IX 146
Djabbara: I 502
al-Djanadiyah: III 455
Djiblah: III 448
Doe, Brian: IV 94
Dowson, V.H.W.: VI 307
Duhur: VIII 72
Durian, Shaikh: VIII 33

East Africa: V 125
East African Coast: V 111

Fadalah family: I 502
am-Fajarah: VIII 33
FAO: VII 28
Far East: VI 309
Fatimah bint Mubarak: I 500
Formosa: V 125
Fu'ad Saiyid: VII 26
al-Furani: VII 45

Gabir al-Gaffari: XI 407
Gardafui, Cape/Ras: V 121
al-Gazali: XI 407, 409
Gazi: V 125
Gershevitch, I.: X 474
Ghail Ba Wazir: VIII 33, 57, 58; IX 152
Ghayl Ba Wazir: *see* Ghail Ba Wazir
Ghazan, Ilkhanid: X 476
al-Ghuraf: VIII 33
Glaser: III 438, 442; IV 95; VII 30–1
Gulf, (Persian/Arabian): I 489–90, 497, 504, 511–2; V 109, 111, 121; VIII 70; X 472

Habban: VIII 59
al-Habib Ali b. Hasan al-Attas: VIII 67
Habshi family: VI 312
Hada: VII 32, 37
al-Haddad: *see* Abdullah al-Haddad
Hadi b. Ahmad al-Haddar: VIII 45
Hadramawt: I 490, 505; III 434, 436, 438–40; IV 96; V 109, 125–6; VI 307–8, 312–3; VII 26, 28–9, 33; VIII 33–4, 43, 46, 48, 49–51, 56, 59, 62–3, 65, 67–70, 72, 74–5; IX 145, 150–1; X 490
Hadur: VII 43
al-Hafah: VIII 55
Hafun: V 118, 121–2
Haid Qasim: VIII 51
Haidarah: *see* Muhammad Ahmad Haidarah
Hainin: VIII 48; IX 150
Hairij: VIII 70
Hais: VII 48, 51
al-Hajar: IX 147
Hajr: I 503
Hallaniyyah: IV 94

Hamagan: X 476
al-Hamdani: III 434, 439, 442–3; VII 31, 33–4
Hamdawh: I 488
al-Hami: II 193; V 109, 125
al-Hanafi, Jalal: I 486, 497
al-Haqul: VII 41
Haraz: VII 53
al-Hasa/al-Ahsa: I 486, 491; X 472, 474, 490
al-Hatim, well: VI 318
Hawrah: VIII 72
Haynin: *see* Hainin
Hazzah: VII 48, 53
Hejaz: VIII 57
al-Hidd, village in Bahrain: I 486, 490–1, 497, 500–2, 507
Higris: IX 150
al-Hijail: VIII 72
Hirab: V 119, 122
Hizyiz: VII 36
Hodeidah: VII 32
Holmes, Frank: X 471
Homfray, J.D.: X 474
Horn of Africa: IV 94
Hinds, Val T.: I 489–90; V 126
Hubaish: III 447
Hubyah: V 119
Hud, prophet and place: VI 307, 309; VIII 51; IX 151
(al-)Hudjariyah: III 449–50, 456–7
Humasiyah: IX 146
al-Hurah, village: I 504
Huraidah: VIII 40, 42–3, 45–9, 65
Hurmuz: VIII 70
Husain, Qadi: VII 45
Husain b. Sa'id: VII 30, 32
Husayn b. Amir al-Mu'minin al-Mansur: XI 407
Husn Falluqah: VIII 68
Husn al-Urr: IX 151
Huwala: I 501
al-Huwali: VII 45

Ibb: III 443, 448; VII 25
Ibn Abd al-Shaykh: VI 312
Ibn Bassal: VII 26–7, 35, 54, 46
Ibn Habfur: *see* Khalifah b. Habtur
Ibn Hajar al-Haithami: VIII 35
Ibn Hisam: XI 424
Ibn Magid: *see* Ibn Majid, Ahmad
Ibn Majid, Ahmad: I 490; V 109; XI 408
Ibn al-Mujawir: I 488; X 473
Ibn al-Murtada: VIII 34
Ibn Yamani: VI 312
Ibzur, name of fish-trap: I 491, 497, 500, 502
Idim: VIII 69, 71
Ihsan Abbas: VII 26
Imam of the Yemen: III 441, 444, 446; VI 312; VIII 34
al-Imrani, Yahya b. Sa'd: VIII 34
Inat: VIII 51, 68, 71–2
India: VII 35
India Office Library: X 478
Ingrams, W.H.: VI 309
Ingrams' peace: IX 151
Iraq: I 489, 506; VI 307; X 477
Iraqi(s): VII 45
Irma: VIII 72
Isa b. Ali Al Khalifah: I 501–2
Isa al-Qutami: V 111, 121–3
Islam: VI 317, 321; VIII 33, 60; IX 146, 151; XI 408
Isma'il al-Akwa', Qadi: VII 25–7, 30, 39
Isma'il al-Jabarti: VII 28; IX 146

Jahanah: VII 37
al-Jahiz, Amr b. Bahr : I 486; X 491
al-Jahmaliyah: VII 35, 53
Jahran: VII 30
al-Janad: VII 45
Java: VI 309
Jawf (Yemen): VII 33–5
Jazi: *see* Gazi
Jazirat il-Sayih, sea-spring in Bahrain: I 490
Jeddah: VIII 57
Jerusalem: VI 316
Jews: I 488; III 458
Jibbali: IV 98
Jidfarat al-Zinji: VIII 58
Jifri family: VI 312
Jizan: IV 94; VII 28–9, 37, 38–9; IX 145–6, 149
Johnstone, T.M.: I 507–8; IV 94–5
Jol: VIII 58–9, 67
Joseph al-Siddiq: VII 40
Ju'aimah: VIII 69
Juba, river: V 123
al-Jubb: V 119
al-Juwali: *see* al-Huwali

Kalifah: *see* Kilifi
Kalifi: *see* Kilifi
Karim Ali al-Urayyid: X 475
al-Kasar: VIII 68–9
Kawlat Na'it: VII 31
Kenya: V 126
Khalaf b. Thani: I 499
al-Karkhi, Husain Hatim: I 486

al-Khadrah: X 475
Khalfawh: I 488
Khalifah b. Habtur: I 501–2
Khami'ah: VI 311
al-Khamilah: VIII 69
Khamis b. Mahamid al-Malhami: IV 95
Khartoum: IX 146
al-Khasha'ah: VIII 63
Khawr Maqta' Tubli: X 475
Khawr Yaya(h): V 120, 123
al-Khazraji, Ali b. al-Hasan: VII 31, 35–6
Khon, Wadi: VIII 68
Khumm: III 444
Khuraybah: VI 311
Kilifi: V 120, 125
Kirzakkan: X 474–5
Kismayu: V 123
Kiyama(h): V 119, 123
Kuhlan: VII 30
Kuwayuh: V 119, 123
Kuraikarah, fish-trap: I 504–6
Kuria Muria Islands: IV 94, 98
Kuwait: I 496–7; V 111

Lahej: III 440; V 122; VII 28–9, 48; VIII 33; IX 145–8
Lahj: *see* Lahej
Landberg, Carlo von: III 433–4, 436, 438; V 123
Lambton, A.K.S.: IX 152
Lamu, island and town: V 111, 120, 123, 125
Lane, Edward: III 434, 436, 439; IV 96–7
al-Lihb: VII 53
Lorimer J.G.: I 488, 511; X 472, 474–5, 489–91
Lower Yemen: VII 27; VIII 34; XI 407
Luqman, Muhammad: III 440, 458

al-Ma'adin, Wadi: IX 147
Ma'bar: V 118
MacIvor, I.: I 509
Madrid: VIII 57
al-Mahazzah, al-Kabir: X 475
al-Mahazzah, al-Saghir: X 475
al-Mahdi Abbas, Imam of the Yemen: III 442; IX 152
Mahfuz Awad Ba Habarah: II 199
Mahfuz b. Said b. Awad Ba-'l-Ra'iyyah: II 201
al-Mahjam: VII 35
Mahrah: I 489; V 121
al-Mahuz: X 475
Maifa': VIII 71
al-Ma'inah: II 193
al-Majraf, quarter of al-Shihr: II 197
Majus: I 488
Maktari, A.M.A.: VII 28; IX 145, 148; XI 408
Malaya: I 497
Malikiyyah/Malichiyyah: X 475
Malindi: V 125
Manakhah: III 443; VII 32
al-Manamah, island: I 504, 506; X 475
al-Manar, mountain in the Yemen: III 443
Manda: V 120,123, 125
al-Ma'qili, palace: VII 35
Maqtari, Abdullah: *see* Maktari, A.M.A.
Marashidah: VI 311
Ma'rib: VII 33–4; IX 151
Markah: V 123
al-Masfalah: VIII 70
Mashrik, the east of the Yemen: III 443, 450, 455
Masilah, of Idim: VIII 71–2
Masilah, Wadi: IX 151
Masilat Al Ba Dhib: VIII 69
Masilat Al Shaykh: VI 311–2
al-Masna'ah: I 489; VI 310
Mati': IX 146
Mattash, barrage: VIII 66
Mauritius: V 111
al-Mawardi: VII 45; XI 407, 409, 424
Mawly, Said H.: VII 38
Mawr, Wadi: VII 48, 51; IX 146, 148–9
Mawshah: VII 36
Mawza': VII 48
Mawzah bint Saif: I 499–500
Maziwah: V 120, 125–6
Maziwi: *see* Maziwah
Mecca: III 444; VIII 57
Medina: III 444
Mi'dan: I 486
Middle Persian: X 474
Mikhlaf (Ja'far): VII 40, 41
Miryam bint Mubarak: I 499
al-Mishqas: VIII 70–1
Mocha: V 109
Modern South Arabian (MSA) languages: IV 98
Mogadishu: V 122–3
Mombasa: I 490; V 120, 123, 125
Morocco: VIII 41, 60
Mtoni: *see* Mutuni
al-Mu'aiyad Dawud b. Yusuf, al-Malik, Rasulid sultan: VII 26, 35
Mubarak b. Fadalah: I 501
Mubarak b. Saif: I 500
al-Mudawwarah: II 193
Mudiyah: VIII 33

Muhammad, the Prophet: III 445–6; V 121; VI 319, 322
Muhammad b. Abd al-Haiy: I 499
Muhammad Abd al-Qadir Ba Matraf: V 109, 111, 122–3, 125–6
Muhammad Abd al-Quddus, Saiyid: VII 33
Muhammad b. Abdullah b. Ahmad Ba Sawdan, Shaikh: VIII 61–2
Muhammad Ahmad Ajlan: II 201
Muhammad Ahmad Haidarah: VII 27–8, 31
Muhammad b. Ahmad al-Shatiri: VII 26
Muhammad al-Akwa', Qadi: VII 27
Muhammad Ali Pasha: VI 312
Muhammad b. Amr: I 504
Muhammad b. Biyat: I 501
Muhammad Haidarah: III 433, 440, 442–3, 445
Muhammad b. Hashim: VI 311
Muhammad Hassan: III 444
Muhammad b. Mahmud: I 501
Muhammad Said Midayhij: II 193
Muhammad b. Umar al-Saqqaf, Sayyid: VI 308
Muhammad Zayd, Shaikh: XI 407
(al-)Muharraq, island in Bahrain: I 486, 490–1, 502, 510; X 478
Muhsin b. Ja'far b. Alawi Bu Numaiy, Saiyid: VIII 63
al-Mujahid Ali b. Dawud, al-Malik, Rasulid sultan: VII 25
al-Mukalla: I 513; III 438; V 109, 122; VIII 58; X 487
al-Mukhainiq, Wadi: VIII 69
Müller, Walter: IX 151
Mulligan, W.E.: I 501
Muraizafan: VII 54
Muruti: V 119, 123
Muscat: V 109
Mutuni: V 121, 126
al-Muzaffar Yusuf b. Umar, al-Malik, Rasulid sultan: VII 26

al-Nabhani: X 472, 489
al-Nabiyy Salih, island: X 475
Najal River: V 122
al-Najadain: VIII 58
Najd (Yemen): VII 35, 52
al-Najdi: V 111
Najran: VII 33, 35
Nakhlah, Wadi: VII 51
Nakhlat al-Hamra: VII 37
al-Naqar: VIII 68
Nashiriyyah: IX 146
Nashwan b. Sa'id: VII 31
Nasr Al/b. Madhkur: I 501–2
Nassur: I 498, 501–2
al-Nawawi: VII 45; VIII 34; XI 409
Norris, H.T.: IX 145
North Africa: VIII 57
Nubah: V 126
al-Nukhr: VIII 71, 72

Oman: I 487, 490, 502; V 109; VIII 72

Pahlawi: *see* Middle Persian
Pate: V 123
PDRY (People's Democratic Republic of Yemen): IX 145
Pemba, island: VIII 45
Perim, island: I 489
Persia: I 506; III 439; VIII 57
Persians: I 487, 501
Plas, C.O. van der: VII 28
Popenoe, Paul: VI 307
Prophet, the, *see* Muhammad: I 489; III 439, 444; V 118, 121, 123, 126; VI 318; VIII 57; XI 409, 424

Qabr Hud: VIII 51; IX 151
al-Qahmah: VII 53
Qahtan al-Sha'bi: VII 27
Qaidun: VIII 58, 59, 75
Qalansiyyah: IV 97, 98
Qalqashandi: VII 35
Qara, tribesmen: IV 98
al-Qarah: VII 30, 31, 58
Qar'ad: V 119
al-Qasasir: I 490
al-Qashar, *see* al-Qishar: II 197
Qasam: VI 312; VIII 69, 71, 72
Qataban: VIII 45
Qatham/Qitham: VI 311
Qatif: I 486, 506
al-Qattarah: X 489
al-Qibli: VIII 72
al-Qishar: II 197, 198, 199
al-Qumli b. Sa'id: VII 53
Quran: I 506; VI 313–7; XI 424
al-Qurtub: VII 36; XI 409
al-Qusaibah: VII 40
Qusay'ar, village: II 193
al-Qutami: *see* Isa al-Qutami

al-Rabadi, group of villages: III 447
Rabi'ah: X 473
Rahaiyam: *see* Abdullah Rahaiyam
Rahayyam, Shaykh: VI 308–9
Rahbat Daw'an: VIII 68

al-Raidah: VII 31; VIII 71
Rakhyah: VIII 72
Ras Asir: V 121
Ras Binnah: V 121
Ras al-Khayl: V 122
Ras al-Khaymah: X 488
Ras al-Khayr: V 118, 122
Rasban: VII 48
Rashad: V 118
Rashid b. Amr al-Fadalah: I 499–500
Rashid Ibrahim Al Khalifah, Shaikh: X 472
Rasulids: VII 27; IX 146
Rawkab: VIII 70
Raydat Abd al-Wadud, village: II 193
Red Sea: I 486
Reinhardt, C.: X 489
Rentz, G.: I 501
al-Ribat: VI 311
Ribat Al al-Zubaidi: VIII 69
Rijam, Wadi: VII 33
Rima', Wadi: III 453; VII 48, 50; IX 146–8; XI 407
Robson, J.: VIII 64
Roe, Sir Thomas: IV 98
Romanus: XI 408
Rossi, E.: III 457; VIII 34, 47
Rudum: VI 307; VIII 58
Ryckmans, G.: III 441

Sabikah bint Amr al-Fadalah: I 499
Sabir, Jabal: VII 35, 41
Sa'd b. Abi Waqqas: XI 409
Sa'd al-Suwayni: IX 145, 150
Sadad: X 475
Sa'dah: VII 31
Sadbah: VIII 72
Sadiq al-Baharnah: X 471, 473
Sah: VIII 72
Sahib al-Mawahib, Imam: IX 146
(al)-Sahlah, park in Ta'izz: VII 35
Sahul: VII 46
Sa'id, of Upper Awlaqi: VIII 74–5
Sa'id Ba Tayi': V 109, 118, 125–6
Sa'id Basha: V 122
Sa'id b. Isa, Shaikh: VIII 75, 76
Sa'id Muhammad b. Abdullah: IV 94
Sa'id b. Sultan: V 111
al-Saidalani: VII 45
Saif al-Islam Ahmad, Imam of the Yemen: III 442
Saiwun: VI 309; VIII 33–5, 46–7, 64–5, 67, 74
Salah Ali al-Madani: X 475
Salih b. Ghalib al-Qu'aiti, Sultan: VIII 59
Salim, S.M.: I 486
Salim Ubayd Bil-Kawr: II 199
al-Salimi: VIII 72
Salman b. Hamad Al Khalifah, Shaikh: I 506; X 471
San'a': III 442–4, 446, 450–1, 458; VII 30–4, 36–7, 39, 43; IX 150, 152; X 476; XI 424
Sar: VIII 69
Sarat: VII 52
Sarih: VIII 70
Saryaqus al-Asfal: VII 36
Saudi Arabia: I 503, 506, 512; VII 28, 39; IX 146, 149, X 474
al-Sawahil: V 123, 126
Sayban: VI 311
Sayhut: V 109, 118, 121
Saywun: *see* Saiwun
Seville: VII 26
Shafi'i(s): VI 308; VII 33
al-Shafi'i: VII 45
Shah, the: I 501
al-Shaharah: III 446; IX 145
Shahri, language: IV 94
al-Sha'ir: VII 41
al-Shaizari: III 434
Shanafir: VI 312
Shanbal: VI 307; VIII 51, 67; IX 151
al-Sahari: VII 36
Sharifah bint Ibrahim: I 506
al-Shawafi: VII 41
Shaylah: V 120, 125
Shelah: *see* Shaylah
Shiah: I 486, 488–9, 495, 506–7, 510–1; III 445; X 473
Shibam: III 434; VIII 33, 68, 69, 70, 71; IX 151
al-Shibami: III 434–5, 438
al-Shihr: II 193, 199; V 109; VIII 58, 68, 70–1
al-Shihri: *see* Ba Faqih al-Shihri
Shii(s): I 488–9, 506
Shilah: *see* Shaylah
Shuhair: VIII 70
al-Shuruqi: *see* Abdullah b. Khamis
Siddiq Ahmad: IX 149
al-Sif al-Tawil (Long Beach): V 119, 122
Silatah: I 499, 503
Sindibar: V 118, 120, 125
Singapore: I 497
Sirr, Wadi: VII 32
Sitrah: I 511; X 475
Smith G.R.: VII 35
Smith, Sidney: III 458
Snouck Hurgronje: III 440; VIII 34
Socotra: IV 94, 98, 99; V 122
Socotri, language: IV 94, 95

Soloman: XI 409
South(ern) Arabia: III 434, 458; IV 94; V 109; VI 308; VII 28–9, 39; VIII 33, 43, 45, 47–8, 60
South-West(ern) Arabia: III 433, 440; VII 33, 72; IX 145
Spain: VIII 60
Sudan: III 436; VII 45
al-Suhul: III 448
Sulayman al-Mahri: IV 97; V 109
Sumuh: VI 311
Sunni(s): I 486, 488, 506–7
Suq al-Thaluth: IX 149
Suraih: *see* Sarih
Surdad: VII 48, 51; IX 146
Surdud: *see* Surdad
Syria: III 439; VII 35
Swahili: I 490

al-Tabari: III 439
Tahir b. Alawi, Saiyid: VIII 75
Tahir b. al-Husayn b. Tahir al-Alawi, Sayyid: VI 308, 310–3
Ta'izz: III 438, 442–4, 456; IV 95; VII 27, 29, 31, 35, 40
Tajir, M.A.: X 473
Tamim: VI 311, 312
Tanzania: V 126
Taribah: VI 309
Tarim: III 436, 439; VI 309, 311; VII 26; VIII 34, 46–8, 50–2, 54, 60–3, 65–6, 68, 70–2
Tetuan: VII 26; VIII 41
Tha'abat: VII 35
Thibi, Wadi: VIII 68–9
Thomas, Bertram: IV 95
Tibbetts, G.R.: V 122
Tihamah: III 446, 448, 450, 453–5, 458; VII 27–8, 30, 35, 39, 42, 45, 48, 50, 53, 56; IX 145–7, 149, 151
Tiwalah: V 119
Trucial Coast: I 490
Trucial States: I 487
Tubli: X 475
al-Tuhaita: VII 36
Tulah: V 109
Tumbatu Island: V 126
Turquh: IX 146

Uhud, raid: V 122
Umar, the second caliph: III 445
Umar Abdullah Hunayn: II 199
Umarah, historian: IX 146
Umbarak b. Said b. Fadalah: I 498
Umm al-Dijaj: V 121, 126
Umm Ijrayri: X 475
Umm al-Khawadir: V 119, 123
Umm al-Sawali, spring in Bahrain: I 490
Umm al-Shu'um: X 475
Umm Sirwal: I 504
Unguja: *see* Unjuja
Ungwana: *see* Formosa
United Nations Development Programme (UNDP): VII 28
Unjuja: V 125–6
Urqub, pass: IX 149
Urwah b. Hizam: I 503
Uthman b. Jami: I 499
Uwal: I 488; X 473

Van den Berg, L.W.C.: VI 309
Vatican Library: VIII 34
Vidal, F.S.: X 488, 491

al-Wadi al-Kabir: IX 146
al-Wadi al-Saghir: IX 146
Wadiyan: X 475
Waqf Department/Office, Bahrain: I 489, 495, 507, 510; X 471
al-Waqidi: XI 424
War Shaykh: V 123
Wasin: V 120, 125
Western Arabia: VII 39
Wilkinson, J.C.: IV 95
Wittek, Paul: III 458
al-Wusta, fish-trap: I 502

Yafi': VIII 33
al-Yafi'i: VI 308
Yahya, Imam, al-Mutawakkil 'ala 'llah: III 443
Yahya b. Saraf: XI 424
Yakut, b. Abdullah, geographer: III 444
al-Yamamah: I 503
YAR (Yemen Arab Republic): IX 145, 152
Yemen: III 438, 440–4, 448, 452–3; VI 307–8, 312; VII 25, 27–8, 30, 32–6, 38–40, 43, 45, 52–3, 58–9; VIII 33–5, 59, 69, 72; IX 145–6, 149, 151–2; X 474, 476; XI 407–8

Zabid, place and Wadi: III 453; VII 27–8, 31, 35–6, 39, 48, 50–1; IX 146, 148; XI 407, 409
Zaidiyah: III 444
Zamzam: VI 318
Zanzibar: IV 98; V 109, 111, 118, 120–1, 125–6
Zayd b. Abd Allah al-Akwa': XI 407
al-Zumayli: *see* Ahmad b. Abi Bakr al-Zumayli